Palgrave Studies in Medieval and Early Modern Medicine

Series Editors
Jonathan Barry
Department of History
University of Exeter
Exeter, UK

Fabrizio Bigotti
Institute for the History of Medicine
Julius Maximilian University
Würzburg, Germany

CENTRE FOR THE STUDY OF
MEDICINE AND THE BODY
IN THE RENAISSANCE

The series focuses on the intellectual tradition of western medicine as related to the philosophies, institutions, practices, and technologies that developed throughout the medieval and early modern period (500-1800). Partnered with the Centre for the Study of Medicine and the Body in the Renaissance (CSMBR), it seeks to explore the range of interactions between various conceptualisations of the body, including their import for the arts (e.g. literature, painting, music, dance, and architecture) and the way different medical traditions overlapped and borrowed from each other. The series particularly welcomes contributions from young authors. The editors will consider proposals for single monographs, as well as edited collections and translations/editions of texts, either at a standard length (70-120,000 words) or as Palgrave Pivots (up to 50,000 words).

Jil Muller
Editor

Automata, Cyborgs, and Mutants

Eccentric Bodies from Humanism to Transhumanism

VivaMente
The Garden of Ideas

palgrave macmillan

Editor
Jil Muller
Paderborn University
Paderborn, Germany

ISSN 2524-7387 ISSN 2524-7395 (electronic)
Palgrave Studies in Medieval and Early Modern Medicine
ISBN 978-3-031-93200-7 ISBN 978-3-031-93201-4 (eBook)
https://doi.org/10.1007/978-3-031-93201-4

Cover illustration: Wellcome Collection

This Palgrave Macmillan imprint is published by the registered company Springer Nature Switzerland AG
The registered company address is: Gewerbestrasse 11, 6330 Cham, Switzerland

Preface

In this anthology, *Automata, Cyborgs and Mutants: Eccentric Bodies from Humanism to Transhumanism*, the reader is taken on a journey through the corridors of history, tracing the evolution of these eccentric bodies from the lofty ideals of humanism to the frontiers of transhumanism. From the ancient myths of automata crafted by skilled artisans, to modern cyborgs that seamlessly integrate man and machine, to speculative mutants that challenge the very essence of what it means to be human, this collective work serves as a testament to humanity's relentless quest for self-understanding and self-transcendence.

As readers delve into the pages of this book, they will be invited to consider profound questions about identity, agency, and the nature of existence itself. What does it mean to be human in an age when technology is blurring the boundaries between the biological and the artificial? How do humans navigate the ethical dilemmas posed by an increasingly intimate relationship with machines and genetic engineering? And what future awaits humans as we venture further into the uncharted territories of transhumanism?

The completion of this collective work has been a collaborative endeavour, made possible by the contributions and support of many individuals and institutions.

First and foremost, we extend our deepest gratitude to the authors whose insightful contributions have enriched the pages of this book. Their dedication, expertise and passion have been instrumental in shaping the diverse perspectives presented in these chapters.

We would like to express our gratitude to the series editors, Dr. Fabrizio Bigotti and Prof. Jonathan Barry, who have generously given their time and expertise to ensure the quality and rigour of this publication. Their invaluable feedback and guidance have been instrumental in refining the content and structure of this collective work.

We are indebted to the Centre for the Study of Medicine and the Body in the Renaissance (CSMBR) and the VivaMente Grant (CSMBR_VIV004.2022), without which we could not have organised an exciting two-day conference, of which this book is the fruitful outcome. The Centre has provided us with the resources, facilities and financial support to undertake this ambitious project. Their commitment to advancing scholarship and fostering interdisciplinary dialogue is deeply appreciated.

Paderborn, Germany
March 2024

Dr. Jil Muller

Contents

Notes on Contributors

Paul Richard Blum is an emeritus professor of Philosophy at Loyola University Maryland in Baltimore. Currently he is a researcher at Palacky University Olomouc. His research is centered on Renaissance Philosophy with excursions into philosophy of mind. His book *Oracles of the Cosmos—Between Pantheism and Secularism* has been published at Schwabe Verlag, in April 2022.

Audrey Borowski is a Leverhulme Early Career Fellow and Isaac Newton Trust Fellow at the University of Cambridge working on the philosophy of artificial intelligence. She received her doctorate from the University of Oxford and is a regular contributor to the Times Literary Supplement and Aeon. Her first monograph Leibniz in His World: The Making of a Savant has been published by Princeton University Press.

Sarah Carvallo is a professor of Philosophy of Science in the Department of Philosophy, at Franche Comté University, France. Her research focuses on the philosophy of medicine from the early modern period till nowadays in order to show how the medical knowledge outlines our anthropology. She actually supervises two funded projects on genomics and vulnerability. Her recent publications deals with vitalism («Stahl et le vitalisme», In: *Équivoques du vitalisme*, Charlotte Morel, Bertrand Nouailles (éd.), Paris, Classiques Garnier Flammarion, 2023), genomics (https://medecine-philosophie.com/index.php/category/valeurs-des-donnees-genetiques-7-2022/), and vulnerability («Prendre soin, restaurer, réparer, de la Renaissance à nos jours», avec A. Lavin, J.Ph Pierron, A Anne Braun eds., Mare Martin, 2022).

Emanuele Clarizio is an associate professor at the Center for Medical Ethics (CEM) of the Ethics Laboratory (EA 7446) of the Catholic University of Lille since September 2019. He hold a Ph.D. in Philosophy from the University of Bari in Italy since July 2015, and then he continued his research through three post-docs: at the LIS laboratory of the University of Paris-Est Créteil, with a project on technological analogy in Descartes' epistemology of the living and its ethical and anthropological consequences; at the Costech laboratory of the University of Technology of Compiègne, as part of the ERDF research project research project "BiobanquePerso. The biobank of Picardie in the challenge of personalized medicine"; at the CNRS, attached to the Institute of Legal and Philosophical Sciences of Sorbonne University, with a project on "The values of life in biobanks". He has edited several books and he is the author of a book on history and philosophy of technology: La vie technique. Une philosophie biologique de la technique (Paris, Hermann, 2021).

Emanuele Costa is an assistant professor of Philosophy at Vanderbilt University. He has received his Ph.D. from Birkbeck College—University of London. His work focuses on the interaction between metaphysics, history of philosophy, and philosophy of religion. He is the author of several essays on Spinoza, Leibniz, Anne Conway, and is currently working on manuscripts regarding Margaret Cavendish and Elisabeth of Bohemia.

Hannah Elmer received her Ph.D. in medieval and early modern history at Columbia University in 2019 and currently holds a postdoctoral research and teaching position at the Leibniz University in Hannover, Germany. She specializes in the cultural, religious, and intellectual history of later medieval and early modern Central Europe. Her current research examines the interplay of religious and naturalistic epistemologies in determining boundaries between life and death. She is also developing work concerning the agency of fire in later medieval and early modern Europe.

Guido Giglioni teaches history of philosophy at the University of Macerata, Italy. His research focuses on the interplay of life and imagination in the early modern period. On this subject he has written and edited several contributions. He is the author of two books, on Jan Baptista van Helmont (Milan, 2000) and Francis Bacon (Rome, 2011).

Thierry Hoquet is a professor of Philosophy of Science in the Department of Philosophy, at Paris Nanterre University, France. His publications focus on different topics such as: (1) the history of natural history and biology; (2) the concept of sex, especially with regard to sexual selection, and the sex/gender binary; (3) the role of words and metaphors in science; (4) the figure of mutants, cyborgs, and other fictional characters. His publications in English include *Revisiting the Origin of Species: The Other Darwins* (Oxford: Routledge, 2018), and an edited volume: *Current perspectives on sexual selection. What's left after Darwin?* (Dordrecht: Springer, 2015). On the topic of Cyborgs: a book chapter: « Cyborgs, Between Organology and Phenomenology: Two perspectives on Artifacts and Life ». In French, two books: *Cyborg Philosophie. Penser contre les dualismes* (Paris: Le Seuil, 2011) and *Les Presque-Humains. Mutants, Cyborgs, Robots, Zombies, … et nous* (Paris: Le Seuil, 2021).

Nicole Howard is a professor of History at Eastern Oregon University with a focus on the history of science and medicine in Early Modern Europe. Her recent book, *Loath to Print: The Reluctant Scientific Author*, was issued from Johns Hopkins in 2022.

Dr. Rahma Khazam is a Paris-based researcher, critical theorist, and art historian affiliated to Institut ACTE, Sorbonne Paris 1 and ENSADLab, Paris. She studied philosophy and art history and received her Ph.D. from the Sorbonne in aesthetics and art theory. Her research spans the fields of modernism, image theory, and contemporary speculative aesthetics, and has been published in edited volumes, exhibition catalogues, and academic journals. Recent publications include: "Art, Science and the Mutant Object", in *Post-Specimen Encounters Between Art, Science and Curating*, Intellect, 2020; "Son et Image: Face au Réel", in *L'écho du réel*, Editions Mimésis, 2021; "Ikonische und spekulative Wende: Von Visualität zu Realität", in *Nach der ikonischen Wende. Aktualität und Geschichte eines Paradigmas*, Kadmos, 2021; "Clement Greenberg's Modernism: Historicizable or Ahistorical? » in *Historical Modernisms*, Bloomsbury, 2021.

Albert Kozik is a research assistant at the Institute of Chinese Studies and a Ph.D. candidate at the Faculty of Culture and Arts of the University of Warsaw. He is the principal investigator for the research grant *King of*

Engineers financed by the Polish Ministry of Education and Science, and collaborator in the *RETOPEA* (*Religious Toleration and Peace*) project funded by the European Union as part of the Horizon 2020 programme. His research interests cover the history of artificial life as well as the early modern cultural and technological exchange between China and the West.

Jessica Lombard graduated from the Special Military School of Saint-Cyr Coëtquidan as a commissioned officer in the French Land Army, and passed her master's degree in philosophy at the École Normale Supérieure d'Ulm and the École des Hautes Études en Sciences Sociales. She defended a thesis in philosophy entitled "Towards an Onto-Anthropotechnic of the Human Sphere. The Question of the Technological Ban through a Phenomenological Approach of Transhumanism", in which she specialized in philosophy of technology and phenomenology. She is currently a University lecturer and researcher at the Chair "Ethics, Technology and Humanities" (ETHICS EA 7446) of the Catholic University of Lille (France).

Jil Muller is an assistant professor at the University of Paderborn and deputy head of the Center for the History of Women Philosophers and Scientists and at Paderborn University in Germany. She received her Ph.D. at the University of Strasbourg, in France, and her book *Soigner l'humain. Péchés et remèdes chez Montaigne et Descartes* has been published in Mai 2022 by Classiques Garnier. Her main research interests include early modern, especially French and German philosophy, history of medicine, moral theories, and Women in early modern philosophy. She has published several articles on Descartes and Montaigne, with a special focus on morality, linked to a medical understanding of the human body and brain.

Thomas A. Murphy is a Career Development Fellow and Tutor in French at New College, University of Oxford, where he teaches and writes on early modern literature and thought.

Michael Stolberg is professor and chair of the history of medicine at the University of Würzburg in Germany. He has published widely on learned medical theory and practice, the patient experience and body history in early modern Europe. His two most recent books are "Learned physicians and everyday medical practice in the Renaissance" (2022) and "Gabrielle Falloppia (1522/23-1562). The life and works of a Renaissance anatomist" (2022).

Charles T. Wolfe is a Professor in the Department of Philosophy, Université de Toulouse-2 Jean-Jaurès. He works primarily in history and philosophy of the early modern life sciences, with a particular interest in materialism and vitalism. He is the author of Materialism: A Historico-Philosophical Introduction (2016), La philosophie de la biologie: une histoire du vitalisme (2019) and Lire le matérialisme (2020), and has edited or coedited volumes on monsters, brains, empiricism, biology, mechanism and vitalism, including (w. D. Jalobeanu) the Encyclopedia of Early Modern Philosophy and the Sciences and (w. J. Symons,) The History and Philosophy of Materialism. He is co-editor of the book series 'History, Philosophy and Theory of the Life Sciences' (Springer).

List of Figures

CHAPTER 1

Introduction: Automata, Cyborgs, and Mutants: Eccentric Bodies from Humanism to Transhumanism

Jil Muller

Since the COVID pandemic, we have come to realize how fleeting human life is and how vulnerable the human body is. It only took a "small" virus to wipe out thousands of lives and put the world on edge. An initially harmless cough and fever quickly turned into a deadly virus that dominated and controlled our world for two years. The political and medical measures were control, isolation, curfews, vaccinations (which were produced at a rapid pace), contact restrictions, and much more that took a toll on everyone's private lives. If some were hoping for science fiction atmospheres and world orders, they were quickly disabused. No one became Will Smith, the loner with a German shepherd, in "I Am

J. Muller (✉)
Paderborn University, Paderborn, Germany
e-mail: jil.muller@uni-paderborn.de

J. Muller (ed.), *Automata, Cyborgs, and Mutants*, Palgrave Studies in Medieval and Early Modern Medicine,
https://doi.org/10.1007/978-3-031-93201-4_1

Legend," or Arnold Schwarzenegger as the Terminator. Instead, we realized that we are still far from the ideal of posthumanism and the immortal human being, and that our bodies are still our limit.

And yet the idea of overcoming the human body as such has been revived, perhaps precisely because it was the weak element in this pandemic. Overcoming the human body means not only upgrading it with technical means, as in the case of the cyborg, or making it more robust and stable (as in the case of the Mutants), but also adapting it to new situations, adapting it so that it can be replaced at will when it is defective or inferior. This is the idea of eccentric bodies, which we can find in transhumanism and posthumanism. Eccentric bodies can take different forms; they adapt themselves or are adapted by men to different environments and changing natural conditions, and they can change their movements and behavior.[1] This idea is not new, but our current research gives it a fresh perspective, presenting it in the form of a cyborg or mutant. Rooted in experiments and theories of automata that date back to the Renaissance, this research seeks to imbue these automata with greater autonomy and a biological dimension.

The aim of this book is therefore to show that the basic ideas of transhumanism and posthumanism are not fundamentally new,[2] that they have their origins already in antiquity, but above all later in humanism, and that they can only be taken up and formulated anew today through the further development of technology. Using the method of reception history (cf. Hans Blumenberg), the book attempts to show that the course for transhumanism and posthumanism was set in humanism itself, and that posthumanism is therefore the natural outcome by overcoming humanism as anthropocentrism.

The objective is to trace the long way to transhumanist or even posthumanist thought, which began in the late Middle Ages, with the first ideas of resurrection and therefore of reanimation. It then passed through the various advances in medicine and the inventions of prostheses and automata in the seventeenth century, reflected not only in philosophy from the Renaissance onwards but also in the art of that period, and it reached its apogee in the thinking of cyborgs and mutants, with contemporary thinkers who imagine a human society that goes beyond its own limits.

Historically and philosophically, there is a continuity from humanism to transhumanism and posthumanism. Current transhumanism has its origins mainly in the English-speaking world and can generally be traced back to the 1960s, when it was very popular in California. The origins of posthumanism, on the other hand, lie more in the mid-1980s and developed out of transhumanism.[3] The central idea of transhumanism was to find a way to prolong human life, delay mortality, and increase mental capacity. The cyborg would be such a life form for transhumanism. Posthumanism focuses on replacing humans with artificial offspring in the form of robots, coupled with artificial intelligence, claiming that there is no fixed nature or a fixed boundary between natural and artificial beings.

In transhumanism, man is still at the center of attention; in posthumanism, man's annihilation is already being sought. Humans are to be replaced by machines, which are to be the carriers of evolution and progress, striving for technical immortality. Transhumanism, on the other hand, sees itself as evolutionary humanism: it wants better health, education, and welfare for the whole of humanity. The aim is to improve the human being as such and thus to correct the "mistakes" of evolution by creating an optimal life form that combines and enhances various biological characteristics so that the human being as such becomes better, morally and intellectually. The idea of the cyborg was to embody this ideal of transhumanism by augmenting human biological characteristics with mechanical and electronic "prostheses."[4]

Posthumanism, however, embraced the fusion of human and machine as an opportunity to transcend humanity—particularly the limitations of the human body—and to challenge traditional dichotomies such as mind-body, male–female, and natural-artificial. For them, the human brain is the neural interface to the computer: the human-machine hybrid was the first step toward a perfect simulation of a person stored in a computer. For posthumanist authors, the supposedly mechanistic mode of operation attributed to the body and mind in the sixteenth and seventeenth centuries makes it possible to compare the human being as a unit with a machine, and even to combine him with a machine. Does this mean that posthumanism is the end of humanism? One thing is clear: the focus of posthumanism is not on the human being as such, but on the economy and the utilitarian character of machines. The goal of posthumanism is immortality, and in order to achieve this, posthumanism also abandons the human being. In order to achieve eternal life, human limits must be transcended to the point where man is no longer human, to the point where his body can be arbitrarily replaced and even overcome.

All these ideas have their origins in humanism. Humanism sought to realize the greatness of man. Reason and human morality were central.

Humanism, at least the one that developed from Cartesian dualism, emphasized a self-regulating and moral reason.[5] In the sixteenth and seventeenth centuries, the focus is on reason, and the aim is to improve people intellectually and morally (just look at the many moral treatises; the spread of philosophical ideas through the invention of the printing press; the emerging process of secularization and thus a profane morality and ethics). The body, as a unit of parts subject to basic mechanistic principles (like in Descartes and Hobbes), plays only a subordinate role in this conception (in contrast to later transhumanism). Although humanism also strives for a healthy body (*mens sana in corpore sano*) that supports reason, the idea that the body is subject to biology and evolution is still very much present. The body, as matter that has its own functioning, has received attention through Cartesian dualism, but it has also been understood as something that humans must accept, live with and heal when it no longer functions. The body is then an element whose functioning needs to be understood so that it can be controlled and/or conditioned, but there is no aspiration to transcend the limits of nature, biology, or the body.

This is not to say that this age did not use external aids to support the body. As early as the sixteenth century, the first attempts were made to support and thus enhance the body with mechanical prostheses. Ambroise Paré, a French military surgeon who was celebrated until the nineteenth century as the pioneer of modern surgery, invented the first mechanical hand prosthesis that could realistically reproduce the specific movements of a hand (opening and closing, grasping, moving fingers, etc.).[6] The idea of humanism was to understand the body as a mechanical element and to reproduce its functioning. This is why it is also the age of the automaton. From Descartes to La Mettrie to Vaucauson, the automaton, as an image of the human body, was very popular because it contained the principle of its movement, making it clear that the human body could move without the soul. This not only supported Descartes's theory of two substances, but later made possible the idea of an arbitrarily interchangeable body.

Automata generally refer to hypothetical beings or entities that operate on the basis of purely mechanical processes, like a machine, without possessing consciousness or subjective experience.[7] They are usually created by humans to mimic the movements of the human body or other mechanical movements. Philosophers often used automata as thought experiments to discuss questions about the nature of consciousness, cognition, and the mind-body problem, as has been the case especially since the philosophy of the Renaissance.[8]

However, automata include not only imitations of human body parts (see prostheses), but also mechanical instruments set in motion by springs, internal weights, etc., such as clocks. We find the clockwork model already in Descartes' idea of man as an automaton in his *Traité de l'homme*.[9] And in the seventeenth and eighteenth centuries, watchmakers, in particular, attempted to reproduce the moving human body. For a long time, the watch was the model for various types of automata, because once it had been assembled and wound by the watchmaker, it moved by itself.

During the eighteenth and nineteenth centuries, automata experienced a golden age in Europe. Intricate clockwork automata, known as "automaton dolls," were crafted to simulate human-like movements. These automata were often dressed in exquisite clothing and performed various tasks, such as writing, painting, playing musical instruments, and even performing magic tricks. One of the most famous automata of this period was "The Turk," a chess-playing automaton designed by Wolfgang von Kempelen in the 1770s.[10] The Turk was later revealed to be an elaborate hoax, operated by a hidden human chess player inside the machine. Nevertheless, it captured the imagination of the public and became a symbol of the potential of machines to exhibit human-like intelligence.

In the twentieth and twenty-first centuries, artists and creators have continued to explore the concept of automata in various forms of media, including literature, film, and digital art. Automata have become a recurring theme in science fiction, raising questions about the nature of consciousness, the boundaries between man and machine, and ethical considerations related to artificial intelligence. In summary, automata have left a significant mark on the history of medicine, philosophy, and arts. They have served as instruments for medical training, prompted philosophical inquiries into the nature of life and consciousness, and inspired artistic expressions of human creativity and technical ingenuity. With the emergence of contemporary transhumanism, the question of automata has changed: automata are now closely related to the study of artificial intelligence and autonomous systems. Today, automata are computer programs or machines capable of performing tasks that normally require human intelligence.

But in the beginning, an automaton was just a machine that moved on its own. This principle of the self-movement of the body was a key element of Cartesian theory and was later taken up by La Mettrie.[11]

For Descartes, all movements in the body were subject to the laws of mechanics. For him, therefore, physiology was a branch of physics in which material events were of a mechanical nature and could ideally also be described mathematically. La Mettrie took up this interpretation of Descartes, but also went beyond Descartes by attributing functionalist and materialist properties to the soul. Descartes' machines are intensive automata based on the clockwork model.[12] La Mettrie's machines, however, can wind themselves up and even read clocks. In Cartesian dualism, the soul was an independent substance that could exist without the body, just as the body could move without the soul. La Mettrie replaced the concept of the soul as a substance with the model of a cerebral mechanism that gave rise to mental activity. This enabled his machines to perform more specific movements than those of Descartes because they were cerebrally controlled and did not merely incorporate the principle of self-motion.

But Descartes' innovation with dualism must not be forgotten. He saw thought as detached from the body and thus indirectly laid the foundations for artificial intelligence. If thought no longer needs a body to exist, then this intelligence can exist outside the body and be used in other ways, which is what we see today with artificial intelligence. The milestones for transhumanism and posthumanism were thus largely set in humanism itself, with Descartes' dualism playing a central role. Descartes' dualism was the perfect foundation for the idea of a machine simulation of man for the posthumanists who came from philosophy[13]: from the hypothetical body machine created by God to posthumanism, where man himself becomes the "watchmaker."

Descartes' dualism enabled posthumanism, on the one hand, to think of reason as separate from the body and therefore in the form of artificial intelligence, and on the other, the body itself as separate from the soul, as a machine, and therefore arbitrarily replaceable. Posthumanists also find support for these ideas in Descartes' doubts about his own body. Descartes never doubts the mind, but he does doubt the body, and therefore any body could be our body. It is just a kind of vessel that functions according to mechanistic rules. Descartes' body can be seen in purely organic terms: it was the Creator Himself who once initiated its movements, and now it continues to move. It does not need a soul. For Descartes, the soul is no longer the principle or the cause of the body's movements.

This leads to two main ideas for posthumanism[14]: "the concrete body has no significance for the existence of a being" and "since the human mind can exist without the body, the continued existence of the human being as a mere 'thinking thing' in the computer is guaranteed for all eternity."[15] For the posthumanists, therefore, the immortality of the human being is guaranteed by the immortality of the mind.

Thus, as we have seen, the following pattern emerges: humanism sought to realize the greatness of the human being; transhumanism focuses on the human body; and posthumanism gives primacy to the mind. Humanism strengthened human reason and morality. The body was taken into account, but it was not the center of attention. Transhumanism, on the other hand, seeks to perfect the body in order to create a longer life. The body is to be technologically enhanced, and there are early attempts to transcend society's gender binary. In posthumanism, the body recedes into the background; it becomes arbitrarily interchangeable and no longer really needed as such. Posthumanism wants the complete abstraction of the human being in a virtual existence, as this is the only way to guarantee the immortality of the human being through the immortality of the mind. Mortality is to be completely overcome technically.

This leads to different goals for transhumanism and posthumanism. Transhumanism[16] is human-centered and individualistic, focusing on human continuity and valuing humanistic values. Posthumanism, on the other hand, goes beyond the human-centered, focuses on relationships and the deconstruction of human identity, and values a paradigm shift. Transhumanism is primarily concerned with enhancing and improving the human condition. The goal is to transcend current human limitations and create "transhumans"—individuals with enhanced abilities beyond what is currently considered normal for humans. Transhumanism also emphasizes individual choice and autonomy and aims to maintain a connection with humanity's evolutionary history while seeking to develop and improve it in the spirit of humanistic values. Posthumanism questions the centrality of humans in the world and explores the possibility of a future in which humans coexist and interact with non-human entities. Posthumanism considers the interconnections and relationships between humans, machines, and the environment. It challenges traditional notions of human identity and subjectivity and represents a paradigm shift in thinking about the future of humanity and technology.

But both agree that humans have been using technology throughout history to improve their lives and the embracing technological advancements is a natural extension of this process. The book takes this very dictum and traces this process through history, from the automaton and humanism to posthumanism. The book is divided into four parts that attempt to trace the process from humanism to posthumanism, not necessarily chronologically, but conceptually. As a result, some chapters follow a certain chronology as ideas develop over time, while others take up similar concepts and show us how they have developed over time.

In Part I, *The "Evolution" of Human Physiology: the First Automata and Humanism*, mechanism and materialism of the sixteenth and seventeenth centuries are at the core. In Chapter 2, Michael Stolberg underlines that physicians of the sixteenth century already, long before the rise of "mechanical philosophy," resorted to a range of mechanical and, in particular, hydraulic notions and images, to explain specific physiological and pathological phenomena in the body. Renaissance anatomists, in turn, devoted considerable attention to the precise anatomy of the arteries and veins and of other vessels and ducts which channeled and guided the flow of blood and spirits and of other fluids and vapors in the body. This new way of thinking has led to major advances in medicine and thus in the health and capabilities of human beings. In Chapter 3, Thomas A. Murphy emphasizes a new understanding of the concept of physiology. The chapter investigates the evolving use of the term *physiologia* ("physiology") in Antiquity and in the sixteenth and seventeenth centuries. In the early modern period, there was a fundamental change in the ways the notions of "nature" and "thought" were conceptualized, which led to a split of these two notions. Understanding the ramifications of this split sheds light on the foundations of contemporary transhumanism. Charles Wolfe, in Chapter 4, questions the concept of materialism and vitalism in the early modern period. Materialism, and its approach to the body, are often presented as "mechanistic": as signifying that the properties unique to organic, living, embodied agents are reduced to or specified as mechanistically specifiable properties that characterize matter as a whole. This is an inappropriate view of the human body: the materialist project, therefore, also focuses on "life" and embodiment, on vitalism, and on the uniqueness of organisms. This chapter, in conjunction with the preceding two chapters, establishes the conceptual framework that enabled early modern philosophers and intellectuals to address the notion of a distinct human nature, whether through its enhancement or transformation.

After discussing the core elements of the sixteenth- and seventeenth-century natural philosophy, i.e., physiology, mechanism, and materialism, the book moves on to the philosopher who combined these elements in his philosophy: Descartes. In Chapter 5, Jil Muller analyses Descartes' famous metaphor that opens the *Treatise of Man*: the human body as a machine created by God. Descartes' dualism allowed him to think about automata and machines and to consider the human body as functioning entirely mechanically, without the involvement of the soul. The chapter offers a precise analysis of the metaphor and of the functioning and role of some other devices and machines in Descartes' philosophy. This helps to understand the development of mechanics in the seventeenth century and the idea of automata, which would be perfected in the following century, among others, by Jacques Vaucanson's automata show. In Chapter 6, Albert Kozik emphasizes the connection between Vaucauson's automata shows and the topography of entertainment in eighteenth-century Paris. Vaucanson consciously referred to the visual language of *fêtes champêtres* and the conventions of garden amusement popular at the time to make his famous automata appear more life-like in the eyes of the beholders. Extending the park's space into the venue where the mechanical shows were staged helped the inventor to skillfully tune the expectations of the viewers to the possibilities of mechanical arts, so that they might easily believe in the ingenious mimetic qualities of the automata. In this way, Vaucanson wanted to make his automata more human-like, so that they could be a model of an improved human being, improved both physically and intellectually.

Part II, *The Foundations for Transhumanism and Posthumanism*, attempts to cross the boundaries of humanist thought into transhumanism and even posthumanism. As we have seen, the ideas of transhumanism and posthumanism are not new; they have their origins in humanism itself, and even, for some reasons, in Descartes's dualism. His dualism is at the center of the first article in this section. In Chapter 7, Richard Blum examines the historical process which enables the idea of robots by analyzing the mind-body problem. Dualism of spiritual or mental functions and the body is the subject of present day panpsychic attempts at establishing monism. The purpose is to make plausible the attempt to describe the mental and the psychic in human affairs by focusing on the physical, that is, the anatomical and empirical workings of the mind, which finds its origins in the humanist thought.

But it is not only in philosophy but also in art and medicine that the first seeds of the ideas of trans- and posthumanism can be found, coming from humanism. In art, the first steps were taken to depict the human body in a distorted way, to separate it from its flesh, its materiality, so to speak. Nicole Howard, in Chapter 8, takes up Giovanni Battista Bracelli's (d. 1649) *Bizzarie di Varie Figure* which showed distorted and fantastical bodies, who were firmly rooted in artistic traditions informed by academic and popular culture of the time. In the postures and faces of his figures, the characters of the *Commedia dell'arte* appear in exaggerated form, from the *giullari* (minstrels) to the *buffoni* (clowns). And through Bracelli's lens, the naturalistic bodies of Jacques Callot's *Varie Figure* are freed from their flesh, then contorted in ways reminiscent of popular puppetry of the time. This is another manner, alongside automata, of representing eccentric bodies, and it opens the way to thinking about mutants and cyborgs.

Sarah Carvallo, in Chapter 9, presents a genealogy of vegetative life, which until the eighteenth century played a major role in medicine and philosophy for understanding living beings but then became instead an indicator of non-human or inhuman life. To this day, the vegetative reference has become less and less important for understanding human life. By focusing on levels of consciousness and neglecting the common forms of vegetative life shared by all living beings, contemporary medicine is setting the stage for transhumanism, which overvalues consciousness and intelligence and consequently claims to emancipate human beings from their bodies.

In Part III, *The Quest for Human Perfection in Transhumanism*, we will address two core principles of transhumanism: increasing human knowledge and prolonging life. One important idea for transhumanism is to enhance human intelligence and cognitive abilities, potentially leading to increased learning capabilities, memory enhancements, and improved problem-solving skills. Chapter 10 by Guido Giglioni hints at this idea by looking at the complex history of the interplay of thinking and tooling within the universe of the human mind by dwelling on the concept of natural organs of knowledge and action. The chapter argues that the action of thinking should be reincorporated into a new ontology of active matter. This perspective is also explored in Chapter 11, focusing on Spinoza. Emanuele Costa explores how far the embodiment, the self-determined ontology, and the anti-anthropocentric stance in Spinoza's philosophy might open the way for the transhumanist paradigm, based on his metaphysics. He argues that a Spinozist account of human nature is beneficial to transhumanism insofar as it provides two crucial components: first, an open ontology capable of capturing change and individualizing the essence of human beings; second, a criterion for the limitation of mutation that operates as a safeguard against the disaggregation of the human whole.

Transhumanists are also interested in extending human lifespans and improving health to achieve a longer and healthier life. They see aging and mortality as problems to be overcome through scientific means. The issues of aging and mortality have been present throughout history, and they remain prominent in the present day. In Chapter 12, Hannah Elmer examines cases of later medieval reanimation in order to consider boundaries of human life, the circumstances in which they are made visible, and means by which they could be crossed. She discusses reanimation and presents the different purposes it might serve and the different ways in which it might occur. This chapter, thus, shows that the idea of allowing people to live longer is not a new idea of transhumanism; it is just being taken up again and implemented in a different way.

Jessica Lombard, in Chapter 13, takes up this fundamental idea for transhumanism: the desire for immortality, and integrates it into Heidegger's Phenomenology. Even if the challenging topics of aging and death are usually seen as psychological, social, or political issues, transhumanist authors and associations envision how the human being could transcend his mortality through technological means, and how the transhuman could be immortal or amortal. However, for transhumanism, mortality remains a biological matter that is not easily overcome. Therefore, in a Heideggerian perspective, man should rethink his existence so that death does not condition his life.

Finally, Part IV, *The Transcendence of the Human: A Posthumanist Dream*, focuses on the idea of annihilating humans and transcending the human by inventing new models of existing beings. Posthumanists envision a future where humans evolve beyond their current biological form, leading to the emergence of posthumans—beings with greatly augmented abilities and potentially different forms of existence. Chapter 14 by Audrey Borowski reflects on a world without human mankind. Deleuze, drawing on the philosophies of Spinoza and Nietzsche, alongside the philosopher Felix Guattari, reimagined a more immanent form of philosophy and speculative anthropology based on "assemblages." Bodies were mutable, plastic and in possession of an infinity of virtualities.

The body, by combining itself with machines and computers, could challenge "common sense," produce unfamiliar visions of reality, and create new worlds and modes of existence which would overthrow and flatten existing hierarchies. Ramah Khazam, in Chapter 15, questions the decentering of the human being. As a branch of speculative realism, object-oriented ontology (OOO) (Harman, Bogost) argues that there is always more to reality than humans are able to grasp. Likewise focusing on decentering the human, the new materialists view the world in terms of continuities, paying attention to the entanglements between the subject and the world beyond it. Both OOO and new materialism put humans and non-humans on a par, but whereas OOO asserts the autonomy of all objects, new materialism upholds relationality and subject-object entanglements. Emanuele Clarizio, in Chapter 16, focuses on the living being which is not only organized but also autonomous. Conceiving the living as autonomous allows the relationship between life and technology to be reversed. This overturning leads from the figure of Automaton to that of Cyborg, which presents a strict continuity between organic and inorganic organs. However, Cyborg is no more autonomous than Automaton: an organism whose functioning is regulated by technology is in fact heteronomous. And finally, Chapter 17 by Thierry Hoquet offers a comparison between *Mutants* and *Cyborgs*. Both figures are analyzed in respect to the concept of "organology"—the general theory of instruments, for which tools are just external organs and organs internal tools. In this tradition, mutants and cyborgs embody two ways to achieve human enhancement: mutants as spontaneously generated new functions and cyborgs as the product of conscious and intentional design. Both cyborgs and mutants imply that "natural" and "artificial" organs have been designed by various instances (divine or human engineers or natural selection) and that they might be improved.

We have seen that in the history of medicine, philosophy, and the arts, there have been various approaches to thinking about or even creating automata, all of which had the basic ideas of today's transhumanism or posthumanism at their core. In the history of medicine, automata have played a role in the development of medical devices and instruments. An important example is the creation of medical automata to simulate human anatomy for teaching and surgical training. These automata were early attempts to provide physicians with realistic models to practice surgical techniques without operating on live patients. Automata have also long been the subject of philosophical reflection. Ancient philosophers such as Aristotle and the Stoics pondered the idea of artificial beings and the potential for machines to mimic human behavior. In the seventeenth century, René Descartes discussed automata as part of his views on animal behavior and the nature of consciousness, raising questions about the distinction between living beings and machines. Finally, automata have had a profound impact on the arts, inspiring creativity and technical innovation. In the Middle Ages and Renaissance, automata were popular with the European nobility as elaborate mechanical toys and entertainment. These early automata were often featured in court performances and festivities, showcasing the skills of skilled craftsmen and engineers. Designed to mimic human actions, automata have a fascinating history that intersects with various fields, including medicine, philosophy, and art, and culminates in today's transhumanism and posthumanism.

The aim of this book is to trace the evolution of the idea of trans- or posthumanism, that began with the ancient desire to prolong life and strengthen human nature, then to perfect it through technology. Through the idea of the automaton and the transhuman, the book raises the question of the human being, its nature and its limits, in the vast field of the humanities, including medicine, philosophy, literature, and the arts. By asking this question again, without prejudice but also without judgment, the book traces the history of a question that philosophy has never stopped asking itself: what are we? What is our nature and our destiny or purpose?

Notes

1. See Léa Barbisan, "Eccentric Bodies: From Phenomenology to Marxism- Walter Benjamin's Reflections on Embodiment," *Anthropology & Materialism* (2017): 1–14, p. 6: "Edgar Allan Poe's short-story *The Man of the Crowd*, to which Benjamin devotes a chapter of his essay on Baudelaire, describes from an overhanging standpoint the gesticulations of anonymous passers-by, who seem to behave like semi-defective automata, mechanically repeating the same gestures. [...] The human physis is trained to resemble the infallible body of the robot, animated by no specific fears or desires, yet always ready to react. Here also the primary eccentricity of the human body is reactivated: its ability to quickly adapt to new tasks relies on the splitting of the organic unity into a series of functioning entities."

2. See Oliver Krüger, *Virtualität und Unsterblichkeit, Gott, Evolution und die Singularität im Post- und Transhumanismus* (Freiburg i.Br./Berlin/Wien: Rombach Verlag KG, 2019).
3. See Krüger ibid., 75–76.
4. See for example: Roberto Manzocco, *Transhumanism—Engineering the Human Condition: History, Philosophy and Current Status* (Cham: Springer International Publishing, 2019).
5. See Rosi Braidotti, *The posthuman* (Cambridge: Polity, 2013).
6. See for example: Jil Muller, "Ambroise Paré and René Descartes on sensation in amputated limbs," in *Descartes and Medicine, Problems, Responses and Survival of Cartesian Discipline*, ed. F. Baldassari (Turnhout: Brepols Publisher, 2023), 81–97.
7. See for example: https://cs.stanford.edu/people/eroberts/courses/soco/projects/2004-05/automata-theory/basics.html; and https://iep.utm.edu/rene-descartes-scientific-method/; but also Giogio Marchetti, "Consciousness: a unique way of processing information," *Cogn Process* 19 (2018): 435–464.
8. See Joseph Campana and Scott Maisano (eds), *RenaissancePosthumanism* (New York: Fordham University Press, 2016).
9. See Jil Muller's article in this book.
10. http://hoaxes.org/archive/permalink/the_great_chess_automaton.
11. See his work on L'homme Machine: Julien Offray de La Mettrie, *L'Homme Machine* [1748] (Paris: Fayard/Mille et Une nuits, 2000).
12. See for example: Bernhard Irrgang, *Posthumanes Menschsein?: Künstliche Intelligenz, Cyberspace, Roboter, Cyborgs und Designer-Menschen—Anthropologie des künstlichen Menschen im 21. Jahrhundert* (Stuttgart: Steiner, 2005).
13. See for example: Francesca Ferrando, *Philosophical Posthumanism* (London: Bloomsbury Academic, 2019).
14. See for example: Janina Loh, *Trans- und Posthumanismus zur Einführung* (Hamburg: Junius, 2020); Stefan Herbrechter, *Posthumanismus Eine kritische Einführung* (Darmstadt: WBG, 2012); Stefan Herbrechter, Ivan Callus, Manuela Rossini, Marija Grech, Megen de Bruin-Molé and Christopher John Müller (eds.), *Palgrave Handbook of Critical Posthumanism* (Switzerland: Palgrave Macmillan Cham, 2022).

15. Oliver Krüger, op. cit., 180.
16. See for example: Newton Lee (ed.), *The Transhumanism Handbook* (Switzerland: Springer Cham, 2019), Robert Frodeman, *Transhumanism, Nature, and the Ends of Science* (New York: Routledge, 2019) and Stefan Lorenz Sorgner, *On Transhumanism*, transl. Spencer Hawkins (University Park USA: Penn State University Press, 2020).

PART I

The "Evolution" of Human Physiology: the First Automata and Humanism

CHAPTER 2

Mechanical and Hydraulic Elements in Sixteenth-Century Medicine

Michael Stolberg

Physicians and natural philosophers in the sixteenth century were still firmly rooted in the traditional theory of the soul. The vegetative, the vital and, in humans, the rational soul with their respective faculties accounted for the bodily movements and for the many complex processes that took place inside the body, from the concoction and assimilation of food to sensory perception and intellect.[1] This theory was far removed from the idea that the body was governed by the laws of mechanics or hydraulics and could be understood as a machine, and the challenge for the new mechanical philosophy in the seventeenth century was to offer a plausible alternative explanation for all those movements and processes in the body which the Aristotelian-Galenist tradition attributed to the workings of the immaterial soul.

According to this tradition, the soul's rule over the body depended on material instruments, however, and, in particular, on the various organs,

M. Stolberg (✉)
University of Würzburg, Würzburg, Germany
e-mail: michael.stolberg@uni-wuerzburg.de

J. Muller (ed.), *Automata, Cyborgs, and Mutants*, Palgrave Studies in Medieval and Early Modern Medicine,
https://doi.org/10.1007/978-3-031-93201-4_2

on the innate heat, and on the spirits. The organs were endowed with peculiar properties and performed specific tasks. The stomach concocted food and drink, the liver generated nutritious blood, the kidneys attracted and filtered the urine, the brain was the site of intellect, imagination, and memory, and so forth. The innate heat with its substrate, the radical moisture, resided as *calidum innatum* in the heart. It was the source of movement and change in the body and enlivened the whole organism. The radical moisture was consumed with time and in the end it could no longer sustain the flame of life of the aging organism, which explained the limited natural life span of humans and other animate beings (which was difficult to account for in mechanical terms).[2] The vital spirits, finally, were generated by the innate heat in the heart and some of them were further elaborated in the brain and transformed into animal spirits. The spirits were taken to be hot and extremely mobile, which allowed them to rapidly convey sensory perception in one direction and the commands of the soul in the other, for example, for the voluntary movement of a limb.

Not only the ability of the spirits to perform their tasks in the service of the soul but also many other processes in the body crucially depended on movement of matter inside the body. In principle, early modern physicians and natural philosophers described the body as permeable and believed that matter could change its place inside the body almost ad libitum, disregarding anatomical barriers like the diaphragm. The crucial life-sustaining movements, however, were taken to be largely confined to pre-established channels. The hot vital spirits moved from the heart through the arteries to the various parts of the bodies they enlivened and back again to the heart to convey sensations. The blood took its much slower course from the liver through the veins to the various parts of the body it nourished. Food and air reached the inside of the body through the esophagus and the trachea, respectively. Last but not least, the body constantly absorbed and produced impure, potentially harmful matter. This matter had to be evacuated through appropriate, predestined excretory pathways, such as the intestines, the ureters and the urethra, the uterus, the pores of the skin, the mouth, the ears, and the eyes.

All these movements could be hindered or totally obstructed, the same way in which the flow of water in brooks, canals, or water tubes could be slowed down or blocked entirely, accidentally, for example by debris, or on purpose, by dams and locks. It is here that mechanical and more specifically hydraulic notions and images could find an important place.[3] As I will argue in this chapter, physicians in the sixteenth century resorted to an impressively wide repertoire of such notions and images, in fact, and some of them were at the very heart of the contemporary understanding of the body and its diseases, in general.[4]

In what follows, I will first outline some major mechanical or, more precisely, hydraulic and pneumatic elements in sixteenth-century physiology and pathology. In the second part, I will then turn to the place of mechanical concepts and images in sixteenth-century anatomy, some of which eventually played a key role in the rise of iatromechanics and the new mechanical philosophy.

Physiology

The use of "mechanical", "hydraulic", and "pneumatic" concepts and images in explaining certain natural operations of the healthy body, goes back to Antiquity.[5] A well-known example, which can be found in Aristotle and other ancient authors already, is the comparison of the lungs with the bellows of a blacksmith. When the bellows—or the lungs—expanded they filled with outside air, which they ejected again when they were compressed. Along similar lines Erasistratos described the movements of the heart. He has been praised as the discoverer of the cardiac valves, which is a highly implausible claim since these valves are easily seen in the hearts of cattle or pigs and any butcher would have known them. His account of the action of the heart, as reported by Galen (De placitis 6.6, Kühn V, 548–50),[6] was detailed, however, and it resembled to some degree that of a water pump that was used to fight fires, as described by Philo of Byzantium, with an alternating opening and closing of valves.[7]

Galen agreed with Erasistratos that the tricuspid valve (between the right atrium and the right ventricle) prevented a movement of the blood from the right ventricle into the vena cava. Erasistratos' findings thus supported Galen's view that the veins which brought the blood to the various parts of the body originated from the liver and not from the heart. By contrast, the valves were built and arranged in such a way that the spirits could be ejected into the aorta and the blood into the lungs. According to Galen, Erasistratos attributed the ejection of the spirits and the blood out of the heart to the compression or contraction of the heart, as in blacksmith's bellows. The valves let the blood and the spirits pass through on their way out but prevented a reflux. Vice versa the valves that could be found where air and blood entered the heart, allowed the movement into the heart, when the heart dilated, but closed when the heart contracted. As we will see, valves and their ability to prevent reflux came to play an even more important role, when Renaissance anatomists identified various new valves in the body. Famously, the discovery of the venous valves paved the way for Harvey's theory of the blood circulation, which, in turn, served as a major building block for mechanical accounts of the body.[8]

An illustrative example for a "pneumatic" rather than "hydraulic" understanding can be found in the Galenist account of certain muscular movements. While muscular movement was commonly ascribed to the ability of muscles to contract, there were two types of muscular movement that, according to Galen, could not be explained in this way: stretching out the tongue and erection (which was also commonly ascribed to muscular action). In both cases, the muscles seemed to expand rather than contract. Galen attributed this expansion in quite literally "pneumatic" terms to an influx of spirits or *pneuma* into the hollow channels or ducts in these organs.[9] Variants of the "pneumatic" theory of erection were still widely accepted in the early modern period, when physicians and vernacular recipe books frequently recommended "inflating" substances and foodstuffs such as peas, beans, and barley to treat male impotence.[10]

A more strictly "mechanical" imagery emerges when some Hellenistic authors compared the physical movements of the body and its limbs to those of a puppet. They found that "neura"—the word from which the modern term "nerves" derives originally referred to the tendons or sinews—were like the strings that were attached to the different parts of a puppet, which was tellingly called a "neurospaston".[11] I have so far not found sixteenth-century physicians, however, who resorted to this imagery.

Pathology

Mechanical and hydraulic elements played an even more important role in sixteenth-century pathology. This was due to the great importance physicians and laypersons alike attributed to "obstructions" of the natural movement of spirits, humors, blood, and excrements in the body and across its borders.

"Obstructions" famously already held a conspicuous place in the Hippocratic "On the sacred disease". Rather than claiming some divine or supernatural influence, the author of the treatise opted for a strictly physical explanation: the characteristic seizures resulted from an obstruction of the efflux of phlegm or mucus through the openings in the base of the skull. As a result, the phlegm or mucus accumulated in the skull—and caused the seizures.[12]

Variations of this explanatory model were still widely accepted in the sixteenth century, when it offered a convenient explanation for hemiplegia as well as for epilepsy. It was common knowledge that the animal spirits which conveyed the commands of the souls to the muscles had to pass through the narrow openings at the base of the skull and that these openings could be blocked by mucus or some other viscous morbid matter.[13] Hemiplegia, as the famous anatomist Gabrielle Falloppia explained to his students along these lines, occurred when the movement of the animal spirits was blocked entirely on the affected side. Epileptic fits, by contrast, were due to a partial obstruction, which the animal spirits sought to overcome, thereby causing the characteristic bizarre movements of epilepsy.[14]

With other diseases the obstruction and its material cause even manifested themselves directly to the senses. A prime example was bladder stones, a complaint we find quite commonly reported in the sixteenth century. Many patients intermittently could not let water and then excreted stones of different sizes, often accompanied by intense pain, which left no doubt that stones had blocked the natural pathways through which the urine had to exit. When, in the course of the sixteenth century, post-mortems of deceased patients came to be performed more commonly, stones were frequently also found in the kidneys, in the gallbladder and the biliary duct, and sometimes also in other parts of the body.[15]

Some post-mortems also brought to light a blockage of larger passages inside the human body, especially by tumors.[16] An early example is Antonio Benivieni's account, of the post-mortem of a man with a hardened, tumorous narrowing of the colon that almost totally blocked the passage of the feces.[17]

In ordinary medical practice, "obstructions" were most commonly diagnosed in solid organs. They could not usually be seen in post-mortems, except when the organ had hardened or shrunk but they were taken to account for numerous illnesses. Frequently the liver was thought to be affected. "Your liver is obstructed", Georg Handsch explained to one of his patients, for example; "this is the root of all your diseases".[18] The ultimate cause of liver obstructions was often found in the stomach:

when it concocted the food insufficiently, in the first step of "concoction", it produced slimy, viscous matter. When this slimy, viscous matter was transmitted to the liver to be turned into blood in the second step of concoction, it could easily obstruct the liver. On top of that, the stagnating, viscous fluid in the liver could putrefy, generating excessive heat, which caused some of the fluid in the liver to evaporate. The loss of fluid made the humors and the blood thicken further and inaugurated a vicious circle. A cold stomach, which poorly concocted the food, in combination with a hot liver that resulted from it, was one of the most common diagnoses we found in case reports and *consilia* at the time.[19]

Inevitably, the blood, which the liver generated from the slimy, sticky matter provided by the stomach and delivered to the rest of the body, was likewise sticky and viscous and prone to clog and obstruct vessels. "I said the stomach does not digest well", Handsch reported how he communicated his diagnosis to a sick woman. "For that reason obstructed blood comes into the vessels und causes pain".[20]

Moreover, just like mud, dirt, and dung could attach to canals and water pipes and clog and block them, the viscous, sticky blood could adhere to the walls of the vessels. Paracelsians could conveniently link this notion to that of "tartar". The term was commonly used for the solid matter that frequently settled at the bottom of wine barrels but it epitomized, in Paracelsian medicine, what Galenist physicians termed "peccant" or "morbific" matter. Tartar in the body could not only form stones in the gall-bladder and in the kidneys, as Johannes Willenbroch, a personal physician with Paracelsian leanings at the Habsburg court, explained, for example. Just like limescale narrowed and ultimately plugged water pipes, the tartar in the human body stuck to the walls of the vessels. Another vicious circle was thus inaugurated: as the bloodflow slowed down, more matter attached to the wall of the vessel narrowing it further, with the ultimate risk of a total obstruction.[21]

An obstruction of organs and vessels by sticky and hardening matter could also interfere with the excretion of harmful, morbid matter. This was particularly important for the understanding of female diseases. Obstructions of the uterus and the resulting reduction or indeed "suppression" of the periods was a very common diagnosis in women in the sixteenth century. Women's health was widely taken to rest on the regular, monthly excretion of impure, morbid matter that constantly accumulated in their bodies, due to the inferior strength of their innate heat and to the fact that they often lacked the physical exertion, which, in men, helped excrete the matter with sweat. When this matter stagnated in and around the uterus—the "cloaca" of the female body, as some writers called it in rather unflattering terms—it corrupted further and could ultimately putrefy, with devastating consequences. The menstrual matter could regurgitate into the veins, mix with the blood and harm other parts of the body. The treatment of an obstructed uterus and of the ensuing reduction or total suppression of the menstrual flow therefore relied on medicines that promoted the evacuation of the menstrual blood through its natural pathway as well as on getting rid of the accumulating morbid matter by other means.[22]

The "suffocation of the mother", a common diagnosis at the time,[23] was also frequently attributed to an obstruction of the uterus, in the sixteenth century. Vapors were thought to ascend from the accumulating corrupt or putrefying menstrual matter. On their rise upwards, toward the diaphragm, the chest, the neck, and the head, they caused the characteristic symptoms of uterine suffocation, such as hard breathing and a feeling of pressure or narrowing in the chest. Some women even lost their voice and could only gesture. Others reported a burning sensation in the heart or felt their heart beating faster than usually.[24] When the vapors reached the brain, the symptoms could be dramatic. We have reports of women who suddenly collapsed in church and fell to the ground during mass, screaming loudly and eventually fainting.[25] Others lay in bed as if petrified or indeed looking dead, and their husbands sought to wake them up in vain.[26]

Anatomy

In the sixteenth century, mechanical and, in particular, hydraulic concepts and images thus already played a major—and widely underestimated—role in explaining pathological processes and especially the obstructions, to which many diseases were attributed. The rise of anatomy to a major field of empirical enquiry in the sixteenth century fostered a growing interest in mechanical or hydraulic elements also in the healthy human body. As Domenico Bertoloni Meli has shown for the seventeenth century, interest in the anatomical structure of the individual parts or organs and the desire to find mechanical explanations for bodily processes mutually reinforced each other, and the synergy grew even more powerful with the systematic use of the microscope.[27] Important first steps into this direction can already be identified in sixteenth-century Galenic medicine.

As we have seen, the sick body, in particular, was described as remarkably permeable but the natural movement of the humors and the vital spirits as well as that of the natural evacuations was believed to be largely confined to specific pathways in the body: the arteries, the veins, the biliary duct, the ureters and the urethra, the intestines, and the pores of the skin, respectively. They channeled the movement of the fluid.

Renaissance anatomists went to considerable lengths to identify the precise course of the numerous vessels in the body. Their findings were also of considerable practical relevance, not only for the diagnosis and treatment of injuries. In the case of the veins, this knowledge was crucial for the choice of a suitable vein for bloodletting, a principal element of therapy in many diseases. There was a long-standing tradition of visual representations of the numerous veins on the surface of the body that could be found and cut. But only a detailed examination of the anatomy of the veins inside the body could ascertain that a given vein really "drained" the area in which the peccant matter was believed to reside. It was therefore not just a minor anatomical detail, when Gabrielle Falloppia corrected the account, which Vesalius had given of the branching of the veins from the upper *vena cava*. Quite possibly based on a drawing Falloppia showed in his lectures, one of his students even illustrated his notes with a little sketch, which showed the anatomy of these veins "secundum Vesalium" in the lower half and "secundum veritatem"—according to truth—in the upper half.[28]

Their investigations also led some anatomists to discover new, unknown vessels that channeled specific fluids and as such could be blocked as well. Thus, in the 1550 s, some 70 years before Gaspare Aselli found the "lacteals"—and has since then been credited with the discovery—Gabrielle Falloppia described vessels in the abdominal cavity that were filled with a whitish-yellowish and somewhat oily, bitter fluid that seemed to flow toward the area of the pancreas and were connected to some "glandulae"—presumably the lymph nodes of modern terminology. Probably around the same time, Bartolommeo Eustachi discovered the thoracic duct through which the content of the lacteals as well as the lymphatic fluid from the lower parts of the body are ultimately discharged into the large veins.[29]

The anatomists devoted particular attention to valvular structures and their hydraulic function in the various organs and vessels. In 1550 s-Padua, medical students not only were shown the cardiac valves, which, the anatomist explained, prevented a reflux of the blood.[30] They also heard about a valve-like structure at the site where the ureters inserted into the bladder, which prevented a reflux of urine toward the kidneys. By 1553, at the latest, Gabrielle Falloppia also showed his students the valve-like structure at the end of the small intestine, in the area of the *caecum*. This "Bauhinian valve", as it is commonly called after the Swiss physician who later rediscovered it, allowed the feces to flow into the colon from the small intestines but prevented their reflux into the small intestines. Falloppia demonstrated the valvular function, by filling in water first from the side of the small intestines and showing how it flowed into the colon and then from the side of the colon, from which it did not pass into the small intestines. This finding was again important also for therapy: clysters, a fairly common means for applying medicines at the time, clearly could not pass further up into the small intestines or indeed the stomach, as the Galenist tradition claimed.[31]

The valves in the veins eventually became a crucial building stone for William Harvey's theory of blood circulation, which in turn called for a major reassessment of Galenic physiology.[32] Harvey probably learned about the valves in Padua from Girolamo Fabrizi d'Acquapendente and may have believed Fabrizi's claim that he discovered them. However, we know from various contemporary authors that Giovanni Battista Canani in Ferrara already showed the valves in the vena azygos in the 1540 s; the vena azygos is a major vein in the thoracic cavity which branches off from (or according to Galenic theory inserts into) the vena cava. Amatus Lusitanus, who worked in Ferrara at the time documented this finding in his widely read *Curationes*.[33] Andreas Vesalius, on his part, reported that Canani told him, when they were both in Regensburg to treat a famous patient, that he had found such valves not only in the vena azygos but also in other veins in the body.[34]

Canani was not the only one. Jacques Dubois, in his introduction into anatomy that was published shortly after his death, in 1555, offered a fairly precise description of venous valves, of a "membrana epiphysis", he had found at the opening of the azygos vein but also in other places, such as the beginning of the great jugular, the brachial, and the leg veins. He compared these structures to the "membranes" that closed the orifices of the heart.[35]

In sum, sixteenth-century anatomists were familiar with valves of different kinds, long before Fabrizi and Harvey. Somewhat paradoxically, this familiarity may have made it more difficult for them, however, to understand the function of the venous valves. The natural flow of the venous blood was "known" to move from the liver to the other parts of the body. But venous valves clearly did not serve to avoid a "reflux" toward the liver. They were actually in the way. Fabrizi himself acknowledged that it was surprising that the free flow of the blood in the veins should be obstructed by these structures. He concluded that these structures did not actually serve as valves. They did not prevent a reflux. They only slowed down the movement of the blood toward the extremities, which, he thought, helped prevent a rapid, sudden influx of large volumes of blood into the extremities, which would make the limbs swell, while other parts of the body would be deprived of their nourishment.[36]

Conclusion

We can identify a whole range of mechanical and more specifically hydraulic elements in sixteenth-century anatomy, physiology, and pathology. We do not have to search for them in occasional references in the works of a few writers. They were ubiquitous. Physicians even resorted to mechanical and hydraulic images at the bedside, in order to explain to the patients and their families what was happening inside the body. Laypersons undoubtedly could relate to these images and notions. They made sense to them. Except for mills, they may not have been familiar with larger "machines", but they knew from daily experience that the flow of water and sewage in a brook or canal could easily be slowed down or totally blocked, that fluids became thicker when they were boiled for some time over a kitchen-stove, and that limestone could adhere to the walls of pots and vessels.

These mechanical and hydraulic notions and images would also have been familiar to the natural philosophers and physicians who began to propagate a new, "mechanical" medicine and philosophy. They offered building material they could use in order to construct new "mechanical" theories and they may well have made "mechanical" explanations of the whole body as a whole more acceptable. It is certainly worth noting, in this context, that many members of the Royal Society, a focal point of the new "mechanical" philosophy were trained physicians.[37]

It also has become clear, however, that mechanical notions and images in sixteenth-century anatomy and medicine referred only to specific processes and structures in the body. They were not meant to provide an overarching model for human (and animal) physiology as such. The same goes for corpuscular theories that later stood at the center of some of the new "mechanical" philosophies of the seventeenth century. Sixteenth-century physicians were familiar with corpuscular notions. The term "atom" was part of the medical vocabulary, if only to describe tiny corpuscles in the urine.[38] By the end of the sixteenth century, Daniel Sennert famously even proposed an atomistic theory of the human body. Sennert's "atomism" rested on the notion of a hierarchy of forms, however, with the rational soul as the highest form in the human body. His particles were "a-tomoi", that is indivisible, only in the sense that there was a limit beyond which the superior form could no longer govern the matter. When they were further divided, the inferior forms or indeed the elements would take over.[39] By contrast, Sennert and Galenists, in general, considered the idea—which they often attributed to Epicurus—as implausible or indeed as blasphemy that the complex faculties of the human (and animal) body could result from the mere movement of little particles. For them, a body that was ruled by mechanical principles rather than by the soul and its instruments was still virtually unthinkable.

Notes

1. See Katharine Park, "The Organic Soul," in *The Cambridge History of Renaissance Philosophy*, ed. by Charles B. Schmitt et alii (Cambridge: Cambridge University Press, 1988), 464–484; Gad Freudenthal, *Aristotle's Theory of Material Substance: Heat and Pneuma, Form and Soul* (Oxford: Clarendon, 1995); Hiro Hirai, *Medical Humanism and Natural Philosophy: Renaissance Debates on Matter, Life and the Soul* (Leiden: Brill, 2011); on the trend towards a more somatic, anatomically based notion of the soul see Fabrizio Bigotti, *Physiology of the Soul. Mind, Body and Matter in the Galenic Tradition of the Late Renaissance (1550–1630)* (Turnhout: Brepols 2019).
2. See Peter H. Niebyl, "Old Age, Fever and the Lamp Metaphor," *Journal of the History of Medicine and Allied Sciences*, 26 (1971): 351–368; Thomas S. Hall, "Life, Death and the Radical Moisture. A Study of Thematic Pattern in Medieval Medical Theory," *Clio Medica* 6 (1971): 3–23; Michael McVaugh, "The *humidum radicale* in Thirteenth-Century Medicine," *Traditio* 30 (1974): 259–283; Michael Stolberg, "Die Lehre vom 'calor innatus' im lateinischen 'Canon medicinae' des Avicenna," *Sudhoffs Archiv* 77 (1993): 33–53.

3. Aristotle explicitly referred to puppets or *automata* in a couple of passages in *Generation of Animals* and *On the Movement of Animals* but compared their self-moving capacity only to specific phenomena such as the movement of the limbs and the transmission of the formative powers of the father to the growing embryo via the semen, which depended on an initial impetus which sparked, through a series of intermediate steps, changes and movements that were more complex than the linear trajectory of a stone that left a throwing hand (see Tiberiu Popa, "Mechanism. Ancient Sources," in *The Routledge Handbook of Mechanisms and Mechanical Philosophy*, ed. by Stuart Glennan and Phyllis Illari (London and New York: Routledge, 2018), 13–25, especially 19–21).
4. If we take "mechanical" to mean "corpuscular", because this was the leading paradigm of the new mechanical philosophy, precursors are hard to find (see, e.g., Daniel Garber, "Remarks on the Pre-History of the Mechanical Philosophy," in *The Mechanization of Natural Philosophy*, ed. by Daniel Garber and Sophie Roux (Dordrecht: Springer, 2013), 3–26.). I see no need to take such a narrow and ultimately teleological approach, however.
5. For an excellent overview see Sylvia Berryman, *The Mechanical Hypothesis in Ancient Greek Natural Philosophy* (Cambridge: Cambridge University Press, 2009).
6. See Rudolph E. Siegel, *Galen's System of Physiology and Medicine* (Basel and New York: S. Karger, 1968), 39–40.
7. See Iain M. Lonie, "The Paradoxical Text 'On the Heart'," *Medical History* 17 (1973): 1–15 and 136–153.
8. See ibid.
9. See *Galen on Problematical Movements,* ed. by Vivian Nutton (Cambridge: Cambridge University Press, 2011), 140 (IV, 24).
10. See Pieter van Foreest, "Observatio XVIII. De coitus diminutione et impotentia coeundi," in ibid., *Observationum et curationum medicarum ac chirurgicarum opera omnia, libri XVI-XXVIII* (Frankfurt: Officina Paltheniana, 1634), 585–589.
11. See S. Berryman, *The Mechanical Hypothesis*, op. cit., 210–211.
12. See Owsei Temkin, *The Falling Sickness. A History of Epilepsy from the Greeks to the Beginnings of Modern Neurology,* 2nd ed. (Baltimore and London: Johns Hopkins University Press, 1971); Mervyn J. Eadie and Peter F. Bladin, *A Disease Once Sacred. A*

History of the Medical Understanding of Epilepsy (Eastleigh: John Libbey, 2001).

13. See Österreichische Nationalbibliothek, Vienna (henceforth: ÖNBV), Cod. 11210, fol. 24r (student notes by Georg Handsch).
14. See ibid. and ÖNBV, Cod. 11205, fol. 119r.
15. ÖNBV, Cod. 11210, fol. 207v.
16. See Michael Stolberg, "Post-Mortems, Anatomical Dissections and Humoural Pathology in the Sixteenth and Early Seventeenth Centuries," in *Pathology in Practice. Diseases and Dissections in Early Modern Europe*, ed. by Silvia De Renzi, Marco Bresadola, and Maria Conforti (New York and London: Routledge, 2017), 79–95.
17. See Antonio Benivieni, *De abditis nonnullis ac mirandis morborum et sanationum causis,* ed. by Giorgio Weber (Florence: Olschki, 1994), 95.
18. ÖNBV, Cod. 11205, fol. 78r.
19. For an overview of the prevailing pathological concepts in Renaissance physicians' ordinary medical practice see Michael Stolberg, *Learned Physicians and Ordinary Medical Practice in the Renaissance* (Berlin: De Gruyter 2022), 121–145.
20. ÖNBV, Cod. 11240, fol. 41r.
21. See ÖNBV, Cod. 11183, fol. 163v.
22. See M. Stolberg, *Learned Physicians*, op. cit., 327–333.
23. The Zwickau town physician Hiob Finzel documented hundreds of cases of "suffocatio uteri" in his practice journal (Ratsbibliothek Zwickau, Ms. QQQQ1, Ms. QQQQ1a and Ms. QQQQ1b); see Michael Stolberg, "A Sixteenth-Century Physician and His Patients: The Practice Journal of Hiob Finzel, 1565–1589," *Social History of Medicine* 32 (2019): 221–240.
24. See ÖNBV, Cod. 11183, fol. 286r and fol. 379v; ÖNBV, Cod. 11207, fol. 113v.
25. See ÖNBV, Cod. 11205, fol. 248r.
26. See ÖNBV, Cod. 11183, fol. 197v and ÖNBV, Cod. 11205, fol. 438r.
27. See Domenico Bertoloni Meli, *Mechanism, Experiment, Disease. Marcello Malpighi and Seventeenth-Century Anatomy* (Baltimore: Johns Hopkins University Press, 2011).
28. See Staats- und Universitätsbibliothek, Göttingen (SUBG), Ms. Meibom 20, fol. 133r.

29. See Gabrielle Falloppia, *Expositio in librum Galeni de ossibus. Huic accesserunt observationes anatomicae eiusdem authoris* (Venice: Apud Simonem Galignanum de Karera, 1570), fol. 72v: "In infima parte hepatis sunt quidam parvi meatus, qui desinunt, ac terminantur im [sic] pancraeas [sic] et in glandulas ibi proximas, qui quidem minimi meatus deferunt quendam succum oleaginosum flavum, et tendentem ad aliquam amaritudinem".
30. See ÖNBV, Cod. 11210, fol. 17v, "quomodo sanguis ex eius dilatatione non refluere non potest".
31. See SUBG, Ms. Meibom 20, fol. 140r; ÖNBV, Cod. 11210, fol. 199v.
32. See William Harvey, *Exercitatio anatomica de motu cordis et sanguinis in animalibus* (Frankfurt: Fitzer, 1628).
33. See Amatus Lusitanus, *Curationum medicinalium centuria prima, multiplici variaque rerum cognitione referta.* (Florence: Laurentius Torrentinus, 1551), 258–260.
34. See Andreas Vesalius, *Anatomicarum Gabrielis Falloppii observationum examen* (Venice: de Franciscis, 1564), 83.
35. See Jacques Dubois, *In Hippocratis & Galeni physiologiae partem anatomicam isagoge*, 2nd ed. (Venice: Valgrisius, 1556), fol. 30r.
36. See Girolamo Fabrizi d'Acquapendente, *De venarum ostiolis* (Padua: Ex typographia Laurentii Pasquati, 1603); Jerome J. Bylebyl, "The Medical Side of Harvey's Discovery. The Normal and the Abnormal." in *William Harvey and His Age. The Professional and Social Context of the Discovery of the Circulation*, ed. by Jerome J. Bylebyl (Baltimore and London: Johns Hopkins University Press, 1979), 28–102.
37. See Harold J. Cook, "Victories for Empiricism, Failures for Theory. Medicine and Science in the Seventeenth-Century," in *The Body as Object and Instrument of Knowledge. Embodied Empiricism and Early Modern Science*, ed. by Charles T. Wolfe and Ofer Gal (Dordrecht et al.: Springer, 2010), 9–32, especially 10–11.
38. See e.g., the chapter *De atomis urinae* in Michele Savonarola, "De urinis," in *Practica canonica* (Venice: Valgrisius, 1561), fol. 102v, and Giovanni Argenterio, *De urinis liber* (Heidelberg: Sanctandrea, 1591), 92, who described a characteristic movement of the particles in the urine "more atomorum".

39. See Michael Stolberg, "Das Staunen vor der Schöpfung: ‚Tota substantia', ‚calidum innatum', ‚generatio spontanea' und atomistische Formenlehre bei Daniel Sennert." *Gesnerus* 50 (1993): 48–65.

CHAPTER 3

From Contemplation to Composition: Translating Physiology in the Early Modern Period

Thomas A. Murphy

A 2016 episode of the science-fiction television series *Black Mirror* entitled "San Junipero" offers an instructive illustration of the mind-body relation as reflected in the popular, tech-obsessed imagination. The story centers on two women, Yorkie and Kelly, who meet in a bar and fall in love in (what appears to be) 1987. As we soon discover, however, San Junipero is a virtual reality simulation, powered by a wearable node attached to the foreheads of its human users. In her non-virtual life, Kelly is an elderly widow, dying of terminal cancer in an assisted living facility; Yorkie is paraplegic as the result of a tragic car accident when she was 21. Shortly after their serendipitous meeting in virtual reality, Yorkie informs Kelly of her intention to "pass over" and become a "permanent resident" of San Junipero. Although she will die through assisted suicide in the hospital,

T. A. Murphy (✉)
New College, University of Oxford, Oxford, UK
e-mail: thom.murphy@new.ox.ac.uk

J. Muller (ed.), *Automata, Cyborgs, and Mutants*, Palgrave Studies in Medieval and Early Modern Medicine,
https://doi.org/10.1007/978-3-031-93201-4_3

her virtual self, uploaded to the (fictional) TCKR Systems servers, will live on forever in blissful San Junipero. Yorkie's longtime wish to "pass over" had been prevented thus far by her family, and so Kelly agrees to marry Yorkie and sign the paperwork allowing her to die. Yorkie, or rather the virtual copy of her consciousness, subsequently tries to convince Kelly to become a permanent resident and to live with her forever in the paradise of San Junipero after death. After some debate between the two over whether San Junipero is in fact a "real" place, Kelly decides to "pass over" also and join Yorkie as a "permanent resident," and the two, the viewer is left to presume, live happily ever after in San Junipero.

As the credits roll, the viewer is presented with some of the most philosophically provocative material in the episode, in which three versions of the same event are shown, occurring nearly simultaneously but on separate planes of reality. As the soundtrack thunders out Belinda Carlisle's 1987 "Heaven Is A Place On Earth," the credits are interspersed with Kelly's medically assisted suicide, the insertion of her digitally copied consciousness into the servers in a massive warehouse, and her return as eternally young lover to enjoy the paradise of San Junipero with Yorkie. As the images from the credits unambiguously inform the viewer, this coastal paradise is made possible by a gargantuan warehouse filled with computer servers. That the idea of spending a subscription-based eternity with one's consciousness stored on a computer server in a hedonistic virtual paradise was hailed by critics as upbeat and optimistic, "bittersweet and moving—even hopeful—rather than horrific" suggests as much about the way transhumanist narratives and the promise of wearables have taken root in popular culture as any inherent optimism of the story.[1] What interests me here, however, is not so much great television but rather how these kinds of narratives can feel become believable. Underlying the glossy costumes and nostalgia-inducing soundtrack is a complex conceptual apparatus, part of which can, and has already, been traced to changing patterns of thought in the early modern period. The aim of the present chapter is to carry the reflection about this paradigm shift beyond a classical Cartesian narrative, into a domain preceding and well known to Descartes himself—sixteenth-century and seventeenth-century medical thought.

* * *

The storyline for the *Black Mirror* episode relayed above could find its philosophical justification in something like Descartes's thought experiment at the beginning of his *Principles of First Philosophy* whereby in seeking certainty, the philosopher discovers that the first thing which exists beyond doubt is the thinking being, subtracted from all material supports:

> Moreover, rejecting in this way all those things about which we are able to doubt in some way, and even assuming them to be false, we easily make the supposition that there is no God, no heaven, no bodies, and that we do have hands or feet or even any body; not however that we who are thinking such things are in fact nothing: for it is contradictory for us to think that this thing which things [*id quod cogitat*], at the very same time it is thinking, does not exist. And accordingly, this thought, *I think, thus I am*, is the first and most certain of all things, which occurs to any person philosophizing in an orderly way.[2]

One could say that writers of Black Mirror take Descartes' thought experiment at face value, reducing the human being to the *id quod cogitat*, "the thing which thinks." Defining the essence of the human being in this way permits a total rupture with the body, something which, in fact, not fully realized or even possible in Descartes's thought.[3] Yet the removal of the "thing which thinks" (which in Black Mirror's transhumanist fantasy is simply incorporeal but not a soul per se) from the body and the ensuring of its survival in a virtual eternity is made possible by a massive technological artifice, a wide array of computers and robots.

If indeed, following Descartes elsewhere, the thinking being is what is considered to have a real, independent existence and the body is more or less a machine, what stands in the way of this being which thinks and declares independence from the body, being transplanted into a different machine?[4] Prominent transhumanist Ray Kurzweil, who shares a similar conception of what constitutes the human being, takes this line of thought to its logical conclusion: "Actually there won't be mortality by the end of the twenty-first century. Not in the sense that we have known it[...]. As we cross the divide to instantiate ourselves into our computational technology, our identity will be based on our evolving mind file. We will be software, not hardware...There will still be hardware and bodies, but the essence of our identity will switch to the permanence of our software."[5] But this non-total mind/body distinction in Descartes, which

Kurzweil stretches well beyond the confines of Cartesian thinking, is itself corollary to a more fundamental conceptual separation—that of thought from nature. It is precisely through externalizing thought from nature that humans are able to become, as so famously stated in the *Discourse on Method*, its "masters and possessors."[6]

The relationship between mind (and by extension, thought) and nature is fundamental to the way science, in the sense of knowledge gathering, is structured. The Oxford historian R. G. Collingwood used the nature-mind dynamic to understand the history of Western science from the Greeks to contemporary times. This is because, he argues, the ancient Greek conception of nature, and subsequent development of the idea of nature in Western thought, necessitates a careful elaboration of a corollary concept of mind. Historically, for the Greeks, mind was generally assumed to exist *in* nature; for the moderns, nature lies *outside* of nature.[7] Much contemporary transhumanist thought rests on this modern conception of nature and mind, even if the import of the shift is in no way limited to transhumanism and is an important part of the dynamics of the so-called scientific revolution of the seventeenth century.

The term "physiology" offers a fitting case study for this inquiry insofar as it is, in essence, composed of two key terms: *phusis* (or "nature") and *logos* (or ordered "thought.")[8] The word, which for ancient Greeks such as Aristotle until the sixteenth century signified by and large what we call today "natural philosophy," was first applied to medicine by Jean Fernel in his 1554 *Medicina*, a historical event which Georges Canguilhem calls the "birth certificate" of physiology and for which there exists a fairly extensive historiography.[9] But while historians of medicine have paid some attention to the originality of Fernel's coinage—or lack thereof, as Vivian Nutton has argued[10]—less emphasis has been placed on the particulars of Fernel's definition of physiology, and the significant modification it undergoes in the seventeenth-century French translation of the physician Charles de Saint-Germain. What Fernel describes as a "natural contemplation concerning the human" (*naturalem de homine contemplationem*), Saint-Germain translates "knowledge of the natural composition of the human" (la cognoissance de la composition naturelle de l'homme). As the present chapter will show, this apparent mistranslation of "physiology" is not accidental but rather reflective of a change in the way the notions of nature and thought are conceptualized in the early modern period. Examining this change in a single instance, though necessarily limited, nevertheless demonstrates the ways in which the early modern period may shed significant light on contemporary thought. Here, we might understand the transhumanist philosopher Nick Bostrom's often cited statement that "Transhumanism has its roots in rational humanism"[11] (itself an egregious and misleading oversimplification of the humanist thought of the fourteenth–seventeenth centuries[12]) as an useful invitation to look more closely at how key terms undergo significant change in the world of early modernity.

Physiology in Antiquity

The Greek word "physiology" (*phusiologia*), as understood in Antiquity, was not used to designate a division of medical knowledge, and the adoption of the term by physicians in the sixteenth and seventeenth centuries requires a robust historical explanation that has been offered elsewhere.[13] For the ancient Greeks, the term refers to theorization about nature, what is now commonly translated by the phrase "natural science" or "natural philosophy." Aristotle employs the term a number of times throughout his works. In *On Sense and Sensible Objects*, for example, he uses "physiology" in reference to the title of a now lost treatise on plants: "This finishes our discussion of taste and flavor; the others affections of flavor have their proper place of inquiry in the *Natural History of Plants* [*ἐν τῇ φυσιολογίᾳ τῇ περὶ τῶν φυτῶν*]."[14] Quite literally, the Greek philosopher refers to a "physiology concerning plants," a theorization of the *phusis* of plants. Elsewhere Aristotle uses the term physiologist (*φυσιόλογος*) to describe one who theorizes about nature, especially in reference to pre-Socratic philosophers such as Empedocles. No doubt, the most famous example of the latter usage occurs in the *Poetics* where Aristotle famously distinguishes between the "natural philosopher" and the poet: "Homer and Empedocles have nothing in common except for the meter. That is why it is right to call the former a poet [ποιητήν] and the other a natural philosopher [φυσιολόγον] rather than a poet."[15] Thus, in spite of the fact that Empedocles composed his work in verse, he should be, according to Aristotle, considered a *phusiologos* rather than a poet. The lasting influence of Aristotle's distinction between the poet and physiologist is apparent in the late antique grammarian Servius' Latin commentaries on Vergil, where on several occasions he distinguishes between what properly belongs to poetry and what belongs to physiology.

A short passage from the Greek biographer Diogenes Laërtius' *Lives of Eminent Philosophers* brings to light another, later, nuance in the term physiology, differentiating it clearly from philosophy, a distinction already implicit in Plato's *Phaedo*: "Zeno and Empedocles studied with Parmenides at the same time, that they then left him, and that Zeno went on to do philosophy [φιλοσοφῆσαι] on his own, while the other continued to study with Anaxagoras and Pythagoras and emulated the latter's dignified way of life and bearing, and the former's study of nature [φυσιολογίαν]."[16] Physiology, perhaps as early as Plato and for a certainty afterward, was used in a way distinct from philosophy proper.

The first appearances of the term physiology in the extant Latin corpus are in Cicero, where he provides Latin translations with some slight variance. In *On the Nature of the Gods*, he speaks of "physiology, that is to say a reasoning about nature" (*physiologiam id est naturae rationem*).[17] Here, the Roman philosopher helpfully offers the Latin *ratio* as equivalent to the Greek *logos*. Elsewhere in the same text, he makes use employs an expression that more nearly transliterates the Greek *φυσιολογία* by way of the Latin phrase *physica ratio* ("reasoning about *phusis*"):

> *Nec vero eae stellae quae inerrantes vocantur non significant eandem mentem atque prudentiam. quarum est cotidiana conveniens constansque conversio, nec habent aetherios cursus neque caelo inhaerentes, ut plerique dicunt physicae rationis ignari*[.]
> And likewise these stars that are called "unwandering" are indeed evidence of the same mind and prudence. And these stars whose daily rotation is appropriate and consistent, neither have ethereal routes nor have routes attached to the heavens, as very many claim who are ignorant of reasoning about *phusis* [*physicae rationis ignari*][.][18]

Remarkably, in the extant Latin corpus, this Ciceronian paraphrase for "physiology" occurs numerous times in the commentaries on Vergil by Servius, a Latin grammarian of the fourth or fifth century C.E. In his commentary on a passage from Book I of the *Aeneid* where Neptune pacifies seas made stormy by Aeolus, he writes:

> *sub poetica licentia physicam quoque tangit rationem. Mare enim dicitur esse Neptunus, quem superius dixit graviter commotum, quia temepstas erat. Nunc ait* placat, *quia iam sedari coeperant maria.*
> [Vergil] also touches upon the explanation of *phusis* [*physicam rationem*] using poetic liberty. For the sea is said to be Neptune, which he previously said had been "stirred up mightily," because there was a storm. Now

> he says *placat* [pacifies], because already the seas had begun to be made calm.[19]

This "explanation of *phusis*" in Servius' commentary seemingly amounts to a paraphrase of the term physiology, parallel to the use of the same phrase in Cicero.

Curiously enough, the Latin term *physiologia* came to have a largely different meaning in the Late Antique and Medieval periods. For Christian commentators, beginning with the Greeks, physiology comes to refer to allegory, notably in the title of the Greek bestiary the *Physiologos*, which there means "allegorist" of nature:

> The three great figures of Alexandrian exegesis masterfully illustrate this ultimate meaning of *physiology* for which the allegorical dimension is part of nature. Philo, who is the first to practice a resolutely allegorical reading of the Old Testament, gives qualifies the commentators and exegetes of the Scripture as *physicoi*; since the essence of nature resides in its meaning, just as the essence of the letter is in its spirit, the very name *physis* (nature) without any other precision sometimes designates "the allegorical meaning" of a thing.[20]

One finds a similar sense of the transliterated *physiologia* in Latin several times in Servius' commentaries, published after the *Physiologos*, where it comes to mean very broadly "the explanation of an allegory."[21]

"Physiology" in the Renaissance

In the sixteenth century, the term *physiologia* had already recovered some of its ancient meaning, but was employed especially in reference to Aristotle's natural science, and no longer, as for the Greeks, with the pre-Socratic philosophers. A 1518 edition of Aristotle's *Parva naturalia*, which set William of Moerbeke's medieval translation in parallel with the more humanistic translations of Johannes Argyropoulos and François Vatable, is entitled *Twenty-Eight Books from the Physiology of Aristotle* (*Ex Physiologia Aristotelis Libri Duodetriginta*).[22] Here, *Physiologia* undoubtedly refers to the "natural philosophy" of Aristotle broadly, as is likewise attested in Robert Estienne's Latin-French dictionary of 1538, where it is defined as "the knowledge of natural things" ("la science des choses naturelles").[23] It is with this lexical background in mind that one should understand the standard historiography on Jean Fernel's coinage of "physiology" as a medical term: "The Paris physician Fernel is usually remembered as the scholar who moulded the recently restored classical medical inheritance into a workable synthesis, in the process inventing the Latin words *physiologia* (in the modern sense), *pathologia*, and *therapeutices*."[24] This alleged "modernity," however, is somewhat misleading.

Fernel's definition of physiology only appears for the first time in his 1554 *Physiologia* (part of the threefold division of his massive medical manual entitled *Medicina*) in the second book, entitled "On the Elements." There, he clarifies what this first division, the *Physiologia*, is intended to accomplish: "In this way, a whole physiology will be drawn together, which establishes a natural contemplation concerning the human being by force of demonstration" (*Sic universa contrahetur physiologia, qu[a]e naturalem de homine contemplationem demonstrationis vi constituit*).[25] The definition of physiology offered here, at first glance, bears resemblance to other conventional definitions of the first part of theoretical medicine. In Johannitius' *Isagoge ad tegni Galeni*, a key text in the *Articella*, a standard medieval and Renaissance medical textbook, the author refers to the first division of theoretical medicine as "the contemplation of things pertaining to nature" (*contemplationem rerum naturalium*), by which he means a consideration of the four elements, qualities, and humors.[26] The resemblance of Fernel's definition of physiology to the conventional definition of the "Natural Part of Medicine" (the title of an earlier version of the *Physiologia*) has led some scholars to question the originality of Fernel's contribution. Vivian Nutton has argued that Fernel's use of the term *physiologia* is little more than a marketable Greekism: "Fernel proclaims at one and the same time his individualism and his adherence to Greek doctrine by using a choice word, *physiologia*, and it is this apparent innovation that has gained Fernel the reputation as the founder of the discipline."[27] While much of the criticism of this traditional historical narrative of physiology's "founding" in the sixteenth century is merited, it is important, nevertheless, to note that Fernel himself makes a bold claim concerning the originality of his own definition. In the 1542 version of the text, he refers to Aristotle as the founder of "physiology," that is to say, "knowledge about nature," the standard use of the term in the first half of the sixteenth century; in the 1554 *Physiologia*, he modifies the text, referring to Aristotle as the founder of philosophy, *not* physiology, and therefore claiming for himself the title of founder of physiology by changing the name of his book from

De naturali parte medicinae to *Physiologia*.[28] The Aristotelian sense of "physiology," as is clear from the previous examples, would have been well-known to Fernel's contemporaries. In separating the term from Aristotle in his 1554 *Physiologia*, Fernel nevertheless stays close to the original meaning of the Greek compound, although he departs from Cicero's phraseology. Fernel offers *naturalis* as equivalent to *physio-* (<*phusis*), *contemplatio* for -logia (<*logos*). The human being (*de homine*) is placed in the middle of the two terms, as the link between the two.

The Latin term *contemplatio* ("contemplation") already occurs as early as the first century in the medical treatise of the Roman encyclopedist Celsus. There, the author, describing the history of Greek medicine, discusses how knowledge of the "treatment of disease and the contemplation of the nature of things [*rerum naturae contemplatio*] began through the same authorities... Hence [...] that many who professed philosophy became expert in medicine, the most celebrated being Pythagoras, Empedocles and Democritus."[29] But the use of the term in Fernel's definition owes a more important debt to Florentine Platonic thought, especially as exemplified by Marsilio Ficino.[30] In the long Latin Neoplatonic tradition, "contemplation" is often used in a precise technical sense, designating specifically thought directed toward eternal things.[31] In the *Platonic Theology*, Ficino uses the term numerous times with this meaning:

> *Quamobrem sicut in nobis quodammodo caducae sunt nutriendi sentiendique natura et complexio corporalis, quae circa caduca versantur et cum caducis animalibus sunt communes, ita immortalis erit contemplandi potestas quae circa immortalia versatur et cum immortalibus est communis, quia non potest operatio eadem nisi ab eadem natura et potentia proficisci.*
> For this reason just as in a certain way in us the nature of nourishment and perception and our bodily complexion, which are concerned with mortal things and are shared with mortal animals, are mortal, in the same way the power of contemplating, which is concerned with immortal things and is common with immortal beings, will be immortal, because the same operation is not able to begin except from the same nature and power.[32]

In the *Physiologia*, Fernel makes clear to the reader that he is using the term "contemplation" in the Platonic sense, when he speaks of its relationship to the other faculties, such as the faculty of judging:

> *Supra has omnes se extollit contemplandi facultas, quae sapientiae rerumque aeternarum studio tenetur. Quisquis autem illius amore et suauitate non*

satis illectus, ad res caducas ratiocinandi vim conuertit, omneque studium ad humanas refert commoditates et vsus, is vice contemplationis, actionis sibi facultatem adhibet.

The faculty of contemplating, which is held by a zeal for wisdom and eternal things, raises itself above all these faculties. However, whoever, not sufficiently seduced by the love for and sweetness of the faculty of contemplating turns the force of his reasoning towards mortal things and carries all zeal towards human benefits and advantages, this person, instead of contemplation, puts to use for himself the faculty of action.[33]

Here, Fernel's apparent reprisal of Ficino's definition of contemplation, as well as the use of similar vocabulary, demands a closer examination of its use in his definition of physiology.

Fernel's translation of *logos* by a highly technical term such as *contemplatio* suggests the utility of adopting for himself the Greek *physiologia* over a rival term like *physica*, the meaning of which is identical in the sixteenth century.[34] A parallel example from the seventeenth-century philosopher Pierre Gassendi's discussion of the term *physiologia* provides an instructive of why one term is to be preferred over the other:

Solet porrò Epicurus hanc partem Physiologiam potiùs quàm Physicam dicere; forte quod vox φυσιολογία definitionem huis Scientiae contineat; cum sit sermo quidam, & ratiocinatio circum rerum naturam, in cuius contemplatione versatur.

Epicurus is moreover accustomed to call this part of philosophy *physiologia* rather than *physica*; perhaps because the word φυσιολογία contains the definition of this Knowledge since it is a certain speech and reasoning about the nature of things, in the contemplation of which [the physiologist] is occupied.[35]

"Physiology" makes explicit the "speech" and "reasoning," or for Fernel, the "contemplation" used to investigate nature. Whether or not Fernel exhibits direct influence on the seventeenth-century philosopher's use of *contemplatio*—Fernel's writing was still widely influential in the seventeenth century and Francis Bacon mentions him by name[36]—Gassendi's discussion of *physiologia* nevertheless elucidates the reasoning for Fernel's preference for the Greek compound.

Having dealt with "contemplation" in Fernel's translation, the equally difficult question of what "nature" and "natural" mean comes to the forefront (a problem as fundamental to the sixteenth century as it is now).

Fernel appears to make light of its difficulty to define in his 1548 philosophical dialogue *On the hidden causes of things* when the physician Brutus asks the aspiring physician, aptly named Philiatros, what precisely *nature* is:

> PH. So, while this morning birdsong restores our spirits, begin the discussion and take on these points for explanation.
> BR. I am beginning as you wish, and, first of all, I ask you whether you are certain nature is some *thing*?
> PH. Those who pass down the teachings of natural philosophy do not doubt that some [kind of] nature exists, from which they even call themselves the *phusikoi* ["natural scientists"]. And this, as if it were most famous and well known among all people, they establish as a first principle of philosophy such that it neither ought nor is able to be demonstrated.
> BR. Have you at any time seen or touched with your hand that well-known nature?
> PH. I do not struggle to see it, if only I am just able to obtain it sufficiently through thinking. So then, continue in a different direction.[37]

The exchange makes clear the problem with "nature," and Fernel's wit is readily apparent to the reader. Philiatros supposes that there must be a thing called nature because there is a group of "natural philosophers" who have named themselves after it. But Philiatros' reply also points to a deeper metaphysical problem. These natural philosophers suppose nature to be, at the same time, both a concept and a physical entity. This not only allows them to equivocate when pressed on what nature is. For them, nature has become simply "a principle of philosophy, which neither needs proof nor can be proved."

The complex relationship between two parts of physiology, "nature" and "thought," is sketched out roughly by Fernel in a chapter from the *Physiologia* entitled "On the Functions of the Mind" (*De mentis functionibus*). There he again states that contemplation is focused on eternal things: "This thing which they call contemplative understanding [*contemplantem intelligentiam*], is that thing which is most greatly and only concerned with universal things, necessary and eternals things, those same divine and difficult things that must be marveled at" (*Ea est quam contemplantem intelligentiam appellant, quae maximè et duntaxat circa est universales, necessarias et sempiternas, quae eaedem admirandae sunt difficiles atque divinae*).[38] But Fernel, following Aristotle, suggests at the same time that understanding (*intelligentia*), from which contemplation emerges, has a material basis founded in the senses. From the material world, the senses receive stimuli that are then, largely though not entirely. The active understanding (*intelligentia agens*) shapes the sensory information (the still material images, or *simulacra*) into something incorporeal, which then remains in the mind. But the border between mind and matter is not neatly separated, and the faculty of understanding encounters the material remains in the images, which persist in retaining something of their material existence: "These notions are not indeed able to penetrate more deeply into the mind and understanding because, although they are without body, they nevertheless reek of a certain impure smell of matter" (*Hae quidem in mentem & intelligentiam altiùs non penetrare possunt, quia tamesti sine corpore sunt materiae tamen impurum quiddam redolent*).[39] The natural world remains the necessary physical support for mental content about the physical world, and it is through the human being that this content is conducted. The human being thus forms the troublesome link between contemplation and nature.

Fernel's definition of physiology (placing each time the phrase *de homine* between *naturalis* and *contemplatio*) confirms the relation between nature and thought established in his chapter on cognition. Rather than a physiology from the point of view of nowhere, Fernel's physiology identifies the object of physiological thinking with the medium through which the thought is conducted. Here, the double sense of the intentionally ambiguous adjective "natural" becomes evident: Physiology, for Fernel, is the highest thought about nature through the human medium and is therefore also *natural* contemplation about the human being. The mind is the physical place in which "thought" and "nature" interact to arrive at the highest form of knowing, that is, the knowledge gained through contemplation. The loop-like quality of Fernel's understanding of cognition, and ultimately of physiology, falls in line with Ficino's discussion of contemplation:

> *Quis autem dubitet contemplationem nostram supernis esse persimilem, cum per eam animus, ut caelestes illi, seipsum et opera sua consideret, investiget quoque supernas causas earumque effectus, item ab effectibus inferioribus per medias causas usque ad causam supremam ascendat, atque vicissim a suprema causa usque ad infimos eventus* circulo*remeet?*
>
> Furthermore, who could doubt that our contemplation is very similar to celestial beings, since our mind through contemplation, just as those

> heavenly minds, considers its very self and its own works, because it also searches out celestial causes and their effects, since it likewise climbs from the lowest effects through the middle causes all the way to the highest cause, and in turn from the highest cause returns to the lowest occurrences *in a circular way*? [emphasis added][40]

Fernel's physiology, far from an anticipation of seventeenth-century functional physiology, is foregrounded by the idea that nature and thought are never fully separable, that matter's stench can never be fully expunged from thought.

Breaking the Loop

Despite being credited as the founder of physiology, Fernel's highly metaphysical definition of physiology never catches on. The physician Jean Riolan the Elder, who published elaborate commentaries on all of Fernel's major works, writes fawningly that Fernel "called Philosophy back from its exile."[41] At the same time, in his commentary on the *Physiologia*, he all but ignores Fernel's definition of physiology, offering in its place: "the knowledge" about "our nature, namely the elements, temperaments, humors, spirits, parts, faculties, and functions."[42] In the 1655 translation of Fernel into French, the physician Charles de Saint-Germain likewise departs significantly from Fernel's definition, translating "thus we will understand all of Physiology, which teaches by the work of demonstration the knowledge of the natural composition of man" ("ainsi on comprendra toute la Physiologie, qui enseigne par l'effet de la demonstration la cognoissance de la composition naturelle de l'homme").[43] The Platonic *contemplatio* is substituted for the material "composition."

Presumably, Saint-Germain's modification of the text (perhaps with the innovations in seventeenth-century anatomy in mind) follows in line with resistance to Fernel's metaphysics, as well as with the increasingly common definitions of physiology in the seventeenth century. Cotgrave's 1611 French–English dictionary provides the following meaning: "that part of Phisicke which treats of the composition or structure of mans bodie."[44] Roberto Poma, writing in reference to the definition of physiology in Saint-Germain's translation, sees in it a potential precursor to William Harvey's functional physiology, insofar as "Fernel places the foundation of medicine in the theory of the healthy functioning of the body and of the mind."[45] But Saint-Germain is a revisionist, and in modernizing the text he translates, he deprives it—or, in his view, perhaps, relieves it—of any treatment of the relationship between nature and thought.

Contemplation is no longer itself *natural* or concerned with nature (Ficino's paradoxical proposition adopted by Fernel). Instead, the "natural composition of the human being" is made the object of human knowledge. By removing "contemplation" from physiology in the seventeenth century—while the roughly contemporaneous counterexample of Gassendi is worth noting—and setting nature plainly on the side of the material body, the translator seems to believe he has fixed Fernel's troublesome definition. In so doing, he opens a path for seeing thought, and by extension the *id quod cogitat*, "the thing which thinks," as something removed from the natural, a separate entity that transcends the material body upon which it no longer depends. Or so the story goes.

Notes

1. See James Poniewozik, "Review: 'Black Mirror' Finds Terror, And Soul, in the Machine," *New York Times*, October 20, 2016.
2. *Sic autem rejicientes illa omnia, de quibus aliquo modo possumus dubitare, ac etiam falsa esse fingentes; facile quidem supponimus, nullum esse Deum, nullum cœlum, nulla corpora; nosque etiam ipsos non habere manus, nec pedes, nec denique ullum corpus; non autem ideo nos qui talia cogitamus nihil esse: repugnat enim, ut putemus, id quod cogitat, eo ipso tempore quo cogitat, non existere. Ac proinde hæc cognitio,* ego cogito, ergo sum, *est omnium prima & certissima, quæ cuilibet ordine philosophanti occurat.* Translation mine. René Descartes, *Prinicipia primae philosphiae* (Amsterdam: Ludovicus and Daniel Elzeviri, 1656), 2. Cf. *Principia primae* philosophiae, AT 8, 6–7.
3. See R. G. Collingwood puts it well: "Descartes knows very well that transcendence must not be pushed to the point of dualism; the two things must be connected somehow; but cosmologically he can find no connexion short of God, and in the individual human being he is drive to the desperate expedient, justly ridiculed by Spinoza, of finding it in the pineal gland…" See *The Idea of Nature* (Oxford: Clarendon Press, 1945), 6–7.
4. See, for example, Descartes, *Le Discours de la méthode* [V], in *Œuvres et Lettres*, ed. André Bridoux (Paris: Gallimard, 1953).

5. Ray Kurzweil, *The Age of Spiritual Machines* (New York: Penguin, 1999), 128–9.
6. See Descartes, *Le Discours de la méthode*, 168: "Mais sitôt que j'ai eu acquis quelques notions générales touchant la physique, et que, commençant, j'ai remarqué jusques où elles peuvent conduire et combien elles diffèrent des principes dont on s'est servi jusqu'à présent, j'ai cru que je ne pouvais les tenir cachées sans pécher grandement contre la loi qui nous oblige à procurer autant qu'il est en nous le bien général de tous les hommes. Car elles m'ont fait voir qu'il est possible de parvenir à des connaissances qui soient fort utiles à la vie, et qu'au lieu de cette philosophie spéculative qu'on enseigne dans les écoles, on en peut trouver une pratique, par laquelle, connaissant la force et les actions du feu, de l'eau, des astres, des cieux et de tous les autres corps qui nous environnent aussi distinctement que nous connaissons les divers métiers de nos artisans, nous les pourrions employer en même façon à tous les usages auxquels ils sont propres, et ainsi nous rendre comme maîtres et possesseurs de la nature."
7. Collingwood maintains that the general view can be confirmed by the fact that those who rejected it "stated this discovery in a way that shows unmistakably how paradoxical it seemed to them and how remote from their habitual... ways of thinking." *The Idea of Nature*, 6.
8. See "λόγος [IV.]," *Liddell, Scott, Jones Greek-English Lexicon* (Oxford: Oxford University Press, 1848).
9. My translation. Georges Canguilhem, "La constitution de la physiologie comme science," *Études d'histoire et de philosophie des sciences* (Paris: Vrin, 2015): 226: "Peu importe ici que Fernel ait de la nature humaine une idée plus métaphysique que positive. L'essentiel à retenir c'est l'acte de naissance, en 1542, de la physiologie, comme étude distincte de la pathologie et la précédant, la pathologie précédant elle-même l'art du pronostic, l'hygiène et la thérapeutique." See also, Laurence Brockliss and Colin Jones, *The Medical World of Early Modern France* (New York: Oxford University Press, 1997), 129.
10. For Vivian Nutton's challenge to the traditional narrative, see "Physiologia From Galen to Jacob Bording," in *Blood, Sweat, and Tears: The Changing Concepts of Physiology from Antiquity into*

Early Modern Europe, eds. Manfred Horstmanshoff, Helen King, and Claus Zittel (Leiden: Brill, 2012), pp. 27–40.

11. Nick Bostrom, "A History of Transhumanist Thought," *Journal of Evolution and Technology*, vol. 14 (2005): 1–25, 2.
12. The idea of a "rational humanism" was notably called into question by Alexandre Koyré, among others. See "L'apport scientifique de la Renaissance," in *Études d'histoire de la pensée scientifique* (Paris: PUF, 1966), 38–47.
13. My Ph.D. dissertation (under revision for publication) devoted a chapter to the subject, which differs in several respects from previous accounts. See Chapter 4: "Jean Fernel's 'Physiology'," in "Translating 'Nature' in Early Modern France" (Ph.D. diss., New York University, 2025). See also, notably, Nutton, "Physiologia From Galen to Jacob Bording."
14. Aristotle, *On Sense and Sensible Objects* [442 a], trans. W.S. Hett (Cambridge, Massachusetts: Harvard University Press, 1957 [1986]), 246–7.
15. Aristotle *Poetics* 1 1447b17–19, *Early Greek Philosophy V*, ed. and trans. André Laks and Glenn W. Most (Cambridge Massachusetts: Harvard University Press, 2016), 582–3.
16. Diogenes Laertius 8.56, *Early Greek Philosophy V*, ed. and trans. André Laks and Glenn W. Most (Cambridge Massachusetts: Harvard University Press, 2016), 338–41. In the *Phaedo* [96a], Socrates mentions his passion, as a youth, for "inquiry about nature" [ἱστορίαν περὶ φύσεως]. See *Platonis Opera*, vol 1., ed. E. A. Duke (Oxford: Oxford University Press, 1995), 151. My thanks to Guido Giglioni for bringing this latter reference to my attention at the 2022 VivaMente Conference in Pisa.
17. Cicero, *De Natura Deorum*, Book I, ed. W. Ax (Stuttgart: Teubner, 1980), 9.
18. Ibid., Book II, 70. Translation mine.
19. Servius, *Commentarii in Vergilii carmina* [*in Aen.* I.142], editio Harvardiana, vol. 2, ed. E. K. Rand (Lancaster, Penn.: American Philological Society, 1946), 86. Translation mine.
20. "Les trois grandes figures de l'exégèse alexandrine illustrent en maîtres ce sens achevé de la *physiologie* pour laquelle la dimension allégorique est une partie de la nature. Philon, qui est le premier à pratiquer une lecture résolument allégorique de l'Ancien Testament, donne précisément le qualificatif de *physicoi* aux

commentateurs et aux exégètes de l' Écriture; puisque l'essence de la nature réside dans son sens, comme celle de la lettre est dans son esprit, le nom même de *physis* (nature) désigne parfois sans autre précision 'le sens allégorique' d'une chose." Translation mine. Arnaud Zucker, "Introduction," *Physiologos: Le bestiaire des bestiaires*, texte traduit du grec, établi et commenté par Arnaud Zucker (Grenoble, Éditions Jérôme Millon, 2004), 23.

21. See, e.g. Servius, *in Aen.* VI.295; VI.893.
22. See *Ex Physiologia Aristotelis Libri Duodetriginta...Francisco Vatablo interprete* (Paris: Henri I Estienne, 1518).
23. "Physiologia," *Dictionarium Latinogallicum* (Paris: Robert Estienne, 1538).
24. Laurence Brockliss and Colin Jones, op.cit., 129.
25. Fernel, *Medicina* (Paris: André Wechel, 1554), 69.
26. See "Johannicius: Isagoge Ad Techne Galieni," ed. Gregor Maurach, *Sudhoffs Archiv* 62. no. 2 (1978): pp. 148–74, 151: *Medicina dividitur in duas partes, scil. in theoricam et practicam, quarum theorica dividtur in tria, i. in contemplationem rerum naturalium et earum, quae sunt contra naturam, ex quibus sanitatis, aegritudinis et neutralitatis scientia procedit, cum ipsae naturales res extra cursum suum procedunt, i. est quando extra cursum naturalem IV humores excrescunt; ex qua occasione et qua significatione aegritudo habeatur.*
27. Vivian Nutton, op. cit., 35.
28. I have documented these modifications to the 1542 text extensively elsewhere. Compare Fernel, *De naturali parte medicinae* (Lugduni: 1551), 189–90 and *Medicina* (1554), 1–2.
29. Celsus, *On Medicine*, Books I–IV, trans. W. G. Spencer, (Cambridge, Mass.: Harvard University Press, 1935), 4–5.
30. For Fernel's indebtedness to Marsilio Ficino, see Hiro Hirai, *Medical Humanism and Natural Philosophy: Renaissance Debates on Matter, Life and the Soul* (Leiden: Brill, 2011), pp. 46–79.
31. The Late-Antique Neoplatonist Calcidius gives a precise definition of *contemplatio*, which is only directed at eternal things: "for the soul engages in contemplation with constant and stable ease, in that contemplation is an unhindered state of attention, whereas sensation arises out of the coupling of the passions of body and soul." *On Plato's 'Timaeus* [Sect. 193], trans. John Magee (Cambridge, Mass.: Harvard University Press, 2016), 422–425.

32. Translation mine. Marsilio Ficino, *Platonic Theology*, vol. 3 [Books IX–XI], ed. James Hankins (Cambridge, Mass.: Harvard University Press, 2003), 94.
33. My translation. Fernel, *Medicina* (1554), 147.
34. The two terms are defined identically by Robert Estienne. See "Physica," *Dictionarium Latinogallicum* (1538).
35. Pierre Gassendi, *Ainmadversiones in decimum librum Diogenis Laertii* (Lyon: G. Barbier, 1649), 159.
36. Bacon mentions Fernel in his *Temporis Partus Masculus*. See *The Works of Francis Bacon Vol. VII*, eds. James Spedding, Robert Leslie Ellis, and Douglas Denon Heath (Boston: Houghton Mifflin), 21. My attention was drawn to the connection between Bacon and Fernel by Richard Serjeantson's article "Francis Bacon and the 'Interpretation of Nature' in the Late Renaissance," *Isis*, 105 (2014): 681–705, 689.
37. I have altered the translation in several places. See Jean Fernel, *De abditis rerum causis* (*On the Hidden Causes of Things*), trans. John M. Forrester (Leiden: Brill, 2005), 124–127.
38. Fernel, *Medicina*, 206.
39. Ibid.
40. Ficino, *Platonic Theology* [Book IX.6], 94, 96.
41. Jean Riolan, *Praelectiones in libros Physiologicos, & De abditis rerum causis* (Paris: Hadrianus Perier, 1602), 4.
42. Jean Riolan, *Universae medicinae compendia* (Paris: Hadrianus Perier, 1598), 3.
43. Charles de Saint-Germain, *Les VII. livres de la physiologie. Composez en latin par Messire Jean Fernel…* (Paris: Jean Guignard le Jeune, 1655), 228.
44. See "Physiologie," *A Dictionarie of the French and English Tongues, Compiled by Randle Cotgrave* (London: Adam Islip, 1611).
45. See Roberto Poma, "Tradition et innovation dans la *Physiologie* de Jean Fernel. L'accord difficile entre expérience et raison dans l'œuvre d'un médecin de la renaissance," *Corpus: revue de philosophie*, n. 41 (2002): 97–118, 98: "Ainsi Fernel pose-t-il l'assise de la médecine dans la théorie du fonctionnement sain du corps et de l'esprit." Translation mine.

CHAPTER 4

Flesh, Soul, and Steel: Aporias of Materialist Embodiment

Charles T. Wolfe

"I beheld the wretch—the miserable monster whom I had created. He held up the curtain of the bed; and his eyes, if eyes they may be called, were fixed on me" (M. Shelley, *Frankenstein*)

"Both children loved her. Once, when Nanny had to be sent to the repair shop, they cried and cried without stopping. Neither their mother nor their father could console them. But at last Nanny was back again, and

The following chapter overlaps with several other of my papers on materialism and embodiment, on mechanism, and on automata, cited in the bibliography. It reflects both certain shifts in my understanding of the issues (ten years ago, I sought to position embodied materialism squarely in opposition to mechanism, something I definitely would not do now), and my response to papers presented at the excellent conference 'From Automata to Transhumans', organized by Jil Muller and Guido Giglioni in Pisa in April 2022. Thanks to Hannah Elmer for useful comments.

C. T. Wolfe (✉)
Department of Philosophy, Université de Toulouse-Jean-Jaurès, Toulouse,

J. Muller (ed.), *Automata, Cyborgs, and Mutants*, Palgrave Studies in Medieval and Early Modern Medicine,
https://doi.org/10.1007/978-3-031-93201-4_4

everything was all right. And just in time! … Nanny came to a halt in front of him …. Two unwinking photocell eyes appraised him, eyes on flexible wire stalks. The stalks moved speculatively, weaving slightly. Then they withdrew" (P.K. Dick, "Nanny")

Our fascination with machines, artificial humans and, half-way between, prosthetically enhanced forms of life, seems to be matched, whether evenly or not, by our fear or loathing toward them—their uncanniness, the feeling they provoke that when we look into their "eyes," the lights are on but "no one is home"…[1] Perhaps this was what the ship captain felt, in the apocryphal story told in all biographies of Descartes,[2] of how the arch-mechanist built an automaton in the shape of a young girl (in some versions of the story, his beloved daughter Francine) and took it aboard ship—but when the ship captain found her, he grew terrified by the "mechanical monstrosity" and her life-like motions, and threw her overboard. Such fascination and fear go back a long way, as Minsoo Kang has studied in his impressive book *Sublime Dreams of Living Machines*,[3] and, for reasons that are not my concern here, seems to pick up speed in the nineteenth century, with works like *L'Ève future*, moving on to Frankenstein (admittedly an entirely flesh-and-blood creation rather than some version of an android), "Metropolis" and so forth. Further, this feeling is also present in many reactions to the mechanical philosophy, particularly when thinkers worried that mechanism might extend beyond its "safe" or "legitimate" province. Thus, the Cartesian Jacques Rohault wrote: "If you currently hold as certain that animals are just mere automata, aren't you afraid that someday you will also believe that humans, too, are just mere machines?".[4]

This fear extended to materialism itself, often in a blurred or confused relation to mechanism. That is, the anguish felt at the possibility that humans might be machines, or better, that increasingly sophisticated machines or machine-organism hybrids might be increasingly indistinguishable from good old-fashioned flesh-and-blood creatures, often spills over into strident denunciations of materialism for reducing the soul and life (indeed, the soul which *is* life on some accounts) to the world of dead,

France
e-mail: charles.wolfe@univ-tlse2.fr

inanimate, senseless, brute matter. Curiously, the attacks on the "stupidity" of matter are matched by the attacks on hylozoism, in which, one might say, matter is *too* intelligent, *too* alive. On the one hand, "There can be no such thing as God if stupid and senseless matter be the first original of all things; and if all being and perfection that is found in the world, may spring up and arise out of the dark womb of unthinking matter" (that is Cudworth[5]); on the other hand hylozoism is "the death of all natural philosophy" (this is Kant; who arguably bridges both views when he writes that "the possibility of living matter cannot even be thought; its concept involves a contradiction, because lifelessness, *inertia*, constitutes the essential characteristic of matter"[6]). If the fault of materialism is either that matter is too inert and unthinking, or that it is too alive and sentient, one might feel that antimaterialists doth protest too much—but that is material for another discussion.

I would like to reflect in what follows on this fear, including the question of whether its two variants I have just sketched out amount to one and the same response, or they are different. In so doing, I will make use of a third term, or perhaps cluster of terms, centering on embodiment and the flesh. Because so much of the hostility to the machine, and to "blind" or "stupid" matter, is based, explicitly or not, on an opposition between merely robotic processing power, and feeling and sensing flesh. That many defenses of the uniqueness and/or sovereignty of life (the living body, the organism, the flesh) are themselves thinly disguised versions of defenses of the soul (whether in Rohault or in Hans Jonas) is rather well known, and I will at most allude to it below. But the basic fact of how strongly many scholars continue to oppose "mechanism" and "body," including in valuative terms, is surprising, at least to me, and motivates the present remarks, among others. How often are we still told about La Mettrie's "mechanistic materialism" (the result of taking rather naïvely and literally the title of his most famous book)? How often do works, particularly in the humanities, paint a "Cimmerian" vision of the Cartesian context in which "as a machine, the body became objectified; a focus of intense curiosity, but entirely divorced from the world of the speaking and thinking subject"?[7]

One problem with such labels and judgments is the irreducibly organic and affective dimension(s) in the body-machine, which in La Mettrie's hands is a very different creature from the iatromechanist creations in which the body is literally reduced to a series of mechanisms, whether with some analogical dimension, as in Descartes (or Steno), or in an attempted non-analogical and fully literal way, when physicians, philosophers or other practitioners claim that bodily processes *simply are* mechanistic processes (rather than being *like* mechanistic processes). Thus the Scottish physician Archibald Pitcairne stated in his 1717 *Elementa Medicinae* that "The cure of every Disease, whether in the Vessels or Fluids, or both, is to be effected only by mechanical Laws."[8] Thus, moving back and forth between iatromechanism as a discourse specifically concerned with medicine and the body, and philosophical mechanism notably of the Cartesian sort, such approaches to the body-as-machine are too often caricatured as somehow blind to embodiment, blind to feeling, blind to affect—whether in the language of "blind scientistic drives," of the fear (as in Rohault) that we turn out to be "mere machines," or of the "death of Nature."[9]

That is, the canonical portrayal of mind/body conceptualizations downplayed and even erased the presence of the affective, the visceral, the reality of "blood and guts" in all its wonder. Culprits usually paraded in front of the tribunal of the return of the affective included Descartes (despite his *Passions of the Soul* and correspondence with Princess Elisabeth which insists so much on the "union" of body and soul: the real Descartes was much more bothered by this union than the "textbook Descartes" is), the Scientific Revolution (which, according to this endlessly recycled story, "cut us off from the world" in the pursuit of the "mastery of Nature"), industrialization, and of course virtualization which gradually plugs our affective "connectors" into disembodied, ghostly forms of self-realization and vaporous satisfaction.

Faced with all this dehumanization, this philosophical Fordism (thinking of Sartre's statement that materialism is a philosophical form of Taylorism/Fordism which decomposes human action into manipulable, mechanistic processes[10]), people call for a return to affect(s)—not at all unlike Carolyn Merchant's denunciation of the "death of Nature" in the Scientific Revolution (at the hands of Descartes and Bacon in particular—a violence, a constraint, a condition in which Nature "takes orders from man and works under his authority"[11]), proclaiming that "The machine image that has dominated Western culture for the past three hundred years seems to be giving way to something new. Some call the transformation a "new paradigm"; others call it "deep ecology"; still others call for a "postmodern ecological world view."[12]

Now, this history of mechanistic alienation (from organismic meaning?) is itself a bit fraudulent—when was the machine ever so divorced from bodies in the plural or the body *in esse*? Deborah Brown has noticed that despite our almost unanimous condemnation, indeed contempt for Cartesian dualism, we can't seem to turn away: "We can't accept the Cartesian mind but we can't seem to avert our gaze either."[13] John Sutton observed wryly that on the anti-Cartesian view, "An earlier enchanted world, criss-crossed by networks of sympathies and antipathies, embracing analogy and suggestion over representation and intervention, traversed by holist herbalists and natural magicians, coupling early bodily realism with organicist ecologism, was sundered and lost with Descartes' blind scientistic drive for the mastery, possession and penetration of nature."[14] This "blind, scientistic drive" is often still associated with the Scientific Revolution (e.g., in work on Margaret Cavendish[15]), with materialism somehow tagging along as a fellow-traveler…

The emergence of modern materialism is often presented as a fiery, ideological outgrowth of the forms of mechanism that emerged in the Scientific Revolution. Now, mechanism itself comes in a variety of forms: for Robert Boyle, the basic properties or "qualities" of things can be exhaustively explained in terms *of*, if not reduced *to* "the motion, size, figure and contrivance of their own parts," with new qualities being produced by "changing the texture or motion" of these basic parts.[16] As regards the body, and thus more macroscopically, Descartes says he'll assume the body is nothing other than "a statue or machine made of earth," to which he adds celebrated analogies with the machinery of fountains and other sorts of clockwork.[17] Some purported mechanists such as Herman Boerhaave, the great Leiden professor of medicine, go as far as claiming that "the nature of the human body is the same as that of the whole of the Universe,"[18] which is an ontologized restatement of the essential claims of mechanism: it is not just that matter itself can exhaustively be defined in terms of shape, size and motion, or that the body should be studied as if it were a mechanistic arrangement of matter—but that the body is itself the same *in essence* as the rest of the universe: mechanical.

If materialism really was an outgrowth of mechanism, a typical example would be La Mettrie's *homme-machine* (he was after all referred to as "Herrn Maschine" in Berlin in his day), which is often described, presumably by authors unfamiliar with his writings or blinded by the prevalence of older interpretive schemas, as simply the Cartesian *bête-machine* extended to humans. This is most often a narrative of progression or decline (depending on whether one is a positivistically inclined historian of the behavioral sciences intent on tracing a line from Descartes to cybernetics via La Mettrie[19] or a moralistically inclined philosopher intent on showing how the triumph of mechanism and/or materialism spelled the death of "meaning and value."[20]) But my point here is that if we consider these narratives from the standpoint of *embodiment*—of our existence as embodied beings—it would appear that we have ended up with an atomistic, reductive, depersonalized way of relating to our bodies, to the fact of our embodiment. This is often decried by theorists who think something was lost at a certain historico-destinal moment of "dehumanization" or alienation (Husserl, Jonas, Merchant). Whether we like it or not, or whether it matches our "phenomenology," in the sense of our experience of what it is to be in a body—the pain, the enjoyment, the ineffable subjectivity and so on—we have ended up with machine bodies.[21] Notice that this is a powerful narrative (comparable to those focusing on "modernity," including their Latourian rebuttal): it is both an account of the early modern period (sometimes pinning its claims on the Scientific Revolution, sometimes on the Enlightenment) and a "just so" account of the world we have come to inhabit since then.

Of course, there have been other responses to the growing complexity of mechanism and its materialistic outgrowths besides desperate appeals to the respect of the sovereignty of the flesh (which can be seen in Merleau-Ponty, or in Hans Jonas, to name some prominent examples[22]—and they are building on a "crisis" narrative found in Husserl and developed in different directions by Koyré, Strauss, Heidegger and others). Notably, some thinkers have celebrated the potential for hybridization between body and machine,[23] which goes well with a historical emphasis on the *heuristic* role of automata such as Vaucanson's duck.[24] In the overlapping space of Renaissance "art and anatomy," Nicole Howard describes powerfully how Giovanbatista Bracelli's *Bizzarie di Varie Figure* decompose the human body, inter alia, into three-dimensional geometric shapes such as pyramids, cylinders, and cubes, creating a strange space of interplay between monsters, puppets, corpses (and "living dead men," as Howard nicely says citing Shakespeare) and automata, in a kind of anti-mereological parcellisation of life into combinatorial parts without wholes. Echoing (perhaps inadvertently) Foucault on prisons, hospitals, factories, and schools, she notes that the bodies anatomized in dissection theaters of the period came to look more and more like the puppets of the period.[25] And mechanistic ontology itself turns out to be mottled, dappled or kaleidoscopic; traversed by non-ontological moments, "machine-based" (in Keller's terms)[26] or even "mechanistic analogy-based." Thus Robert Boyle, chief proponent of "the mechanical philosophy," as he called it, can pass in one sentence from a statement of corpuscularian ontology to an appeal to mechanical analogies, relying on actual machines: "because it [the corpuscular philosophy] explicates things be corpuscles, or minute-bodies, it may… be called corpuscular," i.e., it explains "the phenomena of nature" in terms of "the motion and other affections of the minute particles of matter"; but because these "affections… *are obvious* and very powerful *in mechanical engines*," Boyle adds, "I sometimes also term it the mechanical hypothesis or philosophy."[27] Similarly, Giambattista Borelli sometimes implicitly appeals to a relation between explanations of the body as, e.g., a lever, a more "big picture" sense of mechanism in which there are laws of nature, and somewhere in between, as it were holding the two together, statements like "The movement of the heart can result from an organic necessity as an automaton is moved."[28] We are in the presence of irreducible hybridity and heuristics: that is, rather than a watertight, restrictive ontology of size, shape, and motion, or an unnecessarily complicated theory of the relation between primary and secondary qualities, or even a kind of "machinic"-analogical project, mechanism turns out to be less foundationalist, more open-ended, more diverse.

Early modern mechanism comes in different forms and flavors: it can be ontological, analogical, or heuristic, sometimes in combination (mechanical analogies can be ontologized, and ontological mechanism can sometimes also serve as a heuristic, as when Descartes says he can't see the processes at work in generation, but they must be the same as he observed elsewhere in the physical world[29]). These relations, zones of influence and of overlaps are further complicated by the interplay between programmatic expressions such as "the mechanical philosophy," machines, practical mechanics, mechanical analogies, etc. This shows, at least so I will suggest, the presence of a more pluralistic, "tolerant" ontology in mechanism than is often thought, and additionally that mechanical models, including actual automata, are *not blind to "Life"* but motivated by it as a challenge. Recall that the basic, and intuitive contrast between bodies and machines, metallic and flesh-and-blood entities, whether applied in the history of doctrines in the life sciences ("mechanism versus vitalism") or as a critique of materialism/mechanism, implies that blindness.

One response to this misguided, at times ideological, and definitely inaccurate vision of things would be to examine the diversity of mechanistic and reductionist explanations (something that would overlap profitably both with current "neomechanism" in philosophy of biology, and with some scholarship like Andrault's and Bertoloni Meli's on early modern medical mechanism). However, I suggest a different point: that our intuition, our confidence in opposing "machines" to "bodies" is misplaced in an important way, both as regards machines and mechanism, on the one hand, and as regards our common intuitions concerning embodiment, on the other hand.

Regarding mechanism, my suggestion would be that the concept was never so separate from "life": in Georges Canguilhem's elegant phrase: "Essentially, a machine is a mediation or as mechanics say, a relay (*relais*),"[30] or, we might say, a go-between. Joan Landes, discussing eighteenth-century automata (Vaucanson's but also the Jaquet-Droz's) seems to make a more modest claim when she calls them "devices of Enlightenment," that is, "philosophically animated experimental objects" or "philosophical experiments," "provoking audiences to consider the workings of the body as a machine."[31] I would say that the automata are doing more than serving as thought-experiments (including because the cognitive relation goes both ways, from machine to body and from body back to machine, as has been discussed in relation to Vaucanson's rhetoric), and that "machine," "mechanism," "mechanical analogy" and "mechanistic ontology" are all in need of reevaluation, as described above. Further, the mechanistic approach to the (living) body is not necessarily an enterprise of denial or *elimination* of all vital properties, even if it does involve a degree of *reduction*.

It may be helpful to distinguish reductionism from eliminativism[32] but it is also possible that the distinction in the end does not fit early modern concerns. Contemporary terminology has distinguished between reductionism and eliminativism, both of which have a respectable materialist pedigree. In the classic case of the existence of the soul, eliminativism holds that the soul and all of its properties that have been described and argued over from, say, antiquity and Scholasticism through Swedenborg does not exist and indeed none of these properties are real; thus, what is real would be the brain, or the heart, or the stomach, and so on. Reductionism holds that the soul (to stay with the same example) is indeed not something that exists in any traditional sense; but notice that when La Mettrie says that we really should only use the word to refer to "that part of us which thinks," he is not saying mental faculties do not exist but that we need to rethink what their "seat" is, where they come from, and the extent to which they are independent from the rest of bodily processes, or not. However, he is not suggesting a weaker thesis, which would be that soul/mind might be autonomous in some sense but could be "defined in terms of" bodily processes. Materialist embodiment as discussed here has a commitment to reductionism, but not to eliminativism (although the extent to which this distinction is clearly applicable to the texts at hand is, again, unclear).

Regarding embodiment, I would like to suggest that there is no need to insist on an irreducibly personal—indeed, immaterial—distinguishing our being in a body/our body, from the status of "bodies in general" in a physical sense. Indeed, as I have suggested in earlier work, there was in fact a *specifically materialist sense of embodiment*. That is, materialism was not merely an obsessive reiteration or heightened performance of the "mechanistic" vision of the body, whatever that might be: reducing it to isolated parts or defining it in accordance with general mechanical laws, but in any case, factoring out the rich, fluid, *personal* sense of what it is to be in a body. This is not just because materialists frequently repeated like a mantra that everything that is real is body, or is a body, à la Hobbes.[33] Indeed, trumpeting that "all is body" or that "wherever I read 'soul' I replace it with 'body'" does not amount to a developed concept of embodiment. But what is embodiment?

Diderot, for instance, gives the "all is body" claim a more reductionist tonality, when, in the *Elements of Physiology*, an unpublished work which occupied him during the last two decades of his life, he explains that "the action of the soul on the body is the action of one part of the body on another, and the action of the body on the soul is again that of one part of the body on another" and elsewhere (here as marginal critique of another's work, viz. Hemsterhuis), "wherever I read *soul* I replace it with *man* or *animal*."[34] Similarly an equally representative early modern materialist, La Mettrie, declares in his first philosophical work, the *Natural History of the Soul* (1745) that "he who wishes to know the properties of the soul must first search for those which manifest themselves clearly in the body."[35] Unsurprisingly given the "body language" on both sides of the divide (that is, the attacks on materialists for reducing everything to the body, and the fascination with "what bodies can do" on the materialist side), a word that was used as a synonym for "materialism" in the late seventeenth century, if not a very common one, was "corporealism."[36]

What does it mean to state, "wherever I read *soul* I replace it with *man* or *animal*"? To use a familiar distinction, this is a reductionist rather than an eliminativist position, which should be clear if we contrast it with La Mettrie's statement that "The soul is just a pointless term of which we have no idea and which a good mind should only use to refer to that part of us which thinks."[37] Many such quotations can be adduced, whether they invoke the body or just matter/atoms in general. Briefly put, all materialists must necessarily be reductionists, but this leaves open the (big) question: reduction *to what*? In the name of *which science*? Because, as La Mettrie's statement shows, a reduction to the properties of basic physics is a very different creature from a reduction to the properties existing in medicine, or even chemistry.[38] Similarly, to replace "soul" with "man" is not to replace it with basic physics. The embodied-materialist faced with an entity like the soul can appeal to Epicurean notions of the material soul, to humoral ontology, to the world of affects, drives, and desires. But some might respond, has she not exited the province of materialism then?

The concept of embodiment is a rich and at times fuzzy one, as it can be claimed by work on embodied cognition, by feminist theory, by phenomenology, and by the "body studies" trend in the humanities.[39] One can say that in all this very diverse scholarship and theoretical writing, two features stand out: an emphasis on "our biology" as relevant to an understanding of the mind, consciousness, will, agency, (im)moral behavior, etc.; and an emphasis that we are not just *in* a body but that we have a deep, personal (and on some views, typically in phenomenology, ineffable) relation to our bodily states. We are not just bodies in a physical world, on this view, because we inhabit this world as *subjects*. Thus, the frequent denunciations of those who seek to explain life by the study of corpses[40]; thus the fear that we might be confused with machines.

This fear seems justified when we read some of the classic iatromechanist proclamations, such as the London anatomist William Croone's: "We shall consider the living body to be nothing else but a kind of machine or automaton."[41] Faced with this, we tend to feel that we have some broad intuitive grasp on the issue: a machine is a system of inanimate parts, presumably without a central controller, and certainly without an internal "vital principle." To borrow a phrase from a political context, applied here to Croone (and Pitcairn, and beyond), "The reactionaries tell the truth about the object they love, be it the Indian Office or the Bank of England for Keynes, or constitutional jurisprudence for Rawls."[42] Hence, when a living body—animal or human—is described as being *like a machine* (or "nothing else but a kind of machine or automaton"), we can feel fairly confident about what is happening: the various properties of organic life—the real, basic properties of what it is to be alive and in a body: self-maintenance, goal-directed behavior, and perhaps even intentionality or consciousness—are being reduced to basic mechanical properties. Yet we should notice that even these reductions are less straightforward than we might think: are living, embodied properties being defined as *the properties of machines* as understood at a given time and place, like Descartes' fountains, or as the basic *properties of nature understood mechanically*? It is after all different to say that the heart is *like* a pump, the lungs *like* a pair of bellows, and to state (like Boerhaave) that "the nature of the human body is the same as that of the whole of the Universe."[43]

It is not enough to oppose machine to body, steel to flesh, in terms of even a modified dualism, not least because mechanistic *ontology* is really a set of diverse *ontologies*, and much of the iatromechanist "puppet régime"[44] relies on mechanistic *analogies*—including their relation to "the flesh." The problem is not mixing the ontological and the analogical—as Des Chene showed beautifully,[45] that was part of the reality of early modern mechanism and indeed its power. Additionally, what is flesh or embodiment if we set aside appeals to a transcendentally grounded unity of experience, or an Aristotelian or Stahlian *anima*? Yet even a card-carrying materialist like Diderot insists on the difference between a metal watch and a flesh-and-blood watch (i.e., an organism): "What a difference there is, between a sensing, living watch and a golden, iron, silver or copper watch!"[46] One might say the challenge for the materialist on Diderot's view is to do justice to this difference! But what would the Diderotian materialist say of the cyborg? I suspect he is more invested in the potential and interest of "the flesh" (in the non-sublimated sense of the biological as a whole: molecules, tissue, fibers, spirits, brains…) than in the potential and interest of "the machine" (networks, subroutines, canalizations, ramifications…). The more general question, however, is what happens to the "dialectic" of machine and body I have described so far, once we add the specifically materialist appropriation and transformation of these concepts and debates?

Importantly, a feature of the materialist concept of embodiment that I hope I have made clear is that, unlike the approach to the body that sees it as just so many funnels, pulleys and bellows, or that seeks to establish basic mechanical laws of the body and the rest of nature, the embodied-materialist approach is "*visceral,*" figuratively and literally. (Even Leibniz suggests the possibility, in the *New Essays*, that "something occurs in the soul in response [to] the internal motions of the viscera."[47]) It is a materialism of vital fluids, touch, affects and passions. But second, this approach, if it is to be legitimately qualified as "materialist," necessarily has a *reductionist* component, in the sense of the ambition to explain a higher-level phenomenon X in terms of lower-level processes Y: "where I read *soul* I replace it with *man* or *animal*." Notice, however, that if I am reducing "soul" to "animal," my reducing theory or level of explanation is still something *alive*. The "machine" for La Mettrie or Diderot is much less a clock, watch, or even self-regulating steam engine, than it is a physiologically (and appetitively) determined system, that thinks itself free.

Thereby, even if Diderot is serious when he ponders the difference between machines and organisms, he is no way rejecting categorically that mechanism—mechanistic processes, causal processes consonant with a mechanistic ontology—could give rise to a life-form. Except that (a) he is more concerned with the living world, that of Buffon, Haller, Trembley, Needham et al., and (b) he tends, most if not all the time, to attribute vital properties to all of matter.[48] Additionally, his interest in the complexity of the flesh tends to emphasize what he calls *organization* (structural complexity) rather than some sort of phenomenology of the flesh. In other words, as a careful observer of the evolving medical (indeed, biomedical) fields, Diderot would not have shed a tear at the fact that "Flesh had been written out of medical discourse by the end of the nineteenth century and been replaced by differentiated tissues defined via their specific structure and function."[49]

In contrast, for some prominent contemporaries of Diderot, without even turning back to Cudworth's attacks on the stupidity of matter, "the least plant or insect" cannot be explained in mechanistic terms—the reader will hopefully have recognized Kant. For Kant sharply asks a question he clearly intends as rhetorical: "Are we in a position to say '*Give me matter, I will show you how a caterpillar can be produced*'?" and he adds that we will "sooner understand the formation of all celestial bodies, the cause of their motions, in short, the origin of the entire present arrangement of the world-edifice, than we will come to know distinctly or completely the production of a single herb or of a caterpillar from mechanical grounds."[50] It is not that I am suggesting that

Kant is scientifically or metaphysically mistaken whereas Diderot and La Mettrie are in the right; rather, I have highlighted the ways in which early modern approaches to "the machine" and "the body" are both oppositional and overlapping, hostile and hybrid—and that these "two complexities," mechanism and embodiment, steel and flesh, themselves both permeate the problem of materialism and are transformed by the "triangulation" with materialism itself.

The power of mechanistic analogies is not a "denial of the flesh" (thus by implication, the construct known as "mechanistic materialism" is something of a conceptual monster).[51] In fact, most early modern materialists were deeply concerned with what we would call embodiment—pleasure, the flesh, sensation, organic sympathies, instincts, psychosomatic interconnection, the specific nature of the brain, and so on. And the same applies to mechanism. Soul or mind needn't be denied in favor of brute matter (or body), although the materialist has a variety of strategies at her disposal. One strategy—eliminativism broadly construed—is to deny that there is such a thing as the soul. Irritability and other basic physiological properties would then account for the visible phenomena that we attribute, falsely, to a purely mental agent, a "sailor in the ship" of the body.[52] Another strategy—reductionism of the more mechanistic and less embodied sort—is to say that "soul" and its processes are real inasmuch as they can be assimilated to or explained in terms of basic mechanical laws (d'Holbach: "our minds are subject to the same physical laws as material bodies"[53]).

It is not that the embodied materialist denies that our mental processes are subject to basic laws of physics (although there was very little talk of such laws at the time). Rather, the corresponding form of reductionism she defends, the *embodied* version, when presented with, say, the sense of self, or appetite, or desire, does not look for strictly mechanistic ways of explaining it but rather seeks to embed it in a general "spirited and bloody" account, employing medical, biological, and physiological perspectives on what it is to be such a living agent. Thus, Diderot does not assert that there is no soul, but rather that "I challenge [you] to explain anything without the body"[54]—to which we could add the brain (as he does, briefly, in a discussion of memory and the role of the brain).[55]

If materialism is not merely mechanistic, but instead seeks to articulate an *embodied* account of our mental life (i.e., that which was traditionally accounted for by the soul, including volition, desire, etc.), that doesn't mean it will provide an account of intentionality or "first-person" states of experience that will satisfy everyone. But such states may in fact be nothing other than certain kinds of narratives or projections by the agents themselves: stories we tell ourselves.[56] Granted, some materialists, speaking about action, motivation, and desire, describe us as if we were no better than pigs wallowing in filth, in La Mettrie's provocative image. But where are these "disastrous results" that some speak of, that occur when materialism turns to the inner life, as one Diderot scholar wrote in the 1960s?[57] Such criticisms seem to either resurrect (or perpetuate) the moralistic criticisms of materialism in the seventeenth century, or to operate with a conception of the self (personhood, self-consciousness, etc.) that is itself at best an ungrounded claim, which is then opposed to a "merely passive" definition of matter. Just because the materialist cannot go along with the sacralization of the flesh, or the insistence that there must be a "self," a controller, a *hegemon* without which we are mere "stupid" matter, does not have to mean that materialist bodies are just piles of flesh, mere "aggregates" in the language of the period.

Machines, automata, and mechanistic models of the body are of course *resources* for materialism in the early modern period. This does not make the ensuing forms of materialism especially "mechanistic" (consider La Mettrie, Diderot). But those mechanistic concepts themselves are not as neatly or strictly opposed to "the flesh" as historical actors *and scholars* tend to repeat. This does not change the fact of "uncanniness," if a machine looks at us with "unwinking photocell eyes." Would this uncanniness be tempered if it was an embodied, material, flesh-and-blood agent looking at us, created in a laboratory? No doubt, only temporarily…

Notes

1. Jonathan Smallwood, Merrill McSpadden and Johnathan W. Schooler, "The lights are on but no one's home: meta-awareness and the decoupling of attention when the mind wanders," *Psychon. Bull. Rev.* 14 (2007), 527–533—building on a classic Dennettian intuition.
2. Stephen Gaukroger, *Descartes: an intellectual biography* (Oxford: Clarendon Press, 1995), 1.
3. Minsoo Kang, *Sublime Dreams of Living Machines: The Automaton in the European Imagination* (Cambridge, Massachusetts: Harvard University Press, 2011).

4. Jacques Rohault, *Entretiens sur la philosophie* (Paris: Michel Le Petit, 1671), II, 191. On responses to automata in particular see the excellent collection of papers edited by Aurélia Gaillard et al., *L'automate: modèle, machine, merveille* (Bordeaux: Presses universitaires de Bordeaux, 2012).
5. Ralph Cudworth, *A Treatise concerning Eternal and Immutable Morality*, posthumously published 1731, I, chap. II, § 8. For further reflections on the problem of 'stupid' versus 'intelligent' materialism, see my attempt in "History of materialism and intelligent materialism," in *The History and Philosophy of Materialism*, ed. C. Wolfe & J. Symons (London: Routledge, 2024), 323–333.
6. Immanuel Kant, *Metaphysical Foundations of Natural Science*, III, Prop. 3, Observation; Ak IV: 454; trans. M. Friedman, in Kant, *Theoretical Philosophy After 1781*, eds. H. Allison & P. Heath (Cambridge: Cambridge University Press, 2002), 252; *Critique of the Power of Judgment*, § 73, Ak. V: 394; trans. and ed. P. Guyer and E. Matthews (Cambridge: Cambridge University Press, 2000), 265. See also "Der Hylozoismus belebt alles, der Materialismus dagegen, wenn er genau erwogen wird, tödtet alles" (Ak. II: 330), and the comments in John Zammito, "The Lenoir thesis revisited: Blumenbach and Kant," *Stud. Hist. Philos. Biol. Biomed. Sci.* 43, no. 1 (2012): 120–132.
7. Jonathan Sawday, *The body emblazoned: dissection and the human body in Renaissance culture* (London: Routledge, 1995), 29.
8. Archibald Pitcairne, *Elementa Medicinae* (1717), translated as *The Philosophical and Mathematical Elements of Physick: In Two Books, the First Containing the Theory, the Second the Practice: Compos'd for the Use of All who Study the Art of Medicine* (London: Andrew Bell and John Osborn, 1718), § LXXXVIII, p. 354.
9. Carolyn Merchant, *The Death of Nature: Women, Ecology, and the Scientific Revolution* (New York: Harper and Row, 1980).
10. "materialism, by decomposing man into rigorously defined behaviors like in Taylorism, serves the purposes of the master: it is the master who conceives of the slave as being like a machine," Sartre, "Matérialisme et revolution," (1946) in Sartre, *Situations philosophiques* (Paris: Gallimard, 1990), 127–128.
11. Merchant, *The Death of Nature*, 170. For a useful critique of Merchant's reading of Bacon see Peter Pesic, "Proteus Rebound:

Reconsidering the 'Torture of Nature'," *Isis* 99 (2) (2008): 304–317.

12. Merchant, *The Death of Nature*, 1990 Preface, xvii.
13. Deborah Brown, *Descartes and the Passionate Mind* (Cambridge: Cambridge University Press, 2006), 1.
14. John Sutton, *Philosophy and Memory Traces. Descartes to Connectionism* (Cambridge: Cambridge University Press, 1998), 82.
15. E.g., for Lisa Sarasohn, "Margaret Cavendish used revolutionary scientific ideas and an underlying feminist ideology in her conception of a living universe, infused with motion, and ordered by a female spirit, which could best be understood from the empathetic viewpoint of a female scientist" (Lisa Sarasohn, "A Science Turned Upside down: Feminism and the Natural Philosophy of Margaret Cavendish," *Huntington Library Quarterly* 47(4) (1984): 290). For a very incisive critique of this kind of feminist reading of Cavendish see Deborah Boyle, "Margaret Cavendish's Nonfeminist Natural Philosophy," *Configurations* 12(2) (2004): 195–227.
16. Robert Boyle, *The Origin of Forms and Qualities* (1666) in Robert Boyle, *The Works of the Honourable Robert Boyle*, 6 vols., ed. T. Birch (London: J. and F. Rivington et al., 1772; reprint, Hildesheim: Georg Olms, 1965), III, 13; *Usefulness of Experimental Natural Philosophy*, II (1671), in Boyle, *Works*, III, 427.
17. René Descartes, *Treatise on Man*, AT XI, 120, 130–131, 202.
18. In a 1703 lecture revealingly entitled "On the Usefulness of Mechanical Methods in Medicine": Herman Boerhaave, *De usu ratiocinii mechanici in medicina*, in *Opuscula selecta Neerlandicorum de arte medica*, I (Amsterdam: F. van Rossen, 1703/1907), 46; Herman Boerhaave, *Boerhaave's Orations*, trans. and ed. E. Kegel-Brinkgreve and A.M. Luyendijk-Elshout (Leiden: E.J. Brill/ Leiden University Press, 1983), 96.
19. Jean-Pierre Dupuy, *The Mechanization of the Mind: On the Origins of Cognitive Science*, trans. M. B. DeBevoise (Princeton: Princeton University Press, 2000).
20. Raymond Ruyer, "Ce qui est vivant et ce qui est mort dans le matérialisme," *Revue Philosophique* 116, no. 7–8 (1933): 28–49; Hans Jonas, *The Phenomenon of Life. Towards a Philosophical Biology* (New York: Harper & Row / Dell, 1966).

21. I am not presuming to tell, in a few pages, a (Heideggerian or Harawayian) story of how we have moved from seventeenth-century bodies and machines (and the *mentalités* and perceptual schemes accompanying them, as might be found in a virtuoso analysis like Benjamin's account of the historicity of perception, from painterly to photographic to cinematic eras) to steam engines, Frankenstein monsters, and onto replicants and CRISPR/Cas9 clones. If Ian Hacking couldn't pull that off ("Our Neo-Cartesian Bodies in Parts," *Critical Inquiry*, 34(1) (2007): 78–105), I don't see how I could. Rather, I am interested in shifting conceptual frameworks (involving mechanism and embodiment, mechanistic materialism and embodied materialism—and the flesh).
22. Thus Hans Jonas writes that "the modern life-sciences represent […] a scheme which is essentially the residual of a subtraction, viz. of the zoomorphic elements in experience […] Expatriated from their own birthplace, they are now more homeless in the scheme of science than is Ahasuerus in a world which at least lets him wander in its midst. Each experience, including the scientific undertaking itself, reaffirms the disowned traits by the very act: their claim is bought off by according them a shadowy existence in the no-man's land of 'subjectivity', the convenient Hades of natural science, as if this disposed of them and left for scientific description an untainted 'objectivity' which need no longer answer for them" (Hans Jonas, *Organism and Freedom*, chapter 1, 'Life' and the Scientific Spirit, in *Organismus und Freiheit. Philosophie des Lebens und Ethik der Lebenswissenschaften*, (Freiburg: Rombach, 2009)).
23. Donna Haraway, "A Cyborg Manifesto: Science, Technology, and Socialist-Feminism in the Late Twentieth Century" (1985), in *Simians, Cyborgs and Women: The Reinvention of Nature* (New York: Routledge, 1991); N. Katherine Hayles, "Flesh and Metal: Reconfiguring the Mindbody in Virtual Environments," *Configurations* 10 (2002): 297–320; Victoria Pitts-Taylor, *The Brain's Body: Neuroscience and Corporeal Politics* (Durham, NC: Duke University Press, 2016).
24. See Jessica Riskin, "The Defecating Duck, or, the Ambiguous Origins of Artificial Life," *Critical Inquiry* 29, no. 4 (2003): 599–633; Riskin, ed., *Genesis Redux: Essays in the History and Philosophy of Artificial Life* (Chicago: University of Chicago Press, 2007); Charles T. Wolfe, "Le mécanique face au vivant," in *L'automate:*

modèle, machine, merveille, ed. Aurélia Gaillard et al. (Bordeaux: Presses universitaires de Bordeaux, 2012), 115–138. At the risk of stating the obvious, I am not engaging here with questions like the relation between the growing fascination with automata and the rise of industrial capitalism, e.g., in cases like the invention of the steam engine (on this relation, including in Marx's *Capital*, see Adelheid Voskuhl, "The Android and the Machine: Materialism, Mechanicism, and Industrialism in the Early and Late Modern Ages," *SubStance* 47 no. 3 (2018): 7–26).
25. Nicole Howard, "Neither Comic Nor Corpse: Revisiting Bracelli's *Bizzarie di Varie Figure*," this volume.
26. In her analysis of the so-called 'living instruments' of the 17th-century Dutch inventor Cornelis Drebbel, Vera Keller speaks of "the ways quantitative mechanical philosophy intervened in an already robust, machine-based, but non-mechanical natural philosophy" ("Drebbel's Living Instruments," *History of Science* 48, no. 1 (2010), 42). See also Domenico Bertoloni Meli (here synthesizing numerous scholars' works on the complexities of early modern mechanism, including Des Chene, Garber, Roux, etc.): *Mechanism. A visual, lexical and conceptual history* (Pittsburgh: University of Pittsburgh Press, 2019).
27. Robert Boyle, Preface to "Some specimens of an attempt to make chymical experiments useful to illustrate the notions of the corpuscular philosophy," in *Certain Physiological Essays*, *Works* I, 356.
28. *De motu anim.* II, ch. VI, Prop. LXXIX, *On Movement of Animals,* 282.
29. "The formation of all the parts of the human body [...] is something so difficult that I dare not attempt (to explain it) yet" (Descartes to Elisabeth, May 1646, AT IV, 407); as Dennis Des Chene comments: "Among the phenomena of life, generation offers, along with the apparently reasoned behavior of higher animals, the greatest challenge to a science based on Cartesian principles" (Des Chene, "Life After Descartes: Régis on Generation," *Perspectives on Science* 11, no. 4 (2003), 413).
30. Georges Canguilhem, "Aspects du vitalisme," in Canguilhem, *La connaissance de la vie*, revised edition (Paris: Vrin, 1965 (1952)), 87.

31. Joan Landes, "Vaucanson's Automata as Devices of Enlightenment," *Sjuttonhundratal* 8 (2011), 54–56. For a useful survey of the automata of the eighteenth century, android and other, see Adelheid Voskuhl, "The Android and the Machine," op. cit.
32. See my "Forms of Materialist Embodiment" and work on Descartes by authors such as Stephen Gaukroger, John Sutton, and differently Barnaby Hutchins.
33. Pierre-François Moreau, *Hobbes. Philosophie, science, religion* (Paris: PUF, 1989).
34. Denis Diderot, *Éléments de physiologie*, in Diderot, *Œuvres complètes*, ed. H. Dieckmann et al. (Paris: Hermann, 1975), XVII, 334–335.
35. Julien Offray de La Mettrie, *Traité de l'âme*, I, in *Œuvres philosophiques*, ed. F. Markovits, 2 vols. (Paris: Fayard—"Corpus", 1987), 125.
36. Cudworth describes "Corporealists" as those who "acknowledge no other *Substance* besides *Body* or *Matter*": *The True Intellectual System of the Universe* (London: R. Royston, 1678; Reprint, Hildesheim: Georg Olms, 1977), I, ch. III, § xxx, 135. Coleridge, speaking of his past loyalties to Hartleyan associationism, writes that he had believed in "the corporeality of thought, namely that it is motion" (*Collected Letters*, ed. Griggs, 6 vols. (Oxford: Clarendon Press, 1956–1971), I, 137–138).
37. La Mettrie, *L' Homme-Machine*, in *Œuvres philosophiques*, I, 98.
38. As has been studied notably by François Pépin, Gabriel-François Venel's entry "Chymie" in the *Encyclopédie* is flagrantly a small treatise of *vital* chemistry, obsessively returning again and again to the question of *life*.
39. Gail K. Paster, "Nervous Tension: Networks of Blood and Spirit in the Early Modern Body," in *The Body in Parts*, ed. D. Hillman and C. Mazzio (London: Routledge, 1997), 107–25; Ulinka Rublack, "Fluxes: The Early Modern Body and the Emotions," *History Workshop Journal* 53 (2002): 1–16.
40. In his writings on painting, Diderot recommends painting from corpses (which offended Goethe) and elsewhere he approved of the idea that prisoners condemned to death could be used for scientific experimentation. Authors associated with the 'Montpellier vitalist' school, conversely, tended to criticize the recourse to anatomy (including observations based on cadaverous anatomy).

41. William Croone, *De ratione motus musculorum* (London: S. Hayes, 1664), sect. XXVI, 15.
42. Michael Hardt & Antonio Negri, *The Labor of Dionysus. A Critique of the State-Form* (Minneapolis: University of Minnesota Press, 1994), 19.
43. Herman Boerhaave, "On the Usefulness of Mechanical Methods in Medicine" (*De usu ratiocinii mechanici in medicina*, 1703), trans. in Boerhaave's *Orations*. Translated and edited by E. Kegel-Brinkgreve and A.M. Luyendijk-Elshout (Leiden: E.J. Brill/Leiden University Press, 1983), 96.
44. Richard Westfall thus judged the scientific pertinence of iatromechanism harshly: "Iatromechanism did not arise from the demands of biological studies; it was far more the puppet regime set up by the mechanical philosophy's invasion [...] For the most part, iatromechanism was simply irrelevant to biology. It did not prevent the vital work of detailed observations; it contributed almost nothing to understanding what was seen. Besides the subtlety of biological processes, the seventeenth century's mechanical philosophy was crudity itself. Above all, it lacked a sophisticated chemistry which has turned out to be a prerequisite for a rapprochement of the physical and biological sciences. One can only wonder in amazement that the mechanical explanations were considered adequate to the biological facts, and in fact iatromechanics made no significant discovery whatsoever" (*The Construction of Modern Science. Mechanism and Mechanics* (Cambridge: Cambridge University Press, 1971), 104). For a sophisticated (and convincing) version of the opposite view, see Raphaële Andrault, *La raison des corps. Mécanisme et sciences médicales* (Paris: Vrin, 2016). Andrault has also raised concerns about the status of iatromechanism as a historiographic category, which lie beyond the scope of the present essay.
45. Dennis Des Chene, *Spirits and Clocks: Machine & Organism in Descartes* (Ithaca, NY: Cornell University Press, 2001).
46. Diderot, *Éléments de physiologie*, op. cit., 335. For my early attempts to unpack this idea, see my "Machine et organisme chez Diderot." *Recherches sur Diderot et l'Encyclopédie* 26 (1999): 213–231.

47. Gottfried Wilhelm Leibniz, *New Essays on Human Understanding*, ed. and trans. Peter Remnant and Jonathan Bennett (Cambridge: Cambridge University Press, 1996), II.i.15.
48. See Charles Wolfe, "Sensibility as vital force or as property of matter in mid-eighteenth-century debates," in *The Discourse of Sensibility. The Knowing Body in the Enlightenment,* ed. Henry Martyn Lloyd (Dordrecht: Springer, 2014), 147–170.
49. Mechthild Fend, *Fleshing out Surfaces: Skin in French Art and Medicine, 1650–1850* (Manchester: Manchester University Press, 2017), 12.
50. *Universal Natural History and Theory of the Heavens*, 1755, Ak. I: 230. See also: "it would be absurd to regard the first production of a plant or animal as a mechanical side-consequence [*mechanische Nebenfolge*] of universal laws of nature" (*Only Possible Proof,* 1763, Ak. II: 114). For discussion, see Hannah Ginsborg, "Two kinds of mechanical inexplicability in Kant and Aristotle," *Journal of the History of Philosophy* 42, no. 1 (2004): 33–65.
51. In earlier work I tended to oppose 'mechanistic' and 'vital' materialism too strongly, and sometimes even claimed that there was no such thing as mechanistic materialism, indeed an excessive judgment (I thank John Zammito for making me see this). For some reflection, see my "Varieties of vital materialism," in *The New Politics of Materialism. History, Philosophy, Science*, eds. S. Ellenzweig and J. Zammito (London: Routledge, 2017), 44–65.
52. That the immaterial soul is in the material body like a sailor in a ship is something that Aristotle considers (*De Anima* II, i, 413a5) and Descartes rejects, without mentioning Aristotle, and sounding for all the world like a phenomenologist: "Nature... teaches me, by these sensations of pain, hunger, thirst and so on, that I am not merely present in my body as a sailor is present in a ship, but that I am very closely joined and, as it were, intermingled with it, so that I and the body form a unit" (Sixth Meditation, AT IX, 64).
53. Paul Henri Thiry, Baron D'Holbach, *Système de la Nature ou des lois du monde physique et du monde moral* (London, 1770); reprint of 2nd edition (1781), ed. J. Boulad-Ayoub (Paris: Fayard-"Corpus", 1990), I, xi, 220.
54. Diderot, *Éléments de physiologie*, in *Œuvres complètes*, XVII, 334.
55. See Charles Wolfe, "'The brain is a book which reads itself'. Cultured brains and reductive materialism from Diderot to J.J.C.

Smart," in *Mindful Aesthetics: Literature and the Science of Mind*, *eds.* H. Groth and C. Danta (London: Continuum, 2014), 73–89.

56. See Charles Wolfe, "Diderot and materialist theories of the self," in *The Self and Self-Knowledge in Early Modern Philosophy*, eds. Gábor Boros, Judit Szalai and Oliver Istvan Toth (Budapest: Eötvös Loránd University Press, 2020), 37–52.
57. "Materialism as a working philosophy, used as a tool in the scientific investigation of the material universe, is appropriate and highly effective. Intended for the objective analysis and description of the world of externals, it yields disastrous results when applied to the inner, subjective world of human nature, thought, and emotions" (Emita Hill, "Materialism and Monsters in the *Rêve de D'Alembert*," *Diderot Studies* 10 (1968), 90).

CHAPTER 5

Descartes on Clocks and Automata

Jil Muller

Introduction

The most famous metaphor in Descartes's writings, next to the wax seal, is probably the metaphor that opens the *Treatise on Man*, where Descartes compares the human body to a machine. "I suppose the body to be nothing but a statue or machine made of earth, which God forms with the explicit intention of making it as much as possible like us" (AT XI, 120/ CSM I, 99).[1] The basic form of the human body is thus a statue or construct made of earth and consequently immovable, and yet this machine resembles the human being. At first sight, this seems contradictory: man is the natural moving object *par excellence,* and he is compared to an immovable statue. Is the God-made statue of the *Treatise* a mere simulacrum of the human being, an illusion to deceive the viewer? Descartes adds: "Thus God not only gives it externally the colors and shapes of all the parts of our bodies but also places inside it all the parts required to make it walk, eat, breathe, and indeed to imitate all those of our functions which can be imagined to proceed from matter

J. Muller (✉)
University of Paderborn, Paderborn, Germany
e-mail: Jil.muller@uni-paderborn.de

J. Muller (ed.), *Automata, Cyborgs, and Mutants*, Palgrave Studies in Medieval and Early Modern Medicine,
https://doi.org/10.1007/978-3-031-93201-4_5

and to depend solely on the disposition of our organs" (AT XI, 120/ CSM I, 99). Descartes thus recognizes an external part and an internal part of the machine. The external part is matter, a machine made of earth. The internal part, however, has different pieces with different functions, which proceed from matter and are possible because of the disposition of the organs.

This passage from the *Treatise on Man* reminds us of Descartes's explanation in a letter to Reneri for Pollot of April or May 1638: "[...] suppose that he [a man] was very devoted to the study of mechanics, and had made, or helped to make, various automatons shaped like a man, a horse, a dog, a bird, and so on, which walked and ate, and breathed, and so far as possible imitated all the other actions of the animals they resembled, including the signs we use to express our passions, like crying when struck and running away when subjected to a loud noise" (AT II, 39–40/ CSM III, 99). In general, the actions of the God-made machine and the man-made automaton are the same: eating, breathing, and walking. However, the automaton in the letter seems to have even more functions: Descartes adds crying and running away (which can be associated with instinctive reactions).

A comparison of these two passages reveals two significant discrepancies. Firstly, the artifact created by God is defined as a machine, while that constructed by man is referred to as an automaton. Secondly, the automaton appears to possess a greater number of qualities than God's machine. It seems that Descartes aims to establish a clear distinction between the two concepts of machine and automaton, particularly in terms of their quality and functions. In the *Cambridge Descartes Lexicon*, under the entry "automaton," Dennis Des Chene holds that machines and automata are not the same. "Automaton and machine, though allied, are distinct concepts. A Cartesian machine is a collection of bodies whose operations can be explained on mechanical principles alone: its parts are just parts of extension, and their motions are entirely subordinated to the laws of nature. [...] An automaton, on the other hand, is a self-mover, the proximate cause of certain of its changes."[2] An automaton is a self-acting device; it has the cause of its action within itself, whereas the machine is subject to the laws of nature. But does this really mean, as Des Chene seems to suggest, that the automaton and the machine are intrinsically different and that Descartes uses them as two different concepts?

With regard to the first point, the linguistic distinction between machine and automaton is maintained by Descartes at the end of the

Treatise on Man, yet the contrast seems to be diminished by Descartes himself and therefore somehow contradicts the definition given by Des Chene: “I should like you to consider that these functions follow from the mere arrangement of the machine’s organs every bit as naturally as the movements of a clock or other automaton follow from the arrangement of its counter-weights and wheels” (AT IX, 202/ CSM I, 108). For Descartes, the internal configuration of the machine is analogous to that of a clock or automaton, indicating a probable similarity in operational functionality.

In consideration of the second point, namely that the automaton exhibits a greater number of qualities than God’s machine, it is possible to narrow the affirmation by examining the rest of the letter for Pollot: “Let us suppose that our man had noticed this, and so believed firmly that if there were automatons made by God or nature to imitate our actions, they would imitate them more perfectly, and be incomparably more skillfully constructed than any which could be invented by man” (AT II, 41/ CSM III, 100). In the eyes of Descartes, the artifacts created by God are more sophisticated and more perfect than those created by man. Furthermore, in the *Discourse*, Part V, Descartes reiterates the complexity of God’s creation: “This will not seem at all strange to those who know how many kinds of automatons, or moving machines, the skill of man can construct with the use of very few parts, in comparison with the great multitude of bones, muscles, nerves, arteries, veins and all the other parts that are in the body of any animal. For they will regard this body as a machine which, having been made by the hands of God, is incomparably better ordered than any machine that can be devised by man, and contains in itself movements more wonderful than those in any such machine” (AT VI 55–56/ CSM I, 139).[3] In this context, the term “automaton” is used to refer to a moving machine, without any distinction between the two concepts. This raises the question of whether the terms “machine” and “automaton” are synonymous in Descartes’s writings or whether there is a fundamental distinction between the two concepts, as supposed by Dennis Des Chene. Alternatively, might Descartes be responsible for a conceptual shift?

The human-like machine, as conceived by Descartes in the *Treatise on Man*, is a composite and complex construction, comprising a variety of mechanical functions. These functions can be related to real-existing mechanical machines and procedures, which must operate together in order to imitate the human body. Descartes employs the machine and automaton analogies to enhance the plausibility and intelligibility of his mechanical philosophy. This is also evident in the case of the human-like machine, whereby the invisible is rendered visible through the utilization of a "mechanical" metaphor. Additionally, there are references to a variety of machines and/or automatons, which Descartes employs as analogies to facilitate discourse on the human body.

In order to demonstrate that, as I propose, Descartes considered all automata to be machines, but not all machines to be automata, it is first necessary to examine the concept of a machine in Descartes's writings. This analysis will concentrate on passages that address the function of the human body, as these contain numerous references to machines and automatons. Subsequently, the analysis will focus on specific devices, including fountains, mills, organs, and clocks, and their relationship to the functioning of the human body. Finally, the use of automata in Descartes's writings will be examined to ascertain whether it can be concluded that, for Descartes, automata are machines.

Descartes's Understanding of Machines

In the seventeenth century, mechanical philosophy used different artifacts,[4] machines and constructions to visually explain the human body.[5] As Domenico Bertoloni Meli has underlined: "The pendulum, the spring, the barometer, but also locks and sluices, played an important role in conceptualizing the body, together with notions associated with them, such as periodic motion, elasticity, pressure, and unidirectional flow."[6] These mechanical procedures, systems and functions can be visualized and explained. In his works, Descartes uses the term "machine" nearly 200 times, because the machine metaphor makes the invisible visible and the incomprehensible explainable through the laws of nature.[7] In the *Principles of Philosophy* (art. 203), he explains: "In this matter I was greatly helped by considering artefacts. For I do not recognize any difference between artefacts and natural bodies except that the operations of artefacts are for the most part performed by mechanisms which are large enough to be easily perceivable by the senses [...] The effects produced in nature, by contrast, almost always depend on structures which are so minute that they completely elude our senses" (AT IXb, 321/ CSM I, 288). Because the devices used by man are larger and therefore easily perceivable by the human senses, the artifacts can give "insight" into the functions of natural bodies.

But what is a machine for Descartes? Though Descartes himself does not give any precise definition of a machine, we can formulate a general definition: a machine is (i) "an 'assemblage' or 'aggregate' of distinct functional parts, linked together by their mutual action; (ii) [...] an instrument; (iii) [...] artificial, or at least [...] the paradigmatic machine is an artifact."[8] Or as Daniel Garber explains: "The machine is an artifact, constructed to do some particular piece of work. Considered as an artifact, it has an end, of course; it is constructed by an agent and for some purpose of interest to that agent. This conception of the machine is reflected in the science of mechanics, which is the science of the artificial, the science whose object is the construction and explanation of artifacts built to do things we need to have done."[9] A machine is a mechanical device that acts under the influence of the laws of nature. A machine is created for a certain purpose, and to fulfill or obtain this purpose, the machine, an assemblage of different parts, has to transmit or modify forces, power, and motion, and this in a predetermined manner. In this sense, clocks and water fountains are machines: they are created for a certain purpose and their parts are assembled in a certain way to produce this result.

Louis de la Forge tried to give a definition in his commentary on the *Treatise on Man*: a machine is "[A] body composed of several organic parts, which being united, work together to produce several movements, of which they would not be capable if they were separated. I call organic parts all sorts of simple or complex bodies, which being united together are able to help through their structure, shape, movement, rest, and location, in the production of the motions and functions of the machine of which they are parts."[10] With Gideon Manning, we agree that La Forge says more than Descartes, when using the word "organic." Yet, what is important here is the point that all the parts of the machine have to work together and that there is no "soul" or "mind" in the machine, which controls these motions, nor a principle of self-moving. The motions are produced because the parts of the machine are united.

And La Forge continues his definition: "This being so, not only are clocks and other automata machines, but also the body of man, that of all animals, and the whole universe itself can be considered a machine."[11] We see that La Forge clearly indicates that automatons are machines (but not the contrary), and that, like Descartes, he distinguishes between the two concepts. However, the sentence delivers another interesting idea, which shapes the definition of a machine: the human body and the universe are considered a machine (but not explicitly an automaton).

In the 6th Meditation, among others, Descartes describes the human body as an artifact composed of different parts, thus a machine: "[...] I might consider the body of a man as a kind of machine equipped with and made up of bones, nerves, muscles, veins, blood and skin in such a way that, even if there were no mind in it, it would still perform all the same movements as it now does in those cases where movement is not under the control of the will or, consequently, of the mind but occurs merely as a result of the disposition of the organs" (AT IXa, 67/ CSM II, 58). That the movement is produced through the disposition of the organs is an addition in the French version of the *Meditations*, probably because Descartes had to answer, among others, Arnauld's critique of involuntary movement, i.e., movement not produced by the will or the mind, which means a totally instinctive movement.[12] Descartes writes to Arnauld: "[...] I think the most important point is that, both in our bodies and those of the brutes, no movements can occur without the presence of all the organs or instruments which would enable the same movements to be produced in a machine. [...] Now a very large number of the motions occurring inside us do not depend in any way on the mind. These include heartbeat, digestion, nutrition, respiration when we are asleep and even when we are awake, and also such waking actions as walking, singing and the like, when these occur without the mind attending to them" (AT IXa, 178/ CSM II, 161).

And in the *Principles of Philosophy* in article 188, Descartes likened the universe to a machine: "Up till now I have described this earth and indeed the whole visible universe as if it were a machine: I have considered only the various shapes and movements of its parts" (AT IXb, 310/ CSM I, 279). So not only the human body can be considered a machine, but even the universe.[13] This shows that the concept of device or machine is a very complex one in Descartes's works, and that it must cover different mechanical functions, which account for the different mechanical systems of the body and the earth.

In this context, let us come back to the idea of the human-like machine. In the *Treatise on Man*, it is a very sophisticated construction, to which God has given a greater variety of movements than we can imagine at first sight. They are more sophisticated than the movements we can find in man-made machines because God's power goes beyond human imagination. Descartes affirms that God's creation is more than a "simple" machine: "We see clocks, artificial fountains, mills, and other such machines which, although only man-made, have the power to move of their own accord in many different ways. But I am supposing this machine to be made by the hands of God, and so I think you may reasonably think it capable of a greater variety of movements than I could possibly imagine in it, and of exhibiting more artistry than I could possibly ascribe to it" (AT XI, 120/ CSM I, 99).

The wonderful movements, cited in the *Discourse* too, are beyond our capacity of imagination, so that we have to resort to examples we know (such as clocks, fountains, and mills) and only imagine them as more sophisticated. Consequently, if we want to understand the principles and functioning of the human-like machine created by God, and used as a metaphor by Descartes to explain the internal parts of the human body, then we have to consider automata and machines, the mechanical functioning of which we know well.[14] This mechanical functioning is used by Descartes to speak hypothetically[15] about the internal part of the human body: the part he could not observe or examine in a living human body.[16] It would be advantageous to analyze Descartes's utilization of specific devices, those fountains, mills and organs, in relation to the functioning of the human body, as well as examine the clock metaphor.

Artificial Fountains, Mills, and Organs

In the *Treatise on Man*, Descartes underlines the idea that the human-like machine is a composition of different devices and artifacts, so that we find references and metaphors to fountains, mills, and clocks. One important reference is to the artificial fountains in the royal garden: "Similarly you may have observed in the grottos and fountains in the royal gardens that the mere force with which the water is driven as it emerges from its source is sufficient to move various machines, and even to make them play certain instruments or utter certain words depending on the various arrangements of the pipes through which the water is conducted" (AT XI, 130/ CSM I, 100). This artificial fountain is a very complex device: water does not simply flow through pipes, but by the paths it takes, it operates instruments and emits sounds. Moreover, the fountain is set in motion by the water that emerges from its source with a certain force. The fountain is therefore, on the one hand, a self-mover (the motion is caused by the mere force of the water coming from the source) and on the other hand, its results depend on the disposition of the pipes and instruments (the purpose for which it was created is only produced by combining the individual parts together).

Descartes continues by comparing the neuro-motor system of man to the hydraulic system of fountains: “Indeed, one may compare the nerves of the machine I am describing with the pipes in the works of these fountains, its muscles and tendons with the various devices and springs which serve to set them in motion, its animal spirits with the water which drives them, the heart with the source of the water, and the cavities of the brain with the storage tanks” (AT XI, 131/ CSM I, 100). With this comparison, he can offer a mechanical description of bodily movements. The animal spirits (like the water in the fountains) flow through the human body and trigger reactions, and the heart acts like the source of the water, i.e., it pushes into the body the blood, transporting the animal spirits into the organs, muscles, and tendons. The animal spirits are driven by the pineal gland, which is “the principal seat of the soul” (AT XI, 351–351/ CSM I, 340) and the heart, having the same force as a water source, can be compared to a self-moving principle. On the other hand, Descartes underlines in the first sentence that the nerves are like pipes and the muscles and tendons like the devices and springs which are composed in a certain way to produce a result. This shows that the royal garden fountains seem to be, at the same time, a machine and an automaton.

He even explains respiration by referring to a “water” metaphor. Descartes associates breathing with the mechanical functioning of a mill or a clock. “Moreover, breathing and other such activities which are normal and natural to this machine, and which depend on the flow of the spirits, are like the movements of a clock or mill, which the normal flow of water can render continuous” (AT XI, 131/ CSM I, 100–101). This analogy could hide a dilemma: Descartes speaks of the “normal flow of water” and simultaneously speaks of spirits. In the hydraulic system of a watermill, it is the flow of water that keeps the mill moving. The watermill is thus under the influence of the laws of nature, i.e., the laws of the hydraulic system, and has to be called a machine. In contrast, the spirits are associated with the pineal gland and the soul, suggesting a more complex relationship than that of a simple machine. When the spirits are activated by the pineal gland, their movement is dependent on either the self-moving principle of the soul or they are involuntary movements. Without going into detail, let us just remember that even involuntary movements (without the will intervening) find their starting point, so to speak, in the pineal gland, for all movements are triggered by the animal spirits, directed or not by the will.[17] Consequently, with the watermill metaphor, Descartes can mechanically explain respiration, but again he has to mix the explanation of a machine and an automaton.

Furthermore, the neuro-motor system of man is compared to the organ and its pipes.[18] "You can think of our machine's heart and arteries, which push the animal spirits into the cavities of its brain, as being like the bellows of an organ, which push air into the wind-chests; and you can think of external objects, which stimulate certain nerves and cause spirits contained in the cavities to pass into some of the pores, as being like the fingers of the organist, which press certain keys and cause the air to pass from the wind-chests into certain pipes" (AT XI, 165/ CSM I, 104). This passage not only indicates how the animal spirits are conducted to the different cavities of the brain but even gives an indication for the heartbeat (one of the added functions in the Response to Arnauld). For Descartes, the heartbeat is produced because blood is pushed out of the heart; this push causes the little doors to open to let the blood in and out. The very complex system of the heartbeat, however, makes it impossible to have the four doors open or closed simultaneously: it is the hydraulic pressure in the heart that opens two little doors while closing the other two. In this sense, the heartbeat system depends on the laws of nature that is hydraulic pressure, and is therefore to be considered a machine. On the other hand, blood circulation, including the heartbeat, is a closed system in which the blood continues to move and could therefore be considered a self-moving system, as we have seen before. It can be seen that the intricate human body serves to demonstrate that, in the eyes of Descartes, this automaton is, in fact, a machine. This is due to the fact that the human body is subject to the laws of nature while simultaneously containing an inherent self-moving principle.

The Clock Metaphor

Alongside the metaphors of mills, fountains, and organs, there is one main and very important metaphor in Descartes's texts, namely the clock metaphor. One has to say that the clock came to be widely accepted by civil society at the end of the Middle Ages and during the Renaissance by becoming mechanical. "In the late fourteenth century, the terms *horloge* and *clokke* first appears in English to indicate an instrument of time-keeping, the latter (from Old French cloche, 'a bell') applied in particular to an instrument that sounded the hours or portions of hours by a bell or striking mechanism."[19] However, the clock did not only regulate human activity but even offered through its mechanical system the possibility to track the planets[20] and to measure the pulse, for example.[21] In this sense, clocks were probably the first devices that men embellished to indicate more than just the time. In Venice, in 1512, for example, Poliziano imagined the clock together with an automaton of moving figures.[22] And even earlier, the Strasbourg clock and a clock in Mantua were very ornate. "The famous Strasbourg clock (1352–54) presented to city dwellers a tablet 'showing the body parts and their correlation with the Zodiac for the favorable and unfavorable times for bloodletting', and in 1473 one Bartolomeo Manfredi described a public clock in Mantua that showed the time 'for phlebotomy, for surgery, for making dresses, for tilling the soil, for undertaking journeys and for other things very useful in this world'."[23]

The Strasbourg clock was so famous that even Descartes refers to it, in a letter to Mersenne in 1629: "[…] to say that there were springs and pipes, to express the whole *Pater noster*, like the crowing of the cock in the Strasbourg clock, I find it hard to believe" (AT I, 25). Descartes rejected the idea that an automaton could express the entire *Pater noster*. He was concerned that such a device might be a fraud. However, in the *Treatise on Man*, Descartes acknowledged that certain words can be uttered by artificial fountains. He noted that well-designed pipes and devices can create sounds that resemble words, but that they are not an accurate representation of the entire *Pater noster*.

In the *Treatise on Man*, Descartes uses the reference to the clock when he finally sums up all possible external and internal functions of the human body that are not caused by the will. "I should like you to consider that these functions follow from the mere arrangement of the machine's organs every bit as naturally as the movements of a clock or other automaton follow from the arrangement of its counter-weights and wheels" (AT XI, 202/ CSM I, 108). If the mills, fountains, and organs were employed to elucidate particular mechanical systems within the human body, it becomes evident that Descartes employs the metaphor of the clock to encapsulate these functions within a unified system. In this passage, Descartes presents the argument that, for him, an automaton is a machine, although he does not claim that all machines are automatons.

Descartes refers here to the automaton, or self-moving machine, and associates this concept with the configuration of wheels, which, in conjunction with the laws of nature, facilitate the functioning of the clock.

In an earlier passage of the *Treatise on Man*, however, Descartes spoke of the clock in relation to the functioning of a mill (cf. AT XI, 131), which would indicate that he understands the mechanical functioning of a clock as a self-moving principle, so as an automaton—once set in motion, the clock makes its way, it traces its circle—and as a machine depending on the laws of nature.

In a letter to Morin, on September 12th, 1638, Descartes speaks of a clock as a machine, namely by indicating that its function is a result of the laws of nature: "There is no more problem than if I said that a [wheel] clock shows the time only by the movement of its hands, and that its quality of showing the time is not a more actual or absolute being than its movement, and that this movement belongs to it by its nature and essence, because it would cease to be a clock if it did not have it" (AT II, 367/ CSM III, 121–122). We see that the function of the clock depends on the disposition of its wheels: to trace the circle, the wheels had to be well-arranged and they have to act under the laws of nature. Descartes does not mention the laws of nature in this quote, but he says that the movement belongs naturally to the clock. We understand that he refers to the laws of nature, when we remember the definition he gave in the *Principles of Philosophy*, Part Two, art. 37: "From God's immutability we can also know certain rules or laws of nature, which are the secondary and particular causes of the various motions we see in particular bodies" (AT IXb, 84/ CSM I, 240). Motion, for Descartes, is under the laws of nature.

Consequently, bodies that can be explained physically can also be explained by the laws of nature. A similar idea is mentioned in a passage in the *Principles of Philosophy* (art. 203): "Moreover, mechanics is a division or special case of physics, and all the explanations belonging to the former also belong to the latter; so, it is no less natural for a clock constructed with this or that set of wheels to tell the time than it is for a tree which grew from this or that seed to produce the appropriate fruit" (AT IXb, 322/ CSM I, 288). The clock's functioning can thus be explained by the laws of physics or nature.

Another passage in the *Principles* gives the same affirmation: "For example, if someone walking on board ship has a watch in his pocket, the wheels of the watch have only one proper motion, but they also share in another motion because they are in contact with the man who is taking his walk, and they and he form a single piece of matter" (AT IXb, 80/ CSM I, 236). Descartes continues by saying that the wheels even share the movement of the ship and of the earth turning around the sun, but that man only can have a certain knowledge of the motion that is proper to a single body. Man can only consider with certainty the movement of the wheels and does not notice the other motions in the watch. By asserting that the wheels of a clock can share different motions, Descartes underlines that the motion of a clock depends on the laws of nature, even if all the motions are not visible to man. In this sense, a clock can be considered a machine.

Additionally, we find the same list of "functions," earlier quoted, of the human body as a machine, in article 16 from Part one of the *Passions of the Soul*, in connection with the clock metaphor: "Thus every movement we make without any contribution from our will—as often happens when we breathe, walk, eat and, indeed, when we perform any action which is common to us and the beasts—depends solely on the arrangement of our limbs [...]. This occurs in the same way as the movement of a watch is produced merely by the strength of its spring and the configuration of its wheels" (AT XI, 341–342/ CSM I, 335). Walking, eating, and breathing were in fact the motions Descartes associated with a machine in the *Treatise of Man*, in the letter from 1638 to Reneri—and, in the Response to Arnauld, to explain involuntary movements. This shows that Descartes considers the clock a machine.

In the *Discourse on the Method*, Descartes reuses the clock metaphor while speaking of blood circulation: "[...] I would advise them that the movement I have just explained follows from the mere arrangement of the parts of the heart (which can be seen with the naked eye), from the heat in the heart (which can be felt with the fingers), and from the nature of the blood (which can be known through observation). This movement follows just as necessarily as the movement of a clock follows from the force, position, and shape of its counter-weights and wheels" (AT VI, 50/ CSM I, 136). Not only does Descartes underline once again that the clock system depends on the disposition of the wheels and the laws of nature, but he even gives a complete insight into his method, i.e., he explains how he could deduce the blood circulation from observation and experimentation by referring to his senses.[24] So, by using the clock metaphor, Descartes could express an assumption about the internal function of the human body,[25] and this only by referring to the disposition of the organs. The same idea is expressed in the *Description of the human body*, where Descartes refers to the clock to underline that there is no soul in its internal part that triggers the actions (see AT XI 226/ CSM I, 315).

To conclude, we have seen that when Descartes uses the metaphor of the clock alone, he considers it a machine. However, there are passages, especially in the *Treatise on Man*, which associate the clock with other devices or artifacts and seem to accept it as an automaton or self-moving machine. It is therefore necessary to examine the concept of the automaton as defined by Descartes in order to ascertain whether the clock can be considered both a machine and an automaton, and whether the clock metaphor provides irrefutable evidence that all automatons are machines.

Descartes's Automata

If in the *Discourse*, Descartes associates blood circulation with the clock metaphor, and thus to the functioning of a machine, he did not do so in a letter to Mersenne in June 1637: "For, even if the valves did not close half the openings to each blood vessel, the automaton would nevertheless continue to move necessarily, as I have said" (AT I, 377/ CSM III, 58–59). Here it seems clear that Descartes refers to the self-moving principle to explain blood circulation: respiration and the animal spirits play a role in blood circulation by changing the speed or density of the blood. Nevertheless, the *Discourse* underlines, that this self-moving principle cannot function if there is no exact disposition of the organs and if their interaction does not follow the laws of nature (see for example the blood which responds to the natural principles of the water circuit). This would mean that at least the blood circulation in the human body needs to be explained not only as a machine-like circuit but even as following the principles of an automaton as we have already seen with the watermill metaphor.

In addition, the automaton for Descartes can explain the different movements of the human body, as he explains in the *Discourse*: "[…] distributing the animal spirits to the muscles, make the parts of the body move in as many different ways as the parts of our bodies can move without being guided by the will, and in a manner which is just as appropriate to the objects of the senses and the internal passions. This will not seem at all strange to those who know how many kinds of automatons, or moving machines, the skill of man can construct with the use of very few parts […]" (AT VI, 55/ CSM I, 139). It is possible that Descartes links the self-moving principles to the motions of the muscles and thus to the neuro-motor system. In this case, the automaton self-moving principle, as seen in the fountain system, is linked to the motions of the muscles, and the system depending on the laws of nature, like hydraulic pressure, as in the organ metaphor, is linked to the heartbeat.[26] In the royal garden fountain metaphor, however, the muscles were compared to the pipes, and thus to a machine-like functioning. This means that in order to elucidate the functioning of the human body, as previously already discussed, Descartes employs both the automaton self-moving principle and the machine-like dispositions of the parts and organs.

In Part One of the *Passions of the Soul*, article 6, Descartes once more equates a clock and an automaton: "And let us recognize that the difference between the body of a living man and that of a dead man is just like the difference between, on the one hand, a watch or other automaton (that is, a self-moving machine) when it is wound up and contains in itself the corporeal principle of the movements for which it is designed, together with everything else required for its operation; and, on the other hand, the same watch or machine when it is broken and the principle of its movement ceases to be active" (AT XI, 331/ CSM I, 329–330). We see that in connection with the human body, the self-moving principle and the laws of nature-depending principle work together for Descartes. He no longer distinguishes between an automaton and a machine, which confirms the conclusion we drew after the analysis of the clock metaphor, and even of the other devices. In the more "medical" texts explaining the functioning of the human body, Descartes never distinguishes between a machine and an automaton.

Conclusion

Despite Descartes's apparent acceptance of a conceptual distinction between machines and automata in his mechanical writings, it has been demonstrated that he also asserts that automata are, in fact, machines. This is most evident when we analyze the passages on the functioning of the human body, as in these passages, Descartes is compelled to explain the complexity of the human body by invoking the concept of the automaton. If some muscle movements could be explained in a purely mechanical way, by reference to the functioning of a machine, more complex muscle movements, such as those involved in digestion and respiration, require the metaphor of an automaton.

This is why the metaphor of the clock is an ideal metaphor for conveying the perfection of the machine created by God. It combines the fifteenth-century idea of the self-moving principle associated with the *anima*[27] and the mechanical process programmed in machines for a certain purpose.[28] Probably Descartes, in the *Treatise on Man*, imagined a device which did not exist during his time. With François Duchesneau, we can assume that: "[…] Descartes foresaw a type of automaton of which no artificial model existed in his time, and for good reason, since according to our contemporary categories, it could only be an automaton capable of self-regulation, programmed to perform all vital functions."[29] This is likely the reason why he employs the clock metaphor, as it unites the two principles: the self-moving principle and the principle of following the laws of nature through the disposition of the organs.

Consequently, even if Descartes operates a conceptual shift in the clock metaphor, namely from the "technomorphic sign for the soul as a principle of (self) movement and life to the symbol of a mechanism of the living body and its involuntary movements that functions exclusively according to the natural laws of matter,"[30] we have to admit that the human-like machine, created by God overcomes the purely mechanical side of the clock metaphor, because Descartes does not want to simply explain the "large" movements of muscles and tendons, but also the small movements of organs, such as digestion. Furthermore, one does not forget that the clock metaphor and all the other devices play the role of making visible the invisible in the human body. Thus, even if the devices imitate perfectly the different systems in the human body, they cannot, taken one by one, account for the complexity of man: a kind of construction is needed that brings together all these models. This is the God-made machine invented at the beginning of the *Treatise on Man*, and, for Descartes, it does not deceive us, because in the end, there is no real imitation: the God-made machine is the human body.[31]

Notes

1. The English translation is taken from John Cottingham et al., *The Philosophical Writings of Descartes* (Cambridge: Cambridge University Press, 1985–1991). See Gideon Manning, "Descartes'

Healthy Machines and the Human Exception," in *The Mechanization of Natural Philosophy*, ed. Sophie Roux and Daniel Garber (Netherlands: Springer, 2013), 245: "The 'supposition' or hypothesis about bodies being machines is the touchstone for everything Descartes has to say about the human body in the *Traité de l'homme*. His analogy between God's fabrication and ordinary and familiar machines also serves to clarify and make plausible this fundamental supposition." And p. 246: this leads Descartes to specify "a material, efficient, final and then a formal cause for the human machine. The material cause is 'earth,' a component of the matter theory developed in *Le monde*. The efficient cause is 'God,' whose intention - and here the final and formal causes are introduced - is to create a machine as much like a genuine human being's body as possible."

2. Dennis Des Chene, "Automaton," in *The Cambridge Descartes Lexicon*, ed. Lawrence Nolan (Cambridge: Cambridge University Press, 2016), 46.
3. Concerning the complexity of the God-made machine, see Karen Detlefsen, "Descartes on the Theory of Life and Methodology in the Life sciences" in *Early Modern Medicine and Natural Philosophy*, ed. Peter Distelzweig, Benjamin Goldberg and Evan R. Ragland (Dodrecht Heidelberg New York London: Springer, 2016), 147–150.
4. See Dennis Des Chene, *Spirits and Clocks, Machine and Organism in Descartes* (Ithaca and London: Cornell University Press, 2001), 96: "To call something an artifact is to regard it as (i) made, not eternal; (ii) having a 'design' which was intended as such by its maker in the making of it."
5. See for example the machine analogy from Harvey. Peter Distelzweig, "'Mechanics' and Mechanism in William Harvey's Anatomy: Varieties and Limits," in *Early Modern Medicine and Natural Philosophy*, ed. Peter Distelzweig, Benjamin Goldberg and Evan R. Ragland, op.cit.,., 125–126: "[…] Harvey employs a machine analogy in his articulation of the motions of the heart. However, the *point* of the analogy is that in both there is a quick, coordinated series of motions producing one action. The point is *not* that in both there is a series of motions produced entirely by shape, size, and motions of (rigid) parts, and by their contact. That is, the machine analogy is *not* being employed to make some

kind of 'mechanical philosophy' more plausible. It is not meant to help the reader appreciate the explanatory power of a restricted mechanical ontology," which is, however, the case with Descartes.

6. D. Bertoloni Meli, "Machines of the Body in the Seventeenth Century," in *Early Modern Medicine and Natural Philosophy*, op. cit., 113.
7. See Descartes to Morin, 12 September 1638, AT II, 367–368 (CSM III, 122): "But in the analogies which I employ, I compare movements only with other movements, or shapes with other shapes; that is, I compare things that are too small to be perceived by the senses with other things that can be so perceived, the latter differing from the former simply as a large circle differs from a small one." See also Gideon Manning in his article "Descartes' Healthy Machines and the Human Exception," in *The Mechanization of Natural Philosophy*, op. cit., 249–250: "On Descartes' view, the nature of an ordinary machine, like a clock, always respects the laws of nature. Insofar as such machines frustrate our purposes it is because the laws of nature do not bend to our will. Our challenge when building machines, in fact, is always to realize our purpose within the constraints set by nature and her laws." See also Dennis Des Chene, *Spirits and Clocks, Machine and Organism in Descartes*, op. cit., 4.
8. Dennis Des Chene, *Spirits and Clocks, Machine and Organism in Descartes*, op. cit., 79.
9. See Daniel Garber, "Descartes, Mechanics, and the Mechanical Philosophy," *Midwest Studies in Philosophy*, XXVI (2002): 195.
10. The translation is taken from Gideon Manning in his article "Descartes' Healthy Machines and the Human Exception," in *The Mechanization of Natural Philosophy*, op. cit., p. 239. See also Descartes, *L'Homme* (Paris: GF Flammarion, 2018), 292.
11. Descartes, *L'Homme*, op. cit., 293 (my translation).
12. See Dennis Des Chene, *Spirits and Clocks, Machine and Organism in Descartes*, op. cit., 3.
13. See Daniel Garber, "Descartes, Mechanics, and the Mechanical Philosophy," op. cit., 185: "According to the mechanical philosophy, everything in nature is to be explained in terms of the size, shape, and motion of the small parts that make up a sensible body. In essence, for the mechanical philosophy, the whole world can be treated as if it were a collection of machines."

14. See Daniel Garber, "Descartes, Mechanics, and the Mechanical Philosophy," op. cit., 196: "We can imagine natural things *as if* they were machines, by imagining them *as if* they were artifacts, things constructed for a purpose. To say that a natural object is like a machine is to say that we can consider it as, or as if it had been made by someone for a purpose, God, perhaps, and explain it as such. […] As Descartes put it in a number of places in the treatise, the machine God creates "represents" what happens in the body. That is, we can conjecture that in our natural bodies, the phenomena in question are produced in the same way in which they are in the body of the machine God built to imitate us."
15. See Laurens Laudan, "The clock metaphor and probabilism: The impact of Descartes on English methodological thought, 1650–65," *Annals of Science*, 22:2 (1966): 73–104.
16. See Domenico Bertoloni Meli, "Machines of the Body in the Seventeenth Century," op. cit., 113.
17. For a detailed explanation see Jil Muller, "Montaigne et Descartes. L'imagination dans les passions," *Les Cahiers philosophiques de Strasbourg*, no 48 (2020): 135–164.
18. See François Duchesneau, *Les modèles du vivant de Descartes à Leibniz* (Paris: J. Vrin, 1998), 70.
19. Nancy Mason Bradbury and Carolyn P. Collette, "Changing Times: The Mechanical Clock in late medieval Literature," *The Chaucer Review* 43, no 4 (2009): 351–375 and 355: "These dialogues and proclamations imply a culture in which time-keeping mattered, in which it had become a shared cultural observance and an accepted way of regulating human activity."
20. See Ibid., 358: "For Manzini and his contemporaries, the clock is not merely a mechanism for keeping track of the hours or the planets, but an evocative representation of the forces that link the planets, moon, and stars to human life."
21. See Galileo, S. Santorio or Manzini. See also Jochen Büttner, "The pendulum as a challenging object in early-modern mechanics," in *Mechanics and Natural Philosophy before the Scientific Revolution*, ed. Walter Roy Laird and Sophie Roux, Boston Studies in Philosophy of Science (Dodrecht: Springer, 2008), 227. Vincenzo Viviani offered a purely invented anecdote in Galileo's biography. He wrote that "[…] Galileo discovered a fundamental property of pendulum motion, the pendulum's isochronism, as a student

while observing the swinging of a lamp in the dome of Pisa." For Büttner, "[…] Galileo had to discover its isochronism in order to realize its 'potential as a tool adaptable for time measurement'. Consequently, the first indisputable application of the pendulum as a timing device in medicine, the *pulsilogium*, even though described in a publication by S. Santorio, is usually attributed to Galileo."

22. See Hanns-Peter Neumann, "*Machina Machinarum*. Die Uhr als Begriff und Metapher zwischen 1450 und 1750," *Early Science and Medicine* 15 (2010): 126.
23. Nancy Mason Bradbury and Carolyn P. Collette, "Changing Times: The Mechanical Clock in late medieval Literature," op. cit., 358. See also Hanns-Peter Neumann, "*Machina Machinarum*," op. cit., 131.
24. See Hanns-Peter Neumann, "*Machina Machinarum*," op. cit., 156: "Descartes versucht nachzuweisen, dass der organische Körper keiner unmittelbaren Beteiligung der Seele oder des Bewusstseins bedarf, um seine Tätigkeit zu verrichten. […] Nachdem er die Mechanik des Herzens und der Arterien beschrieben hat, führt er in einem eingeschobenen Resümee seiner Methode und Argumentation noch vor der Schilderung des Blutkreislaufs die Uhr-Metapher ein."
25. See Laurens Laudan, "The clock metaphor and probabilism: The impact of Descartes on English methodological thought, 1650–65," op. cit., 77: "He suggests that we imagine the world on the analogy of a watch, whose face is visible but whose internal construction is forever excluded from view. In such a case, the most we can say about the mechanisms of the watch is conjectural opinion, not infallible knowledge. We can propose mechanisms for how the internal parts of the watch might be arranged, though we can never, *ex hypothesis*, get inside to see if we are right." See also *Principles of Philosophy*, AT IXb, 322.
26. See also Dennis Des Chene, *Spirits and Clocks, Machine and Organism in Descartes*, op. cit., 93–94: "Perhaps the only way for Descartes to distinguish organs from the whole animal in the causal order is to regard the machine of the *whole* body as the only entity of which one may say that it moves itself, while the organs do not. […] The body is capable of restoring, through its cycles of

motion, the initial conditions under which certain motions—digestion, respiration, the pulse—are bound to occur. In particular, it brings into itself the fuel consumed by the *feu sans lumière*. No organ alone does so. The heart, lungs, stomach and so forth can in that sense be said to be subordinate to the whole body in their operation."

27. See Hanns-Peter Neumann, "*Machina Machinarum*," op. cit., 127.
28. Ibid., 158 and 160.
29. François Duchesneau, *Les modèles du vivant de Descartes à Leibniz*, op. cit., 71. See also Mirko D. Grmek, "A survey of the mechanical interpretations of life," in *Biology, History, and Natural Philosophy*, ed. Allen D. Breck and Wolfgang Yourgrau (New York: Plenum Press, 1972), 181–195.
30. Ibid., 162.
31. See Dennis Des Chene, *Spirits and Clocks, Machine and Organism in Descartes*, op. cit., p. 68: "The *Traité de L'Homme*, as we have seen, is framed as a conceit, an extended comparison. The standard by which the comparison is judged is that of successful *simulation*. God is supposed to have made statues as much like us in the disposition of their parts as possible, statues that, but for the absence in them of reason, might well deceive us as the paintings of Apelles deceived their onlookers. […] Yet the simulation proposed in *L'Homme* is a means of conveying the truth about the body. There is, in the end, no illusion. Divine machines imitate nothing: they are us, in every action that does not require thought."

CHAPTER 6

Blurring Boundaries: Eighteenth-Century Garden Entertainment and the Role of the Beholders in Jacques Vaucanson's Automatic Performances

Albert Kozik

Introduction

Around the mid-eighteenth century, Parisian audiences across different social groups had access to a broad selection of amusing activities to choose from.[1] Since the late Middle Ages, the annual city fairs (*foires*) had become stable centers of popular entertainment, with people

This study is a product of a research project entitled *King of Engineers: Stanisław Leszczyński's Theatre of Automata in the Lunéville Gardens* financed by the Polish Ministry of Science and Higher Education as part of the 'Diamond Grant' programme in the years 2017–2024.

A. Kozik (✉)
University of Warsaw, Warsaw, Poland
e-mail: a.kozik@uw.edu.pl

J. Muller (ed.), *Automata, Cyborgs, and Mutants*, Palgrave Studies in Medieval and Early Modern Medicine,
https://doi.org/10.1007/978-3-031-93201-4_6

crowding at Saint-Germain or Saint-Laurent not only to purchase goods of varying prices and qualities but also to watch theater performances—especially *opéras-comiques*—and acrobatic spectacles. They could also enjoy exhibitions of rare and exotic products or animals,[2] and demonstrations of ingenious mechanical devices, all carried out in an atmosphere of wonder and mystery.[3] Technological shows and displays of machines were especially captivating to eighteenth-century audiences. The constant presence of machines in the entertainment of the period was, of course, not limited to places frequented by the lower strata of Parisian townsfolk. In fact, wealthier members of the bourgeoisie and even the aristocracy enjoyed them just as much, especially following the spread of the Enlightenment-driven interest in science.[4]

It was, in particular, the figure of the automaton that often featured in the mechanical shows favored by all circles of Parisian society. The city fairs provided a plethora of automatic performances, including figures playing musical instruments, portrait-drawing sculptures, animated paintings, and more.[5] For those wealthier, better educated, and actively interested in the dynamic scientific life of the period, numerous *salon* events accompanied by lectures on mechanical philosophy were staged by skilled engineers and clockmakers. Among them, the most renowned was Jacques Vaucanson (1709–1782). The exhibition space that he rented in the late 1730s in the Hôtel de Longueville, situated in the very heart of Paris—between the Louvre and the Palais des Tuileries—immediately became a famous destination thanks to the automata displayed there: the *Digesting Duck*, the *Drum Player*, and the *Flute Player* (see Fig. 6.1), three ingenious machines that sparked great interest among the Parisian bourgeoisie.[6] While the poor continued to admire mechanical shows at the fair of Saint-Germain, just across the bridge and a few streets away, the wealthier members of French society enjoyed Vaucanson's automata in a splendid *hôtel particulier*.

This distinctive topography of mechanical entertainment in eighteenth-century Paris, characterized by a strong presence of automatic shows in places visited by all levels of society, allowed the vogue for similar performances to spread and accelerate with time. The upper strata were certainly aware of the types of mechanical shows that one could see in market squares and fairs (not to mention churches, where automata had long been in use).[7] However, their need for a more intellectually sophisticated framing of such events was only met by the engineers such as Vaucanson,

Fig. 6.1 Hubert Gravelot's engraving from the 1737 *Mécanisme du flûteur automate* depicting Vaucanson's automata

who masterfully combined his inventions with the prevailing fascination with mechanical philosophy.

This chapter will center on the celebrated *Flute Player*, one of Vaucanson's automata presented to the Parisian public as part of his mechanical performances. The aim will be to demonstrate how these shows extended well beyond the *salon* at the Hôtel de Longueville, connecting the venue to the topography of Parisian garden entertainment, which Vaucanson deliberately considered when directing his mechanical spectacles. Eighteenth-century automata have long been recognized as a telling example of how philosophy, mechanical arts, and medicine influenced the perception of the 'natural' and the 'artificial' in the early-modern period.[8] Many scholars have focused on the way various automatic performances were staged for their audiences, with Vaucanson's demonstrations of his 'musical' androids often cited as a perfect illustration of carefully orchestrated mechanical shows meant to both surprise and educate the public.[9] Nevertheless, much less attention has been dedicated to the way these automatic shows were connected to the topography of entertainment in eighteenth-century Paris as well as the conventions of garden amusements prevalent at the time. Drawing on written and visual sources from the period, I will attempt to demonstrate that these two topics are, in fact, crucial to understand the character of Vaucanson's performances and the illusion of life that they were meant to create.

Vaucanson's Automata and the Topography of Entertainment in Eighteenth-Century Paris

In 1737, Vaucanson rented an exhibition space in the Hôtel de Longueville to display the first of his mechanical sculptures constructed for his Parisian audience—the *Flute Player*.[10] From 1739, this statue was accompanied by the *Drum Player* and the *Digesting Duck*, all of which attracted considerable attention from the Parisian public, fueling the engineer's fame as an ingenious constructor capable of imitating the movements and functions of both human and animal bodies.[11] The publicity given to these performances soon piqued the interest of the French *philosophes* and the members of the Académie Royale des Sciences, who invited Vaucanson to deliver a presentation on the statue and demanded that he explain the mechanism propelling the automaton, an opportunity for the engineer to furnish an authoritative interpretation of his own inventions.[12] The lecture, describing the *Flute Player* as a scientific tool for understanding the intricacies of human anatomy, was soon printed—in 1738—under the title *Mécanisme du flûteur automate* (*Mechanism of the Automatic Flute Player*) and sold at the Hôtel de Longueville during subsequent mechanical shows.[13]

In modern scholarship, Vaucanson is often described as a person who artfully combined his engineering skills with a good grasp of what could be referred to in modern parlance as market research.[14] Given the prevalence of the mechanical interpretation of the world in the period, it is little surprise that he attempted to write his machines into the context of one of the most fashionable philosophical discussions of the time, promoting the automata as scientific instruments rather than just wonderful toys meant to amuse the audience.[15] Vaucanson's spectacles were, of course, tailored to the expectations and needs of the wealthier and better educated members of the Parisian bourgeoisie and catered to their desires and tastes, contrasting with the atmosphere of the nearby fairs at Saint-Germain, where mechanical devices were exhibited predominantly for the pleasure of those less affluent. Yet the close proximity of such places of popular entertainment was certainly not the only element of local topography that had a bearing on the shape of Vaucanson's mechanical shows.

In the course of the eighteenth century and with the development of the enlightenment movement, scientists and engineers quickly learned to tailor their performances to meet the needs of a broader public, increasingly interested in science and technology. The curious mix of amusement and scholarly inquiry that often accompanied demonstrations of new inventions or recently discovered phenomena, much like in the case of public experiments in electricity or optics, has long been a subject of academic study, with a rich body of scholarship now available on the topic.[16] Vaucanson belonged to the same tradition of thinking about science as a source of entertainment and pleasure, and his automatic sculptures were undoubtedly designed to appeal to the aesthetic tastes of the beholders.[17] This was partly why, as it is often noted in his biographies,[18] Vaucanson patterned his *Flute Player* on the marble *Shepherd* (*Berger-Flûteur*, also referred to as *Faun*, see Fig. 6.2) executed in 1709 by Antoine Coysevox (1640–1720) for Louis XIV.

The visual connection between the two figures is evident (compare Fig. 6.1), even though Vaucanson's automaton was made of painted wood[19] and dressed in a slightly different attire, one more appropriate for *salon* entertainments. The composition of the two statues, as well as gestures they made, corresponded to each other, leaving no doubt that this visual association was absolutely clear to the public visiting the Hôtel de Longueville to view Vaucanson's mechanical sculpture. Numerous sources from the period highlight the similarity between the two works, explicitly identifying Coysevox's *Faun* as a direct source of inspiration for the *Flute Player*. For example, Bernard de Fontenelle (1657–1757), whose brief text lavishing praise on Vaucanson was published in the *Mécanisme du flûteur automate*, unequivocally acknowledged the *Faun* as the source of inspiration for the machine:

Fig. 6.2 Antoine Coysevox, *Berger-Flûteur* (*Faune*), 1709, marble sculpture. Musée du Louvre (originally displayed in the Tuileries gardens)

> Having heard Mr. Vaucanson's lecture, which included a description of a wooden sculpture modeled after Coysevox's marble *Faun* and playing a transverse flute, on which it performs twelve different melodies with precision deserving public admiration—a show that a great number of the Académie's members had a chance to witness—the Académie concluded that this machine was extremely ingenious.

> L' Académie ayant entendu la lecture d'un mémoire de M. de Vaucanson, contenant la description d'une statue de bois, copiée sur le Faune en marbre de Coysevaux, qui joue de la flute traversiere, sur laquelle elle exécute douze airs différens, avec une précision qui a mérité l'admiration du public, et dont une grande partie de l'Académie a été témoin; elle a jugé que cette machine étoit extrémement ingénieuse.[20]

Even at the very end of the century, a Jacobin journalist, François-Xavier Pagès (1745–1802), mentioned the marble sculpture in his retrospective reflections on Vaucanson's automaton:

> There is hardly anyone who has not heard about the automatic flute player made by Vaucanson around the year 1738. It is a wooden sculpture, five and a half feet tall, modeled on Coysevox's *Faun*, which can be seen at Tuileries. It is sitting, much like the original, on a rock and placed on a square pedestal four and a half feet tall and three and a half feet wide. This sculpture, which seemed, under Coysevox's chisel, to play the transverse flute, performs now, thanks to Vaucanson's care, twelve different melodies with a stunning precision.
> Il n'est presque personne qui n'ait entendu parler du flûteur automate, fait par Vaucanson, vers l'an 1738; c'est une statue de bois, de cinq pieds et demi de haut, copiée sur le faune de Coysevox, qu'on voit aux Tuileries, assise comme lui sur un bout de rocher, et placée sur un piédestal carré, de quatre pieds et demi de haut, sur trois pieds et demi de large. Cette statue qui, sous le ciseau de Coysevox, paroît jouer de la flûte traversière, exécutoit réellement, grace aux soins de Vaucanson, douze airs différens, avec une précision très-étonnante.[21]

Interestingly, Pagès clearly regarded the two statues as complementary, as if Vaucanson had not only imitated the original sculpture, but also amplified it, well in line with the early-modern idea of *aemulatio*.[22] What was already an almost-perfect copy of the human body crafted 'under Coysevox's chisel' became a fully animated work, bringing art much closer to nature.

For this effect to be as powerful as possible, Vaucanson had to select an artwork that would be easily recognizable to the audience. Why, then, did he choose Coysevox's *Berger-Flûteur*, a work of a renowned and universally acclaimed artist, for sure, but only one among many potential options? The very topography of Paris and the surroundings of the Hôtel de Longueville may shed some light on his decision. The venue that Vaucanson chose for his performances was, as previously mentioned, an *hôtel particulier* located in the very heart of the city. The famous Turgot map of Paris,[23] published in 1739 (see Fig. 6.3), around the same time the engineer initiated his shows, provides an overview of the 'Ville' (the area now referred to as the 'Rive Droite', or the right bank of the river), with the Hôtel de Longueville (marked in dark blue) positioned exactly between the Vieux Louvre and the Palais des Tuileries. The prestigious location of the building chosen by Vaucanson is evident—positioned between the two royal residences, the *hôtel* was an ideal venue to exhibit the automata and could easily draw crowds from the middle and upper classes of the capital. The *Flute Player* had actually made its first public appearance in 1737 at the fair of Saint-Germain,[24] and thus moving it to the Hôtel de Longueville underscored Vaucanson's aspirations and marked a shift from popular to a more élite type of entertainment. Perhaps more significantly, however, the venue was in close vicinity of the gardens of the Tuileries palace (marked in dark red in Fig. 6.3) where Coysevox's *Berger-Flûteur* was located at the time.[25]

When Vaucanson first arrived in Paris—his biographers often emphasized—he was mesmerized by the city's grandeur and intellectual atmosphere. Coming from the provincial city of Grenoble, he was elated by the perspective of living in the kingdom's capital, a place of endless opportunities and inexhaustible sources of inspiration. For instance, the *Éloge de M[onsieur] de Vaucanson* (*Eulogy for Mr. Vaucanson*), published

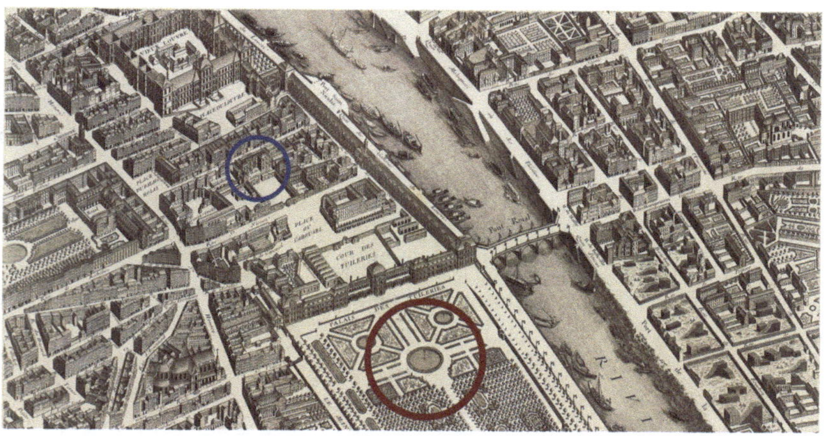

Fig. 6.3 A section of the 1739 Turgot plan of Paris

after his death in 1782, describes one of these early moments of creative infatuation that the engineer experienced in Paris:

> Soon, he left the province to come to Paris where he saw—with joy difficult to picture—that the machine of his *Samaritan* was precisely the one he had imagined in Lyon. This similarity taught him that his taste for mechanics came with some talent, a fact of which he was not aware before. He thus dedicated himself to mechanics with the kind of eagerness that a true hope for success can add to a great passion. A few days later, a statue of a flute player decorating the gardens of the Tuileries appealed to his imagination and he was struck by the idea to play some melodies by means of a similar statue that would imitate the moves of a flute player.
> Il quitta bientôt la province pour venir à Paris, et vit avec une joie, qu'il est difficile de se peindre, que la machine de la Samaritaine étoit précisément celle qu'il avoit imaginée à Lyon; cette conformité lui apprit, ce qu'il ignoroit encore, que son goût pour la mécanique étoit accompagné de quelque talent, et il s'y livra avec toute l'ardeur qu'une juste espérance de succès pout ajouter à une grande passion. Quelques jours après, la statue d'un flûteur qui orne le jardin des Tuileries, plut à son imagination, et il se sentit frappé de l'idée de faire exécuter des airs par une statue semblable, qui imiteroit toutes les opérations d'un joueur de flûte.[26]

The romanticized narrative of Vaucanson's arrival in Paris certainly contributed to the widely accepted image—enthusiastically promoted by local intellectual circles—of the city as the country's unrivaled cultural center. The class context of such perception notwithstanding, there can be no doubt that the engineer indeed saw Coysevox's statue in person. The Tuileries gardens had been open to Parisian bourgeoisie since the seventeenth century, with access formally reaffirmed in 1667, following Charles Perrault's (1628–1703) suggestion that they should be available to everyone for entertainment and *promenades*.[27] As a result, the gardens were effectively transformed into a public park.[28] Few were denied entry (mostly those inappropriately dressed),[29] and thus the Parisian bourgeoisie would often enjoy leisurely walks along the Tuileries' alleys and other fashionable spots in the city, a social practice that later became a favorite activity of the nineteenth-century *flâneurs*.[30] Virtually anyone visiting the park in the 1730s and 40s would have had the opportunity to see and admire Coysevox's *Shepherd*.

Vaucanson's choice of this particular sculpture was therefore by no means accidental. His decision was motivated largely by the universal accessibility of the garden where it was displayed, a place certainly familiar to most members of his intended audience. The performances organized at the Hôtel de Longueville were naturally embedded in the social practices of the beholders, who could compare the automatic *Flute Player* with Coysevox's marble original while strolling through the gardens, either before or after the show. As can be clearly seen on Turgot's plan of the city, the park was within a walking distance from the Hôtel de Longueville, no more than ten minutes on foot away. This proximity allowed for a specific theatrical experience in which the visits to the Tuileries became an extension of Vaucanson's automatic shows, as those who had a chance to see the *Flute Player* would be reminded of the event every time they saw the *Shepherd*.

Points of Comparison: Extending the Garden Space

The observation that the shows at the Hôtel de Longueville were intentionally integrated into the topography of Parisian garden entertainment in the 1730s and 40s, and that the Tuileries served as an extension of Vaucanson's automatic performances, may find confirmation in many other elements of the engineer's plan to captivate his audiences. To attract visitors, various visual representations of the automata were distributed in brochures and posters, advertising the shows and evoking their atmosphere. One of the most important eighteenth-century depictions of Vaucanson's inventions appeared in the *Mécanisme du flûteur automate*—an engraving shown earlier (see Fig. 6.1), executed by the renowned engraver François Vivarès (1709–1780) and based on a drawing by Hubert Gravelot (1699–1773). This image, printed on the booklet's frontispiece, later served as the basis for other illustrations of the machines. It presents a highly curated scene with the figures set on high pedestals, yet another indication of the engineer's desire to emulate marble sculptures. In fact, had the beholders not known that the picture depicted automata, they might have mistaken them for lifeless statues. The theatrical character of the scene is suggested by a curtain unfurling above the figures, subtly highlighting the performative aspect of the display, carefully arranged by Vaucanson. Yet, it is the background that truly sets the tone of the scene—a picturesque view filled with trees and shrubs, opening behind the pillars and inviting the beholders to imagine themselves in a secluded garden pavilion rather than a *salon* located in an *hôtel particulier*.

It is unlikely that the automata were actually displayed against a background of greenery—surrounded by densely built-up Parisian streets, the Hôtel de Longueville offered no such view, unless an artificial backdrop of painted fabric was hung behind the automata. While this is not entirely unlikely, such scenography was not necessary for the audience to feel as though they were part of a garden concerto, since the visual link established between the *Flute Player* and Coysevox's *Shepherd–Faun* was certainly sufficient to evoke associations with garden entertainment. The engraving featured in the *Mécanisme du flûteur automate* visually captured the general atmosphere of these performances, proving that leisurely activities typical of park spaces were indeed the intended context for Vaucanson's shows. The engraving effectively extended the space of the salon into a garden, just as the use of Coysevox's *Shepherd–Faun* as a model for the *Flute Player* transported the beholders to the nearby garden of the Tuileries.

Vivarès's engraving is not the only depiction of the three automata that highlights a similar connection between garden leisure and the mechanical *salon* shows. Much later, when the machines were known and celebrated across Europe, and after Vaucanson sold them to a new owner, the glovemaker Pierre Dumoulin (d. 1781), posters based on the original illustration from the *Mécanisme du flûteur automate* appeared in many places where the performances were organized. Although the machines were removed from their original environment of Parisian *salons* and toured across the continent, the details in such depictions clearly retained the visual links and associations with the practices of garden entertainment. For some audiences in other European countries, Coysevox's *Shepherd–Faun* might not have been immediately recognizable, or perhaps not recognizable at all. This prompted Dumoulin—as the new organizer of the shows—to find an alternative way to suggestively evoke the leisurely activities of parks and gardens. In one such poster, originally printed for a 1746 show of the automata in Frankfurt and now preserved at the Bibliothèque Nationale de France (BNF RESERVE QB-201 (100)-FOL, see Fig. 6.4), the machines are depicted similarly to Vivarès's engraving. However, a few details and subtle changes make this particular depiction especially interesting as far as garden entertainment is concerned.

Here, the lush greenery visible in the illustration from the *Mécanisme du flûteur automate* was replaced with a less overgrown landscape, though still clearly referring the beholder to open-air *promenades*. The curtain

Fig. 6.4 A 1746 poster from Frankfurt advertising Vaucanson's automata. Bibliothèque Nationale de France, BNF RESERVE QB-201 (100)-FOL

remains an important element of the image, continuously reminding the viewers that they are invited to participate in a theatrical event. And yet a closer scrutiny of the scene reveals several new elements added to the original composition of Vivarès's engraving. The pedestals on which the sculptures are set feature a series of three reliefs, each representing a different garden scene (see Fig. 6.5). At the very center, beneath the *Digesting Duck*, a faint outline of a building—possibly a cottage or a picturesque folly in the shape of a ruined tower—can be seen, surrounded by trees and bushes. On the left, where the *Flute Player* stands, the panel depicts a group of dancing people, a fitting scene for the actions performed by the musical automaton. Similarly, on the right, a pair can be observed strolling through a park, with a small figure sitting beside them, presumably a faun—a representation accompanying the mechanical *Drum Player*. All these figures were meant to transport the beholders into an imaginary garden space, a perfect setting for *fêtes galantes* and *fêtes champêtres*, both exceedingly fashionable at the time. The faun additionally echoed Coysevox's *Shepherd–Faun*, helping to preserve the specific context of the sculpture.

It was therefore through a number of conscious references to common eighteenth-century practices of garden entertainment that the two subsequent owners and exhibitors of the automata—Vaucanson and Dumoulin—staged and advertised their performances. Although the figures were, at the time (and still often are in modern scholarship), situated within the discourse of mechanical philosophy, well in line with the engineer's own intentions, it must be noted that the automata were advertised and demonstrated not only through the conventions of *salon* shows, but also by invoking typical park amusements. The general consistency in the style of these performances, in which the beholders were captivated by

Fig. 6.5 A close-up of the pedestals supporting Vaucanson's automata as represented in the 1746 poster

both the ingenious character of the mechanisms and the pleasing associations with garden spaces that they could visit after the show—at least in the case of the original venue, the Hôtel de Longueville—suggests that French parks should be regarded as a natural aesthetic context for Vaucanson's inventions.

It is little wonder that Vaucanson desired to evoke the charm of garden pleasures during his mechanical performances. Since the sixteenth century, following the rediscovery of ancient technological knowledge, automata had become an essential feature of luxurious parks across Europe, with places like Tivoli, Pratolino, and Saint-Germain-en-Laye astonishing visitors with animated sculptures. Even well into the eighteenth century, in 1742, when the shows at Hôtel de Longueville had already begun, Stanisław Leszczyński, Duke of Lorraine and Bar, commissioned a spectacular theatre of automata to his residence at Lunéville, a place often visited by French *philosophes*, including Voltaire, an admirer of Vaucanson.[31] A garden space was commonly associated with mechanical shows, not only because of the automata themselves, but also due to the countless experiments in physics or anatomical displays staged there.[32] When the creator of the *Flute Player* appealed to his audience's familiarity with the topography of the Tuileries, he was certainly invoking the old tradition of technological demonstrations held in parks and gardens, spaces that offered an ideal setting for exploring the boundaries between the natural and the artificial. While highly entertaining and intellectually stimulating, such events were distinct from mechanical shows and displays of machines held in Parisian fairs, as they drew on more intellectual and sophisticated conventions of technological presentation, setting Vaucanson's automata apart from toys meant for popular amusement. The Tuileries, in particular, also evoked associations with the famous Salle des Machines, a renowned venue for mechanical entertainment.[33]

Nevertheless, there might have been yet another, even deeper reason behind Vaucanson's insistence on connecting his automata to garden entertainment. The very form of the original *Berger-Flûteur* executed by Coysevox is, in fact, crucial to understanding the character of the engineer's performances, especially those involving the *Flute Player*. This is because, although the inventor liked to stress that the automaton was meant to demonstrate the workings of a musician's physiology, the statue was actually a double imitation—of both the human body and Coysevox's sculpture. The Parisian audiences attending the shows at the Hôtel do Longueville were undoubtedly mesmerized by the mastery with which the machine replicated human movements. However, their amazed reactions might have been partly a result of their lowered expectations shaped by the association with a marble sculpture. If the direct point of comparison was a familiar lifeless chunk of stone, it became much easier for Vaucanson to refer to the centuries-old *topoi* of works of art coming to life, such as the myth of Pygmalion or the tale of the Corinthian maid,[34] both of which were perfectly familiar to the beholders. For much of the performance, it was not exactly a scientific imitation of the human physiology that the viewers were looking at, but rather a moving marble sculpture, an inanimate object suddenly given the gift of movement.

The visual representations of Vaucanson's automata analyzed above, particularly the posters advertising the shows, promised a theatrical experience rooted in the setting of parks and gardens, rather than a lesson in mechanical anatomy that some could expect based on the engineer's own claims. Eighteenth-century conventions of displaying machines, which often included cross-sections and detailed close-ups of specific structural elements, were still evolving when Vaucanson began exhibiting his inventions. Yet even then, it should have appeared clear to the audience that the engineer was not exactly offering courses in engineering. It was possible to peek inside some of the statues and observe the mechanism, the posters hastened to explain, but only at a higher price, and even then, the construction of the automaton was not fully explained to the viewers. Instead, Vaucanson's booklet was meant as the most important source of detailed information about the devices. Through a series of aesthetic manipulations as well as a carefully crafted nexus of references to known sculptures, cultural practices, and popular scientific discourses, it was easier to convince the public of the ingenious nature of the inventions. The success of this strategy becomes evident if the controversy with the *Digesting Duck* is taken into account—what appeared, at least from the outside, as a perfectly manufactured copy of the animal's digestive system to the public in the 1730s and 40s was later revealed to be a clever deception,[35] demonstrating that much of the fascination came from the performative character and aesthetic context of the shows.

Conclusion

All the methods Vaucanson employed to make his mechanisms appear as lifelike as possible were undoubtedly designed to meet the expectations of his audiences. The engineer was acutely aware of their cultural background, interests, and social aspirations, prompting him to carefully curate the context of his shows. It is therefore crucial to put emphasis on the social and demographic aspect of these performances—what Vaucanson offered was markedly different from what could be experienced during other mechanical shows at, for example, Parisian fairs, because it appealed to a very specific set of cultural practices of upper echelons of society, such as visiting the Tuileries garden. The close proximity of the park allowed the engineer to extend the performance to the natural surroundings of the *hôtel* and invite the spectators to imagine Coysevox's well-known sculpture of the faun—which they could admire either before or after the mechanical show—while observing the automatic *Flute Player*.

As a result, the persuasive visual illusion of Vaucanson's performances did not rely solely on the fact that the automaton was a replica of the human body, but rather on the idea that it was a copy of a marble sculpture. This approach lowered the spectators' expectations regarding how lifelike the machine would appear once brought to life. When there was a need to convincingly blur the boundaries between the real and the artificial, it was also necessary to skillfully tune the expectations of the beholders to the possibilities of mechanical arts, guiding them to believe in what they could see and to imagine what they could not. Vaucanson adopted this strategy on many occasions, consciously referring to concrete visual tropes of eighteenth-century garden entertainment, such as *promenades*, *fêtes champêtres*, and *fêtes galantes*. These careful manipulations of the context in which the sculptures were exhibited were necessary primarily because the engineer never achieved any perfect imitation of either the human or the animal body—in fact, it has never been created to this day. He therefore knew that it was only through careful staging that the shows would convince the public of his incontestable success in this endeavor.

The process of creating artificial life is as much cultural as it is scientific, and it cannot be detached from the dynamics of social and aesthetic life in which it is carried out. In Vaucanson's case, it was a set of concrete references that governed the enthusiastic reception of the automata, underscoring the importance of the audience's participation in eighteenth-century mechanical spectacles. Beholders performed an essential role during such shows, becoming both receivers and co-creators of the events. Moreover, on a deeper level, they defined the scope of cultural and intellectual references on which Vaucanson could rely when he presented his inventions publicly. In seeking the appropriate venue and fashioning an aesthetic framework for his machines, Vaucanson had to balance his ambitions of showcasing himself as a skillful engineer blurring the boundaries between the real and the artificial with the expectations of his audience. By appealing to their cultural associations and familiar practices, he could compensate for the inherent imperfections of his inventions. Shifting the beholders' attention to the nearby space of the Tuileries gardens was one ingenious strategy to achieve this, and it largely contributed to the general appeal of Vaucanson's shows.

Notes

1. For an overview of different forms of popular entertainment in eighteenth-century Paris see the excellent study by Robert M. Isherwood, *Farce and Fantasy: Popular Entertainment in Eighteenth-Century Paris* (Oxford: Oxford University Press, 1986).
2. Louise E. Robbins, *Elephant Slaves and Pampered Parrots: Exotic Animals in Eighteenth-Century Paris* (Boston: John Hopkins University Press, 2002), 68–99.
3. Robert M. Isherwood, "Entertainment in the Parisian Fairs in the Eighteenth Century," *The Journal of Modern History* 53, no. 1 (1981): 24–48, 44–45.
4. Sofia Talas, "Physics in the Eighteenth Century: New Lectures, Entertainment, and Wonder," in *Laura Bassi—The World's First Woman Professor in Natural Philosophy: an Iconic Physicist in Enlightenment Italy*, ed. Luisa Cifarelli and Raffaella Simili (Cham: Springer, 2020), 64–67; Michael R. Lynn, "Experimental Physics in Enlightenment Paris: The Practice of Popularization in Urban Culture," in *Science and Spectacle in European Enlightenment*, ed. Bernadette Bensaude-Vincent and Christine Blondel (Aldershot: Ashgate, 2008), 65–72.
5. Robert M. Isherwood, "Entertainment in the Parisian Fairs," op.cit., 43–44.

6. Jessica Riskin, "The Defecating Duck, or, the Ambiguous Origins of Artificial Life", *Critical Inquiry* 29, no. 4 (2003): 599–633, 600–601.
7. Jessica Riskin, "Machines in the Garden," in *Renaissance Futurities: Science, Art, Invention*, ed. Charlene Villaseñor Black and Mari-Tere Álvarez (Berkeley: University of California Press, 2019), 19–40, 19, 21–22.
8. Jürgen Barkhoff, "Perfecting Nature—Surpassing God: The Dream of Creating Artificial Humans Around 1800," in *Science, Technology, and the German Cultural Tradition*, ed. Christian Emden and David Midgley (Oxford: Peter Lang, 2005), 39–55, 41–44.
9. Jessica Riskin, "The Defecating Duck," op. cit., 601–602.
10. Ibid., 599.
11. Kara Reilly, *Automata and Mimesis on the Stage of Theatre History* (Basingstoke: Palgrave, 2011), 81–86.
12. Adelheid Voskuhl, *Androids in the Enlightenment: Mechanics, Artisans, and Cultures of the* Self (Chicago: University of Chicago Press, 2013), 30.
13. Jacques Vaucanson, *Mécanisme du flûteur automate présenté à Messieurs de l'Académie royale des sciences, par M. Vaucanson, auteur de cette machine…* (Paris: chez Jacques Guerin, 1738). The title page of the book makes it clear that it was "sold in the room of the said automata" ("se vend dans la salle de dite figures Automates"), that is, at the Hôtel de Longueville.
14. For more on this phenomenon see Liliane Hilaire-Pérez, "Technology as Public Culture in the Eighteenth Century: the Artisans' Legacy," *History of Science* 45, no. 2 (2007): 135–153, 143–146.
15. David M. Fryer and John C. Marshall, "The Motives of Jacques de Vaucanson," *Technology and Culture* 20, no. 2 (1979): 257–269, 269.
16. See for instance Ciara Murphy, "Shocks and Sparks: Participatory Electrical Performances in the Enlightenment Period," in *Theatre, Performance, and Analogue Technology: Historical Interfaces and Intermedialities*, ed. Kara Reilly (Basingstoke: Palgrave Macmillan, 2013), 161–182; Christopher Baugh, "Philippe de Loutherbourg: Technology-Driven Entertainment and Spectacle in the Late Eighteenth Century," *Huntington Library Quarterly* 70, no. 2 (2007): 251–268; Simon Schaffer, "The Show that Never Ends: Perpetual

Motion in the Early Eighteenth Century," *The British Journal for the History of Science* 28, no. 2 (1995), 157–189.

17. Simon Schaffer, "Enlightened Automata," in *The Sciences in Enlightened Europe*, ed. William Clark, Jan Golinski, and Simon Schaffer (Chicago: Chicago University Press, 1999): 126–165, 136–137.
18. Kara Reilly, *Automata and Mimesis*, op. cit., 83; Jean-Claude Heudin, *Les créatures artificielles: Des automates aux mondes virtuels* (Paris: Odile Jacob, 2008), 56; Daniel Cottom, "The Work of Art in the Age of Mechanical Digestion," *Representations*, no. 66 (1999): 52–74, 52.
19. Yanyu Wang, "Jacques de Vaucanson (1709–1782)," in *Distinguished Figures in Mechanism and Machine Science: Their Contributions and Legacies, Part 4*, ed. Marco Ceccarelli and Yibing Fang (Cham: Springer, 2022), 15–46, 24.
20. Bernard de Fontenelle's description of Vaucanson's shows in Jacques Vaucanson, *Le mécanisme du flûteur automate* (Paris: chez Jacques Guerin, 1738), 18.
21. François-Xavier Pagès, *Cours d'études encyclopédiques rédigé sur un plan neuf*, vol. 4 (Paris: chez Artaud, an VIII [1799]), 292–293.
22. See Howard D. Weinbrot, "*An Ambition to Excell*: The Aesthetics of Emulation in the Seventeenth and Eighteenth Centuries," *Huntington Library Quarterly* 48, no. 2 (1985): 121–139, 127–128.
23. Antoin E. Murphy, *The Genesis of Macroeconomics: New Ideas From Sir William Petty to Henry Thornton* Oxford: Oxford University Press, 2009), 133.
24. Jessica Riskin, „The Defecating Duck," op. cit., 599.
25. Pierre Edouard Démenieux, *Coysevox (1640–1720)* (Paris: Nadaud et Compagnie, 1882), 98–99.
26. "Éloge de M. de Vaucanson," in *Histoire De L'Academie Royale Des Sciences – Année 1782* (Paris: de l'Lmprimerie Royale, 1782), 157.
27. David R. Coffin, "The *Lex Hortorum* and Access to Gardens of Latium During the Renaissance," *The Journal of Garden History* 2, no. 3 (1982): 201–232, 209.
28. Vandra Costello, "Public Spaces for Recreation in Dublin, 1660–1760, *Garden History* 2, no. 35 (2007): 60–179, 169.
29. David R. Coffin, "The *Lex Hortorum*," op. cit., 209.

30. See Laurent Turcot's excellent study "The Rise of the *Promeneur*: Walking the City in Eighteenth-Century Paris," *Historical Research* 88, no. 239 (2015): 67–99.
31. Minsoo Kang, *Sublime Dreams of Living Machines: the Automaton in the European Imagination* (Cambridge, MA: Harvard University Press, 2011), 110.
32. See *Gardens, Knowledge, and the Sciences in the Early Modern Period*, ed. Hubertus Fischer, Volker R. Remmert, and Joachim Wolschke-Bulmahn (Cham: Birkhäuser, 2016).
33. Sophie Lachapelle, *Conjuring Science: A History of Scientific Entertainment and Stage Magic in Modern France* (New York: Palgrave Macmillan, 2015), 118.
34. For an overview of the myths and their importance in cultural history see Robert Rosenblum, "The Origin of Painting: A Problem in the Iconography of Romantic Classicism", *The Art Bulletin* 39, no. 4 (1957), 279–290.
35. Jessica Riskin, "The Defecating Duck," op. cit., 609.

PART II

The Foundations for Transhumanism and Posthumanism

CHAPTER 7

Robots and the History of the Mind–Body Problem

Paul Richard Blum

Present-day panpsychism is one attempt at overcoming the mind/body dualism that is inherent in thinking about subject, or mind, vs. object, or body, as common to physicalism and materialism in the history of philosophy of mind. To illustrate this, I suggest reading Karel Čapek's theater play on robots, the phrenology of Franz Joseph Gall and Johann Gaspar Spurzheim, the Renaissance physicalists Gregor Reisch and Juan Huarte, and the father of the anatomy of brain functions, Claudius Galen. Robots are the attempt at re-constructing the mental. And thinking of the mental in terms of perfect construction as it originated in phrenology yields two apparently distinct results: either panpsychism or the invitation to build robots, both of which—on a closer look—embody the mind-body antinomy.[1]

P. R. Blum (✉)
Loyola University Maryland, Baltimore, MD, USA
e-mail: PRBlum@loyola.edu

Palacký University Olomouc, Olomouc, Czechia

J. Muller (ed.), *Automata, Cyborgs, and Mutants*, Palgrave Studies in Medieval and Early Modern Medicine,
https://doi.org/10.1007/978-3-031-93201-4_7

Dualism and Monism in Present-Day Panpsychism

Godehard Brüntrup points out that panpsychism is supposed to be a monism, in which the duality of the psychic and the material is solved by claiming that there is only one kind of thing, which still encapsulates both physical and mental properties.[2] The physical part of things is that which can be understood in the research of natural sciences. Therefore, a thorough investigation into the physical must reveal its internal properties that are some sort of mental. The solution offered by panpsychism is this: the intrinsic nature of brain cells explains the possibility for the mind to emerge.[3] Modern panpsychism struggles with two questions, namely, how is it possible for particular things to be known, and secondly, how is it possible for there to be knowledge at all? The theory tries to explain the existence of the understanding mind by finding seeds of understanding, of "the mental," in the physical things that elicit the understanding from the understood. It should be noted that the project repeats the conundrum of any monism and dualism. The rule is: every dualism can be transformed into a monism—in two ways. Pun intended.[4]

In the following, we will look at examples of dealing with the physical vs. psychic interpretation of mental events. The sources selected could easily be replaced by others. Also, each of the sources has garnered a variety of interpretations; the ones offered here serve the purpose to show the systematic problem of physicalism and (pan)psychism over history. Furthermore, the terminology is inevitably flexible and avoids most "isms" of current debate, since terms and meanings have changed over centuries of continual struggling.

Robots Losing Their Minds

As an example, let us take a look at Karel Čapek's theater play *Rossum's Universal Robots* (1920). Čapek was trained in philosophy and specialized in pragmatism. The term "robot" was invented by the author's brother, Joseph[5] in allusion to a Slavic word for "work," indicating that these machines were created to work and only to work. In the introductory scene we learn that the original idea had been to create "real people, real Indians or real professors or real idiots." Such had been the plan of Rossum's father, but the engineer turned that into "making robots that would be a living and intelligent workforce."[6] So, the original idea of creating an artificial human meant to create humans with the same deficiencies that real humans have. This original idea was transformed into the idea of creating working machines for profit. Nature is played against technology: robots are exceeding the sophistication of natural products, and to the same extent they are deficient in comparison to human beings. Robots don't reflect; neither upon themselves nor upon humans. But in this play, there is a crack in the system because these robots have every now and then fits or cramps that make them stop and lose their mind.[7] In other words, despite the intentions of creating an intelligent machine, what came out was something with an unconscious sense of irregularity that doesn't fit the program. The "Robot cramp" is explained as "just some technical disorder." The technical "disorder" is evidently the positive outcome of the mechanic creation: the better the more alive.[8] These robots approach the qualities of human beings, namely, to be "besides" their expectable functioning. There seems to be a "what it is like to be" a robot, and that is unknowable.

Franz Joseph Gall's Toolbox

Noteworthy for our purposes is one character in the play, the physiologist Dr. Gall. The author most likely chose this name after the physiologist of the early nineteenth century, Franz Joseph Gall (1758–1828).[9] He was one of the leaders in research to allocate mental and emotional functions to specific places of the human brain, reflected in the skull. Hence this school of anatomy was named phrenology. Gall's main thesis about the human brain teaches that everything involved in thinking and perceiving is located in the brain, and that the brain is the condition of the possibility for perceiving anything from outside, so that it is not the external impacts that create sensations, but rather the capability of the human brain that produces sensations in quasi mechanical ways. This is to be extended to the life of a human being as part of the animal world.[10] Physiology and psychology overlap. In a letter in 1798 to a friend, Gall gave a summary of his brain theory. Objections against any anatomical interpretation of mental functions that invoke human spontaneity and freedom are answered as follows:

> Those who want to persuade themselves that our qualities are not innate derive them from education. But do we not behave in the same passive way in every case, whether we have been formed in a certain way by innateness

> or by education? In this objection one confuses the concepts of abilities and inclinations – of mere disposition – with the mode of action itself. Even animals are not simply subordinate to their abilities and inclinations without any volition. [...] Man, however, [...] is endowed with the feeling of morality and with clear consciousness, and so on. With these weapons man fights against his inclinations.[11]

The distinction between education and inborn qualities is not relevant; rather, what counts is agency. Gall's response further extends the mental so that it includes animals. The case in point is volition. Hence, he describes whatever is not reducible to the corporeal and anatomical as modes of action that are based on the physiological conditions. Even though there are many "capacities" that go beyond the animal and physical and constitute human self-reference ("fighting against inclinations"), the basis is common with animals.

Paying attention to robots, we notice that one way of imputing something mental on the physical brain is to grant the same for those beings that are non-human and yet capable of spontaneous actions. If the mental is by nature not distinct from the anatomical, then the mental must be present in what traditionally was categorized as non-mental. Such a description perfectly matches the idea of robots as entirely physical and yet human-like.[12]

The next question to be addressed, then, is that of the actual founding of mental acts (like volition or speculation of "an independent being")—in standard philosophical terms: is the human mind dependent upon the body? Gall has this explanation: "The talents and inclinations have their seat, their foundation, in the brain."[13] The arguments for this thesis belong to the field of anatomy: wherever the brain is damaged, specific mental operations fail. Furthermore, biological life exists without a brain; therefore, the brain must have specific functions, which explains why with increasing psychic versatility (as given in the variety of animals) the size of the brain also increases.[14]

Gall's basic assumption is this: there is a continuum from primitive life to highest speculation that can be detected and even reconstructed by means of physiological anatomy. Jokingly he claims to start with plain fibers and to invent nerves; from there he proceeds to animals and sense-tools. With his box of power tools (*Vorrath von Kraftwerkzeugen*) he builds bugs and fish, dogs for the ladies, and horses for the dandies, and so on up to theologians and scientists.[15] Basically the program Čapek's engineer Rossum originally had in mind: real Indians, real professors, or real idiots. Such a description of nature—programmatically mechanical and physical and at the same time ironic—promises to explain both why men fight for women and why theologians hardly abandon their convictions.

A student and coworker of Gall's was Johann Gaspar Spurzheim (1776–1832). He joined him in propagating phrenology, that is, the study of the mind in the biological and specifically anatomical framework. Spurzheim points to the powers within the human mind; his understanding of anatomical givenness of faculties or potentials of the mind in phrenology has some similarities with the theory of the innateness of ideas.[16] But against John Locke he stated: "No sensation, no idea, no principle, is innate. Sensations and ideas of external objects follow from external impressions, and these being accidental, ideas of them cannot be innate; but the faculties which perceive impressions, and conceive ideas, are innate."[17]

For our question regarding the origin and critique of robots, Spurzheim is an example of the wavering between purely physical explanations of mental operations ("external impressions") and the admission that the physical has to be met by some power ("faculties") that antecedes the impressions and thus, while still localized in the brain, is of a higher order. The antagonism of the spiritual and the material (as it also drives modern panpsychism) is not dissolved but only reformulated. In a programmatic statement he declares all dual relations as exchangeable:

> All the functions of man which take place with consciousness, are ascribed to the mind, and constitute animal life. They may be divided into two orders, a division admitted from the remotest antiquity, and known under the names Soul and Spirit; – moral and intellectual faculties; – understanding and will; – heart and head. – I shall design them by feelings and intellect, or by affective and intellectual faculties.[18]

This series of binaries allows for making empirical claims and referring to non-empirical powers or faculties. The ideas are not innate, in this Locke was right, but the physiologist cannot argue without some sort of innateness of the powers. This reasoning, by the way, earned Spurzheim inclusion in a recent anthology on philosophy of mind.[19]

Brain and Soul in Reisch and Huarte

We looked into phrenology because of the reference to Gall in Čapek's play. Another inspiration for giving this name to the physiologist of the robot factory may have been the ancient scientist Claudius Galenus (129–216 AD). In fact, most of the physiological research into the human mind goes back to this Greek scholar. Galen taught that the brain is the source of physiological movements, that it is the seat of sensation, as it receives all the sensations and organizes them into the object of understanding. But before we discuss Galen, let us trace his method in the Renaissance that transmitted it into the modern era.[20] A very good example is accessible in the textbook by Gregor Reisch (1470–1525), *Margarita philosophica*, an encyclopedia of all areas of knowledge. Reisch explains the operation of the human brain by assigning certain parts of the brain to certain functions.[21] The fundamental point is the anatomical allocation of cogitative and sensitive functions within the brain. There are five interior senses, including the "common sense," which collects the sense data; then something like imagination and fantasy that process these images; then the estimative and the cogitative, which is the power of the mind to make statements about sense data, which are then forwarded to the memory—all located physically in the brain.[22] In this tradition the brain is master of understanding, including judgment and memory, and this is the basis of investigating human thinking in physical terms; it is a logical consequence of a purely anatomical and physical view of the human mind. It should be taken into account that Reisch also served the expectations of the metaphysical and theological understanding of human nature. The psychic functions are certainly located anatomically in the brain, while the mind is independent of the corporeal condition, being created in the image of God (Genesis 1:26–27).[23] The metaphysical and the physical approaches meet in the human intellect. However, since we are looking for the source of material and inspirations toward a physical understanding of human agency, we may also identify in this contradiction the root of the ongoing mind-body paradoxes.

Toward the end of the sixteenth century, a Spanish philosopher took up the same idea, with the same purpose of educating students. Juan Huarte (1529–1588) published his "exam of the minds of people with regard to their capability to understand the sciences," in which he pursued a clearly anatomical and medical understanding of the human mind.[24] For that purpose he refers to Galen.

> The wisdom of the sensitive soul also depends on the temperament of the brain [...]. The means Galen had to observe and know by eye sight the wisdom of the sensitive soul was to take a kid as soon as it was born, which, placed on the ground, began to walk as if he had been taught and told that the legs were made for such use, and after this he shook off the moisture that he had from his mother, and lifting the foot he scratched behind the ear, and placed before him many bowls with wine, water, vinegar, oil, and milk – after having smelled them all – he consumed only the milk. [...] Hippocrates had great reason to say that the souls were wise without having had a teacher.[25]

Huarte here reports that Galen made inferences from observations of an animal to prove that through behavior it is evident that the functioning of the animal brain is, so-to-say, wired. In the reference to Hippocrates, we have the ambiguity of mind-body dualism in a nutshell. For that statement can mean, first, that certain functions are built in this soul anatomically or physiologically, so that the animal can perform certain functions that look like sense and judgment. But it can equally mean that certain understanding is impregnated in the souls ("the souls are wise") independently of the body in which the soul is located, which would be the Neoplatonic interpretation of the human soul and human understanding, including innate ideas. This is the same ambivalence we observed in phrenology. In other words, with the tradition invoked by Huarte, the Hippocratic-Galenic medicine, the mind-body problem as the endeavor to explain psychic phenomena physiologically has a solid foundation. Therefore, it is worth looking into Galen's theory of brain and mind, the quasi-origin of the medical and physiological approach to the mind-body relation.

Galen Between Anatomy and Panpsychism

For our purposes, the most interesting work is Claudius Galen's discussion of "The Doctrines of Hippocrates and Plato." In this work, he explains his anatomical discoveries in terms discussed by philosophers and other theorists. The concept of ventricles in the brain that will pervade the history of brain anatomy is constitutive to Galen's theory.[26] After describing the cutting of the ventricles, he weighs the options between whether the soul is incorporeal or corporeal: what resides in the ventricles—where the senses gather in the "pneuma"—is the soul (*psyche*). "[...] if the soul is incorporeal, the pneuma is, so to speak, its first home [*oiketerion*]; or if the soul is corporeal, this very thing is the soul."[27]

Here we appear to have the clear option to view the senses as the dwelling place of the incorporeal soul, so that there is no need to ponder the essence of the soul insofar as it is factually and empirically manifest in those senses in the brain. This then would not make much difference to the alternative that identifies senses and soul. But Galen continues:

> It is better, then, to assume that the soul dwells in the actual body of the brain, whatever its substance [*ousia*] may be, [...] and that the soul's first instrument for all the sensations of the animal and for its voluntary motions as well is this pneuma.

Considering physical impairments of the pneuma and their restauration, which does not entail the death of the animal, Galen concludes:

> Yet if the pneuma were itself the substance of the soul, the animal would immediately die along with the escape of the pneuma. Thus it is reasonable that this pneuma is produced at the ventricles of the brain [...] and that the pneuma is, as I said, the first instrument of the soul. [...] Now the pneuma in the arteries is and is called vital, and that in the brain is called psychic, not in the sense that it is the substance, but rather the first instrument of the soul resides in the brain, whatever its substance [*ousia*] may be.

Galen's is decidedly an anatomical perspective, and, indeed, these statements are based on his detailed descriptions of anatomical operations he had performed and taught in his work on the functions of the human body parts.[28] In this work, however, he is intent on relating what is spiritual to what is empirical. Therefore, his considerations are exemplary for the mind-body problem. One key is Galen's repeated wavering regarding the substance of the soul. Galen suspends judgment about the substance or essence of the soul ("whatever its *ousia* may be").[29]

However, there is one aspect in Galen's physiology of mental operations that is very important for the relation between the external reality and the mental operations. Supposing that all "ideas" come from the external objects impacting the bodily organs, which then process them in the way that can be anatomically demonstrated, there still appears to be some kinship between the material objects and the physiological senses. For the correlation between perceived and perceiving, Galen considers the effect of light-filled air.

> When it has been illuminated by the sun, it is already an instrument of vision of the same description as the pneuma coming to it from the brain; but until it is illuminated it does not turn into a sympathetic [*homoiopathes*] instrument by virtue of the change effected in it by the outflow of the pneuma.[30]

The senses and their objects are "sympathetic," i.e., of germane conditions, whether considering perception in the nature of the senses or in the quality of the objects. The senses and the objects share the same essence as the condition of the possibility to perceive at all.

> But what difficulty is there in supposing that the sunlight is sensitive [*aisthetiken*], much as the pneuma in the eyes that is brought in from the brain is clearly seen to be? For it is luminous [*photoeides*]. And if we must speak of the substance of the soul [*psyches ousias*], we must say one of two things: we must either say that it is this, as it were, bright and ethereal body [...] or that it is (itself) an incorporeal substance [*asomaton ousian*], and this body is the first vehicle, by means of which it establishes partnership [*koinonian*] with other bodies. We must say then that this (pneuma) itself extends through all the brain, and that by partnership with it the optical pneuma becomes luminous.[31]

For the question of the naturalness of mental operations, what do we learn here? The kinship or partnership (*koinonia*) of the senses with the sensible is a plausible consequence from the anatomical analysis of mental operations like sensing.

Physiology sets the framework within which there has to be a commonality encompassing subject and object. If perception is physically possible because the object carries its properties to the senses, and if the senses, branching out from the brain, share the properties of the perceived objects, then there is no essential difference between the perceived and the perceiver. Then the "partnership" between mind and object is all there is. This also entails that questions about the ontological difference between soul and body are moot because, first, the soul is either a body or the body is the instantiation of the soul's capabilities, and second, whatever we want to know about the soul is accessible and intelligible by way of its operations. Focusing on the "partnership" or "kinship" we may say: The soul and the pneuma perceive exactly that which is soul-like in the objects. In other words, panpsychism is a viable option. And yet another consequence is possible when adopting Galenic physiology of the mind:

investigation of the physical nature of mental operations makes it thinkable to construct subjects that have all the necessary, anatomic and psychic properties and functions typical of what otherwise is taken to be living and thinking beings. Or, to put it the other way around: the concept of a robot is that of a living being without real life.

Conclusion

In this investigation from Čapek backwards in time via phrenology and Renaissance medicine to Galen, it becomes evident that psychic phenomena have been researched and understood in anatomical and physical terms. This research by definition and purpose struggles with dualism while trying to reduce the elements of the psychic side to a minimum. We observed that any such attempt can switch over into the opposite. In the same way as the characters of the theater play start to doubt their sensations and feelings while the robots become aware of their agency, in the same way the physical explanation of psychic processes again and again acknowledges the existence of the psychic in spite of the physicalist method.

From this we can draw a number of interesting conclusions. First of all, there has been a strong tradition from ancient medicine into the twentieth century that has explained mental operations and psychic functions in physical terms. Robots are designed by the playwright as a proof in case. The creation of robots is based on the claim that intelligent operations of human beings can be emulated or reconstructed by material mechanisms. Hence the question must always have been: where are the limits to creating thinking machines? One answer is: aesthetics and emotions, including love and violence. These are operations of the human mind that are not built into the physical structure of the human animal. Phrenology was an extreme success of attempts at explaining mental processes and psychic phenomena in physical terms. If phrenology is right, then the laws of nature and psychology coincide. Understanding human feeling aims at medical research: physiology investigates psychological behavior and variations.

Hence resurfaces the question: where exactly is the borderline between the mental and the corporeal? If these realms are not separate or if the distinction between the psychic and the physical is only a matter of convenience in research, then it is unclear how behavioral, emotional, or intellectual mistakes are possible. Where does nonsense come from? In

Čapek's play the issue is expressed as the "Robot cramps," in Galenic medicine it is investigated around failures of psychic organs. For, if the mental is a result of the physical operations and the physical operations are in line with the psychic situation, then nothing is wrong. On the other hand—so we may ask now—what is it that the mind "takes" from the objects, if not the understanding? The same question can be asked the other way around: if these two realms are factually and positively separate, how do they coordinate? Most of present-day discussion in philosophy of mind is concerned with this coordination of the physical and the psychic. Panpsychism within the philosophy of mind is a perfect mirror of phrenology and robotics: instead of thinking of the mind as rooted in the physical to the effect that it explains mental processes, this new approach asks about the roots of the psychic in the material by projecting the psychic into the material, so that the human mind gathers from the physical world whatever was already there.

Notes

1. This study is a result of the research funded by the Czech Science Foundation as the project GA ČR 21-17059S "Pantheism and Panpsychism in the Renaissance and the Emergence of Secularism."
2. Godehard Brüntrup, "Emergent Panpsychism," in *Panpsychism: Contemporary Perspectives*, ed. Godehard Brüntrup and Ludwig Jaskolla (New York: Oxford University Press, 2017), 48–71, 51: "Thus panpsychistic monism is a 'dual-aspect monism,' claiming that there is only one kind of thing but it features physical and mental properties. [...] In the contemporary debate about panpsychism, the mental part is usually constituted by the intrinsic properties of things (the quiddities), whereas the physical part is constituted by the relational-structural properties."
3. Ibid., 57–59. An overview of traditional monism/dualism/pluralism solutions to the mind-body problem in William R. Uttal, *The Psychobiology of Mind*, [1978] (Hove, UK: Psychology Press, 2014), chap. 2 "History of the Mind-Body Problem."
4. Jonathan Schaffer, "Monism," in *The Stanford Encyclopedia of Philosophy*, ed. Edward N. Zalta, Winter 2018 (Metaphysics Research Lab, Stanford University, 2018), https://plato.stanford.edu/archives/win2018/entries/monism/: "There are many monisms."

5. Dominik Zunt, “Who Did Actually Invent the Word ‘Robot’ and What Does It Mean?” (2013), http://web.archive.org/web/20130727132806/; http://capek.misto.cz/english/robot.html.
6. Karel Čapek, *R.U.R. Rossum's Universal Robots. Kolektivní drama o vstupní komedii a třech aktech* [Collective drama in a prelude and three acts] (Praha: Aventinum, 1920), Intro. Scene, 12. The Czech text has: “Myslel, že udělá skutečné lidi, tedy snad nějaké nové indiány, docenty nebo idioty, víte?” English from Karel Čapek, *R.U.R. (Rossum's Universal Robots) A Play in Introductory Scene and Three Acts*, trans. David Wyllie (Adelaide: University of Adelaide, 2014), (this version has no pagination), https://web.archive.org/web/20150914184847/http://ebooks.adelaide.edu.au/c/capek/karel/rur/complete.html.
7. Čapek, *R.U.R.*, Intro Scene, 23.
8. The unintentional transition from machine to human is a frequent theme in robot stories. See, for instance, Paul Richard Blum, “Robots, Slaves, and the Paradox of the Human Condition in Isaac Asimov's Robot Stories,” *Roczniki Kulturoznawcze* 7, no. 3 (2016): 5–24.
9. Biographical and scientific background in Stephen Tomlinson, *Head Masters: Phrenology, Secular Education and Nineteenth-Century Social Thought* (Tuscaloosa: University of Alabama Press, 2005), chap. 3.
10. F. J. Gall, *Sur l'organe des qualités morales et des facultés intellectuelles, et sur la pluralité des organes cérébraux*, Sur les fonctions du cerveau et sur celles de chacune de ses parties [...] 2 (Paris: Chez l'Auteur [etc.], 1822), 72: Summary: “Que les perceptions et la conscience n'existent que dans le cerveau; que, sans le cerveau, aucune impression venue du dehors et aucune irritation née dans l'intérieur ne peuvent produire de sensation; que le cerveau est exclusivement l'organe des fonctions de la vie animale; que tous les phénomènes que nous offrent les zoophytes, tous ceux que nous observons dans les systèmes nerveux différens du cerveau, ne doivent point être regardés comme des phénomènes dus à la faculté sensitive et à la spontanéité animale, mais doivent être attribués uniquement à l'irritabilité.” Unless cited otherwise, translations are mine. Present-day versions of Gall's theories are known, among others, as neuroconstructivism; Denis Forest, “Neuroconstructivism: A Developmental Turn in Cognitive Neuroscience?,”

in *Brain Theory: Essays in Critical Neurophilosophy*, ed. Charles T. Wolfe (Houndmills: Palgrave Macmillan, 2014), 68–87.

11. F. J. Gall, "Schreiben über seinen bereits geendigten Prodromus über die Verrichtungen des Gehirns der Menschen und Thiere," *Der neue teutsche Merkur*, no. 27 (1798): 311–332, 315: "Die sich überreden wollen, unsere Eigenschaften seyen uns nicht angeboren, leiten sie von der Erziehung her. Haben wir uns aber in jedem Falle gleich leidend verhalten, ob wir durch Angebornheit oder durch Erziehung auf gewisse Wiese geartet worden sind? In diesem Einwurf verwechselt man die Begriffe von Fähigkeiten und Neigung, von bloßer Anlage, mit der Handlungsweise selbst. Selbst die Thiere sind nicht schlechterdings ohne alle Willkühr ihren Fähigkeiten und Neigungen untergeordnet. […] Der Mensch aber hat außer den thierischen Eigenschaften, Sprachfähigkeit und die ausgedehnteste Erziehungsfähigkeit. […] Er hat Sinn für Wahrheit und Irrthum, für Recht und Unrecht, für Vorstellungen eines unabhängigen Wesens; das Vergangene und die Zukunft können seine Handlungen leiten; er ist mit dem Gefühl von Sittlichkeit und mit deutlichem Bewusstsein gehabt, u.s.w. Mit diesen Waffen tritt der Mensch gegen seine Neigungen in Kampf." The *Prodromus* was never published. —This letter is, approvingly, quoted in Michael L. Anderson, *After Phrenology: Neural Reuse and the Interactive Brain* (Cambridge, Mass.: MIT Press, 2014), xiii f. There, repeated references to Gall.
12. For modern approaches to robotics see, for instance, John Symons and Paco Calvo, "Computing with Bodies: Morphology, Function, and Computational Theory," in *Brain Theory: Essays in Critical Neurophilosophy*, op. cit., 91–106. The questions discussed here precede systematically those discussed in present-day 'robotics' as, e.g., in Tim Crane, *The Mechanical Mind: A Philosophical Introduction to Minds, Machines and Mental Representation*, 3rd. ed. (London: Routledge, 2016), 149 f.
13. Gall, "Schreiben", op. cit., 317: "Die Fähigkeiten und Neigungen haben ihren Sitz, ihren Grund, im Hirne."
14. Ibid., 317f.
15. Ibid., 328f.; 328: "Narren und Weise, Vestalinnen und Sultane, Dichter und Geschichtenschreiber, Gottesgelehrte und Naturkündiger u.s.w." On the transition from analysis to construction, including emotions, cf. Igor Aleksander, *How to Build a*

Mind: Towards Machines with Imagination (New York: Columbia University Press, 2001).

16. J. G. Spurzheim, *Phrenology, or, The Doctrine of the Mental Phenomena*, vol. 2 (Boston: Marsh, Capen and Lyon, 1835), 78: "The innateness of the faculties must also be admitted, because there is a direct relation between their manifestations and a certain organic apparatus."
17. Ibid., 2:78.
18. J. G. Spurzheim, *Phrenology, or The Doctrine of the Mental Phenomena*, vol. 1 (Boston: Marsh, Capen and Lyon, 1832), 130.
19. "Introduction" to "Innateness and Modularity" in *Philosophy of Mind: Classical Problems / Contemporary Issues*, ed. Brian Beakley and Peter Ludlow, 2nd ed. (Cambridge, Mass: The MIT Press, 2006), 685.
20. The authors discussed here are not featured in Petros Bouras-Vallianatos and Barbara Zipser, eds., *Brill's Companion to the Reception of Galen* (Leiden: Brill, 2019).
21. Gregor Reisch, *Natural Philosophy Epitomised: A Translation of Books 8–11 of Gregor Reisch's "Philosophical Pearl" (1503)*, trans. Andrew Cunningham and Sachiko Kusukawa (Farnham: Ashgate, 2010), 203: The editors explain that there were competing traditions of this kind of assignment as regards the details.
22. Ibid., 205; and Gregor Reisch, *Margarita philosophica totius philosophiae rationalis, naturalis et moralis …* (Friburgi: Schott, 1503), lib 10, tr. 2: De potentiis animae sensitivae, cap. 31; text fol. Hi v, illustration fol. Hii r. The key terminology in Latin is: "Horum organa in substantia cerebri subtilissimis secernentur pelliculis: quae primum totum cerebrum tribus distinguunt ventriculis."
23. Reisch, *Natural Philosophy Epitomised*, op. cit., 225; Reisch, *Margarita philosophica*, op. cit., lib 11, tr. 1: De potentiis animae intellectivae, cap. 3, fol. Hviii r. This duality has been thoroughly pointed out by Richard Oosterhoff, "Review of *Natural Philosophy Epitomised: Books 8–11 of Gregor Reisch's Philosophical Pearl (1503)* Translated by Andrew Cunningham and Sachiko Kusukawa," *Aestimatio* 8 (2011): 149–61.
24. For a historic contextualization see Javier Virués-Ortega et al., "A Systematic Archival Inquiry on Juan Huarte de San Juan (1529–88)," *History of The Human Sciences* 24 (2011): 21–47,

https://doi.org/10.1177/0952695111410929. For the educational context see Paul Richard Blum, "Psychology and Culture of the Intellect: Ignatius of Loyola and Antonio Possevino," in *Cognitive Psychology in Early Jesuit Scholasticism*, ed. Daniel Heider (Neunkirchen-Seelscheid: Editiones Scholasticae, 2016), 12–37. Cf. also Guido Giglioni, "Reading Galen's De naturalibus facultatibus in the Early Modern Period," in *Galen and the Early Moderns*, ed. Matteo Favaretti Camposampiero and Emanuela Scribano, International Archives of the History of Ideas Archives internationales d'histoire des idées (Cham: Springer International Publishing, 2022), 9–35.

25. Juan Huarte de San Juan, *Examen de ingenios, para las sciencias, donde se muestra la differencia de habilidades que ay en los hombres, y el género de letras que a cada vno responde en particular* (Baeça: Montoya, 1575), fols. 45r-v: "Tambien la sabiduría de el anima sensitiva, depende de el temperamento del celebro [...] El medio que tuvo Galeno para contemplar y conocer por vista de ojos la sabiduría del anima sensitiva fue tomar un cabrito luego en naciendo; el qual puesto en el suelo començo a andar (como si le vuieran enseñado y dicho, que las piernas se hablan hecho para tal uso) y tras esto se sacudio de la humidad superfina que saco de la madre: y alçando el pie se rasco tras la oreja, y puniendole muchas escudillas delante con vino, agua, vinagre, aceyte, y leche (después de haberlas olido todas) de sola la leche comio. [...] (gran razon tuvo Hipp[ócrates] en dezir que las animas eran sabias sin aver tenido maestro)." Cf. Fabrizio Bigotti, "'A prima constitutione foetus': La teoria dell'ingegno come facoltà naturale in alcune fonti mediche del tardo Rinascimento.," *Journal of History of Medicine: Medicina nei Secoli: Arte e Scienza* 25, no. 2 (August 2013): 349–394, 369 f. et passim.

26. For the anatomy of brain, spirit, pneuma, and nerves from the Presocratics up to Descartes into the twentieth century see K. E. Rothschuh, "Vom spiritus animalis zum Nervenaktionsstrom," *CIBA-Zeitschrift* 8, no. 89 (1958): 2949–2980, Galen pp. 2951f. Cf. Marielene Putscher, *Pneuma, Spiritus, Geist: Vorstellungen vom Lebensantrieb in ihren geschichtlichen Wandlungen* (Wiesbaden: Steiner, 1973), Galen 15–20 and passim; contains a helpful chronological survey of treatises on pneuma up to the late eighteenth century.

27. This and the following two quotations are from Claudius Galen, *On the Doctrines of Hippocrates and Plato*, ed. Phillip De Lacy, [Greek-English], Corpus Medicorum Graecorum V, 4, 1, 2: On the Doctrines of Hippocrates and Plato, Books VI–IX (Berlin: Akademie-Verlag, 2005), bk. VII 3, Kühn 606 f. Page references are to the standard edition of Galen, *Opera omnia*, ed. Carl Gottlob Kühn, in this case vol. 5, Leipzig: Knobloch, 1823.
28. "De usu partium corporis humani," book 8, in Claudius Galenus, *Opera omnia*, ed. Carolus Gottlob Kühn, Greek and Latin, vol. 3 (Lipsiae: Knobloch, 1822), 609–683. For Galen's brain sections see "A physiology of the brain" in Julius Rocca, *Galen on the Brain: Anatomical Knowledge and Physiological Speculation in the Second Century AD* (Leiden: Brill, 2003), 171–237; Teun Tieleman, "Galenus over de zetel von het verstand: Medisch experiment en filosofische traditie [Galen on the seat of the mind: medical experiment and philosophical tradition]," *Gewina* 18, no. 3 (1995): 24–36 [230–242]. On the materialist aspects cf. Charles T. Wolfe, "Galen's Contribution to the History of Materialism," in *Galen and the Early Moderns*, ed. Matteo Favaretti Camposampiero and Emanuela Scribano, International Archives of the History of Ideas Archives internationales d'histoire des idées (Cham: Springer International Publishing, 2022), 175–191.
29. Numerous quotations of this sort from various works in Julien Devinant, *Les Troubles psychiques selon Galien: Étude d'un système de pensée* (Paris: Les Belles Lettres, 2020), Partie 1. Approche théorique: Le matérialisme galénique en question. On Galen's wavering regarding the essence of the soul Pierluigi Donini, "Psychology," in *The Cambridge Companion to Galen*, ed. R. J. Hankinson (Cambridge: Cambridge University Press, 2009), 184–209.
30. Galen, *On the Doctrines of Hippocrates and Plato*, bk. VII 7, Kühn 642.
31. Ibid., bk. VII 7, Kühn 643f. It would be worth researching whether Girolamo Fracastoro's theory of sympathy was inspired by Galen: Girolamo Fracastoro, *Liber unus, De Sympathia et Antipathia rerum. Item, De Contagione et Contagiosis Morbis, et eorum curatione, Lib. III* (Lugduni: Tornaeus, 1554): chapters 12–14 discuss the sympathy of the soul and the senses, citing also Galen.

CHAPTER 8

Neither Comic nor Corpse: Revisiting Bracelli's *Bizzarie di Varie Figure*

Nicole Howard

In 1943, a rare book and print collector named Lessing J. Rosenwald—president and chairman of Sears, Roebuck, and Company—announced his intention to donate the entirety of his rare book and art collection to the Library of Congress and the National Gallery of Art. The former would receive Rosenwald's 2600 rare books, while the National Gallery was given 22,000 rare prints and drawings.[1] Among these was a small book—its fifty pages were roughly 4 ½ × 6 inches—entitled *Bizzarie di Varie Figure* by the Italian Renaissance artist Giovanni Battista Bracelli. It was one of only two complete copies known in the world, the second being in the Bibliothèque nationale de France, and it contained some of the most confounding and entrancing images of the seventeenth century. This book, like the artist who produced it, is almost a footnote in the history of the Renaissance, but Rosenwald saw in it something essential to the period. In 1968, he commissioned a work by the artist Paolo Boni,

N. Howard (✉)
Eastern Oregon University, La Grande, OR, USA
e-mail: nhoward@eou.edu

J. Muller (ed.), *Automata, Cyborgs, and Mutants*, Palgrave Studies in Medieval and Early Modern Medicine,
https://doi.org/10.1007/978-3-031-93201-4_8

Homage to Bracelli, a collograph made from twelve progressive proofs that were created using a metal collage plate. Clearly, the *Bizzarie* that most of the world had overlooked made an impression on Rosenwald, as it does on anyone who happens across the images today.

To study Bracelli—a self-described pittore fiorentino—is to engage with a style and creativity we often associate with our own century. The Museum of Modern Art classifies his work as Fantastic Art or Surrealism, giving no immediate indication that he lived in Florence in the early seventeenth century among some of the most renowned artists of the time. His images—left to us in a few incomplete books—are enigmatic and arresting, all absurd objects and angular torsos, tossing the viewer off balance. Bracelli's etchings—the eighty-eight of them remaining—are true objects of speculation.

As others have noted, a simple search for Bracelli yields a thicket of misinformation partly owing to the fact that he shared his name with a Genoese painter of the same period. It is clear that he was a member of Florence's Accademia del Disegno from 1619 to 1635 and that he studied under Jacopo da Empoli (c.1551–1640).[2] He spent several years in Livorno working on decorations for the cathedral there, after which he proceeded to Rome where he produced an undated series of etchings, *Figure con instrumenti musicali e boscarecci*. It consisted of thirty images depicting individuals with contemporary instruments: a hurdy-gurdy, a trombone, and a cornet. Historians of music consider these etchings some of the most accurate representations of seventeenth-century instruments. A figurative alphabet, etched in 1632, was published in Naples and resembles work by contemporaries who also sought to contort the human body into letter forms, Peter Flötner's 1534 Menschenalphabet being the earliest example of these.[3]

Bracelli's best-known work, however, is the *Bizzarie di Varie Figure*, from 1624. Published in Livorno, it was a collection of fifty etchings that, from the outset, vexed observers. The images have been described as Mannerist or Baroque in style. They are idiosyncratic and sometimes monstrous. They were dismissed by the art historian Kenneth Clark as "mere stunts" meant to alter the purity of the prevailing mannerist style.[4] I would give them more credit. Generally discussed in the context of northern Italian etchings, the *Bizzarie* was a striking tangent along the curve of Renaissance art, a diagonal that skimmed the more mainstream currents of etching and engraving seen in the works of Annibale Carracci or Jacques Callot. The Dadaist Tristan Tzara marveled at the "sparks of familiarity, mirrors strolling among truths and lies" that Bracelli had imagined more than three centuries earlier.[5] Tzara saw in Bracelli a nascent modernism incongruent with the seventeenth century, something ahead of its time that presaged twentieth century disillusionment. But such a view limits our understanding of this work, for while the human forms Bracelli etched may speak to our modern condition, we should not ignore their resonance in the Italian Renaissance. Bracelli, in fact, sat at the intersection of several important Renaissance artistic currents. We see in his work the influence of two Renaissance spaces. The first is the anatomical theater, where bodies were publicly anatomized and broken down into their constituent parts. The second is the puppet theater, where bodies were assembled, imbued with motion, and animated as living beings. To look at the *Bizzarie* as a reflection of these theatric traditions allows us to see Bracelli's figures as more than unusual or playful forms but as philosophical reflections on the early modern body.

Bracelli's Work and the Influence of the Stage

Anyone who has encountered the contorted, nearly disembodied figures of Bracelli senses the abrupt break from traditional humanist renderings. Angular and deconstructed, fractured and bent, Bracelli's images tear a rent in our shared vision of mannerism and rupture early modern norms of composition. What inspired these images has long been the subject of debate among art historians. Typically, after acknowledging their uniqueness, scholars point to the influence of Renaissance festival designs, an influence traced through the work of Guilio Parigi, his father Alfonso Parigi (1535–1590), and Bernardo Buontaleni (1531–1608), all of whom were actively engaged in set design for the Medici in Florence. A closer look at their work reveals their clear aesthetic influence on Bracelli.

Take, for example, Guilio Parigi's etching The Argonautica: Ship of Hercules, Guilded by the Austrian Eagle [Fig. 8.1].[6] This rendition of the Argonauts was reinterpreted in 1608 by Remigio Cantagallina (1582–1656), one of Parigi's pupils, on the occasion of Cosimo de' Medici's wedding to Archduchess Maria Magdalena of Austria [Fig. 8.2].[7] We can see the decorative bodies—six cupids—that line the side of the hull, each marking the place of an oarsman. There are also the carved torsos of three dignitaries in the aft and numerous other faces and heads adorning the ship's sides. This design was subsequently taken up by Bracelli, who would have been familiar with the motifs, and transformed into corporeal-nautical hybrid, with bodies interlocked and layered upon one another to form the hull, larger figures comprising the masts, and the extended forms mimicking the serpents and dogs in fore and aft [Fig. 8.3]. The only materiality that remains is the fabric of the sails—tied up against yards—and the oars. Bracelli took a well-known image and imbued it with a hyperbolic emphasis on the body. In doing so, he inverted the ideas of another well-known Renaissance artist, Giuseppe Arcimboldo (1526–1593), who famously painted portraits of European monarchs comprised of fruits, vegetables, flowers, and even books.[8] But where Arcimboldo's works presented audiences with a visual paradox—what seems like fresh vegetables is actually, if viewed holistically, the Holy Roman Emperor Rudolf II—Bracelli takes the opposite tack. What appears on first glance to be a ship is not a ship at all but a curious arrangement of bodies. The viewer needs a moment to readjust. The bodies, suggests Bracelli, are simply materials used to construct, and yet his images play off Arcimboldo's ideas of fantasia—understood as "the inventive imagination"—creating bodies that are entirely recognizable but performing the tasks of inanimate objects.[9]

Other etchings in this vein focus on seascapes or cityscapes and point to the influence of both Bernardo Buontalenti (1531–1608) and Jacques Callot. Buontalenti was a major figure in the Medici court, having acted as their primary architect for sixty years.[10] He also built the stage in the Uffizi palace, designed the Boboli Gardens, and served as a military engineer. It was in this latter capacity that he received a commission from Ferdinand de' Medici to improve the fortifications for the city and port in Livorno, the very fortifications Bracelli creatively etched for the *Bizzarie* series. Buontalenti was known for his Neoplatonist approach to design and architecture, with its emphasis on proportionality and geometric design [Fig. 8.4]. This aesthetic seems to resurface in Bracelli's renderings, as we will see shortly, but Buontalenti also adhered to the Platonic idea that man was a microcosm within the macrocosm of the urban space, a notion echoed in Bracelli's humanoid representations of Livorno [Figs. 8.5–8.6].[11] On the left he depicts the port of Livorno, surrounded by walls, as a reclining body. Crenellated parapets sit at the elbows, and a natural hillock forms a raised left knee. As he did throughout the *Bizzarie* series, Bracelli reduces the horizon to a series of basic horizontal lines which comprise water and a single line to reflect land. Looking at the plans Buontalenti drafted for the port of Livorno, we see his use of a

Fig. 8.1 The Argonautica: Ship of Hercules, Guilded by the Austrian Eagle https://collection.mcnayart.org/objects/9246

traditional pentagonal plat, considered at the time the perfect shape for a fortress. This idealized urban form was implemented in the construction of the Fortezza da Basso in Florence, the Palazzo Farnese in Caprarola, and in other citadels throughout the Italian peninsula.[12] One can almost place Buontalenti's plat against Bracelli's human Livorno to locate the five sides, but as with the other images, he does so by morphing that ideal city into a scarcely recognizable body. In an alternate city-scape, Bracelli positions the viewer at the base of Livorno, looking up through what would be the crotch of the city. The five points of his earlier etching remain (head, arms, and legs) but now we are given a unique view, the left knee is no longer elevated and the horizon is reduced to a single line. He has mapped the macrocosm of urban space onto the microcosm of man, rendering the body inanimate again.

Fig. 8.2 Remigio Cantagallina, Battaglia navale rapp. in Arno per le nozze del sermo principe di Tosc. Image courtesy of the Redwood Library and Athanaeum

Street Theater, Puppets, and Monsters

In addition to the artistic and architectural influences of northern Italy, we can also see in Bracelli's figures the influences of Florentine Street theater. Scholars have pointed to the structural parallels between the *Bizzarie* etchings and Callot's *Caprici di varie figure*, a series he completed in 1618. Consider two examples, Callot's etching on the left and Bracelli's on the right [Figs. 8.7–8.8]. The frame in both is void of any background and the pairs are symmetrically placed, each a figure in contrapposto, weapons at hand. In the figures themselves, however, a chasm opens up as Bracelli assembles bodies from metal cylinders and hoops, stripped of any attire, nearly androgynous. One might posit that the figure on the right is the female, with visible navel and implied breasts, and her counterpart with the staff is the male, but Bracelli has largely veiled the cues that would alert a viewer to gender.[13] It is worth recalling the dialogue in Ben Johnson's play Bartholomew Fair when the Puritanical character

Fig. 8.3 From the Bizarrie di Varie Figure

Busy addresses the Puppet Dionysus: "my main argument against you, is, that you are an abomination: for the Male, among you, putteth on the Apparel of the Female, and the Female of the Male." But the puppet ardently refutes the charge: "It is your old stale argument against the Players, but it will not hold against the Puppets; for we have neither Male nor Female amongst us."[14] This notion of puppets leads us to another useful comparison between Callot and Bracelli, one that traces its roots to the popular Commedia dell'arte tradition.

Throughout Callot's *Caprici* we see characters that most Italians of the day would have easily recognized: Pantaloon, the amorous old man who made his living as a merchant; Scapin the Zani, the country bumpkin or trickster; the swaggering Captain; and the lover Amoroso. Each came through in Callot's work with a dynamism and motion that forged new ground in printmaking.[15] Minstrels and clowns pervaded Commedia dell'arte but it is the mountebanks—itinerant doctors and pharmacists—that are of particular interest. The ciarlatano reflected the

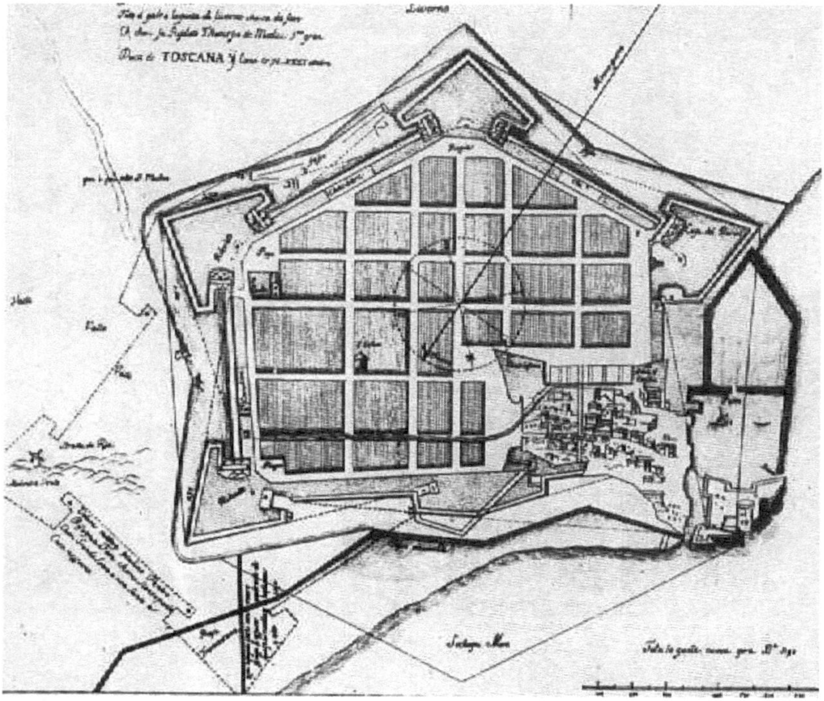

Fig. 8.4 Buontalenti's plan for the port of Livorno

Fig. 8.5 Bracelli's representation of Livorno

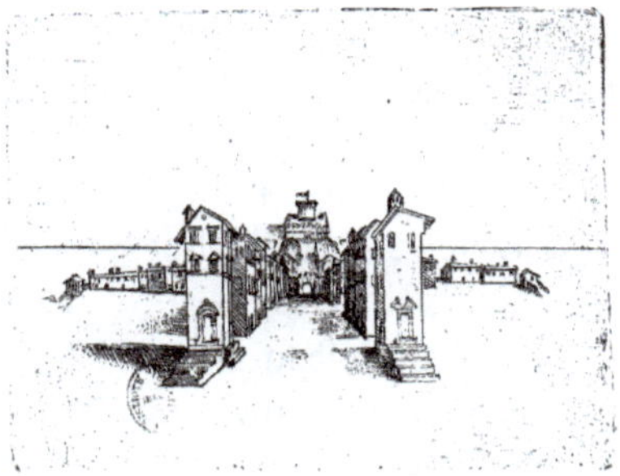

Fig. 8.6 Bracelli's representation of Livorno

Fig. 8.7 Callot's and Bracelli's *Bizzarie Source* Image by Callot

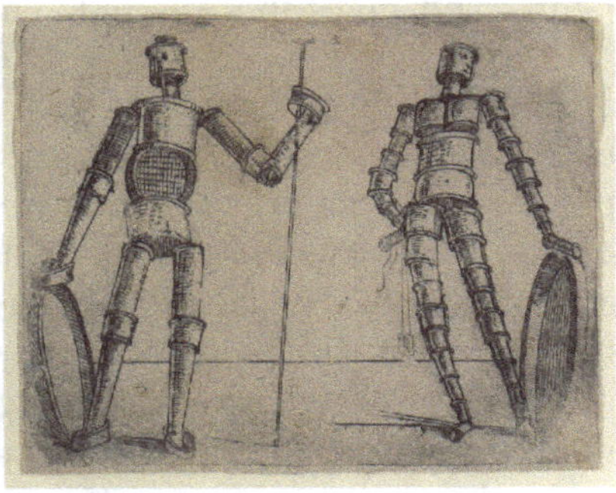

Fig. 8.8 Callot's *Caprici* and *Bracelli*'s *Bizzarie Source* Image by Bracelli

typical pretender who used his verbosity and sly speech to draw people in to his faux medicine, while the cerretano roamed about peddling his quackery through a deceptive theatricality.[16] These two characters would ultimately be collapsed into the single category of "charlatan" but both were well-known theatric characters in the late sixteenth and early seventeenth centuries. They were itinerants who appeared in a town, mounted a make-shift stage (thus, mountebank), and peddled their tonics and cures in dramatic fashion. In Shakespeare's Comedy of Errors there is a character called Pinch:

> a hungry, lean-faced villain,
> a mere anatomy, a mountebank,
> a threadbare juggler and a fortune teller,
> a needy, hollow-eyed, sharp-looking wretch,
> a living dead man.

The mountebank is reduced to a theatric character, "a mere anatomy," a concept which echoes throughout the *Bizzarie*. Our threadbare juggler, reduced by occupation to a hollowed out being, becomes the living dead, the very definition of Bracelli's images. In a similar vein there is Callot's 1622 work Balli de Sfessania, twenty-four etchings that depicted pairs of posturing street performers. Though the work was issued in Nancy, art historians believe the designs were made while Callot was still in Florence, and copies of the work were sent back to Florence after being printed.[17] Scholars note that the Balli figures are distinct from those that appear in Commedia dell'Arte. Rather, Callot uses them to present figures and poses in a carefully choreographed display of etching skill. These characters he created would have been instantly recognizable to both the wealthy and commoners, and not just from live performances of wandering minstrels, but also through the translation of their dramatic personae into puppet theater.

The tradition of puppetry is particularly important when we look at Bracelli's figures, which simulate the movement of articulated limbs hanging from strings. Indeed, early modern audiences referred to these puppets shows as "motions," and puppets themselves came to be called motions as well.[18] Sicilian craftsmen were especially renowned in this period for their marionettes (Opera de Pupi), but puppets from Rome, Milan, and Naples were all considered exemplary, with limbs and torsos and heads animated through a system of silk cords and rods.[19] There were also planchette puppet shows, where a string linked puppets on a board to a musician's leg. As he played, his leg pulled the string to effect movement in the puppets, giving them the appearance of dancing in time.[20]

Looking at Bracelli's etchings in the *Bizzarie*, we see the construction of figures from assorted household goods, manufacturing materials, or more abstract shapes, but all of them carry the appearance of marionettes. There is a figure constructed from a system of pulleys and ropes, standing opposite a plated figure with a serpent that mirrors the strings of its counterpart. Many figures bend at the hips and knees in a stiff manner, all sharp angles with no natural curves. Some appear to be wooden but at times constructed improperly, in one case with the bizarre addition of a second, horizontal figure growing out of the one that stands erect [Fig. 8.9]. Bracelli seems to be drawing on the tradition of puppetry to create a visual jest, a parody of the plurality of our individual selves. In another plate he varies this theme, flattening the characters who seem conjoined, all head and torso with no legs. An error in morphology renders the figures incapable of standing. Arms are raised in pleading while torsos at the opposite end attempt a futile locomotion. The result borders on what Elena Lazzarini described as "the restlessness of the human body; a body filled with internal and external tension; a body that seeks to understand and overcome its limitations..."[21] In this assessment, Lazzarini draws on early modern ideas of the grotesque and monstrous, pointing to the work of the Florentine humanist Benedetto Varchi, whose 1560 work explored the generation of monsters.[22] Varchi categorized monsters into multiple groups, considering some as retributive—those whose bodies bear marks of divine wrath) and others deviant—with bodies that differ from the normal course of nature. Within these groups he further identifies perfect and imperfect bodies, the former reflecting a harmony and symmetry typically found in bodies, but with appearances outside the norms. The latter, imperfect bodies he considered deformed beyond recognition.[23] In his treatise he described a double-monster born at the gates of Prato some years before: "They were two females joined [conguinte] & stuck together, one toward the other in such a way that half the chest of one along with that of the other made up a single chest, & thus they formed two chests, one joining up with the other..."[24] It is hard to look at Bracelli's figures and not see traces of Varchi's monsters. They retain a symmetry and an intrinsic harmony that does not repulse the viewer so much as it challenges and amuses.

By the early seventeenth century, Varchi's ideas were being further developed in the context of scientific and medical advancements. In 1616, less than a decade before Bracelli published his *Bizzarie*, Fortunio Liceti (1577–1657) published his own work on monsters, *De monstrorum*

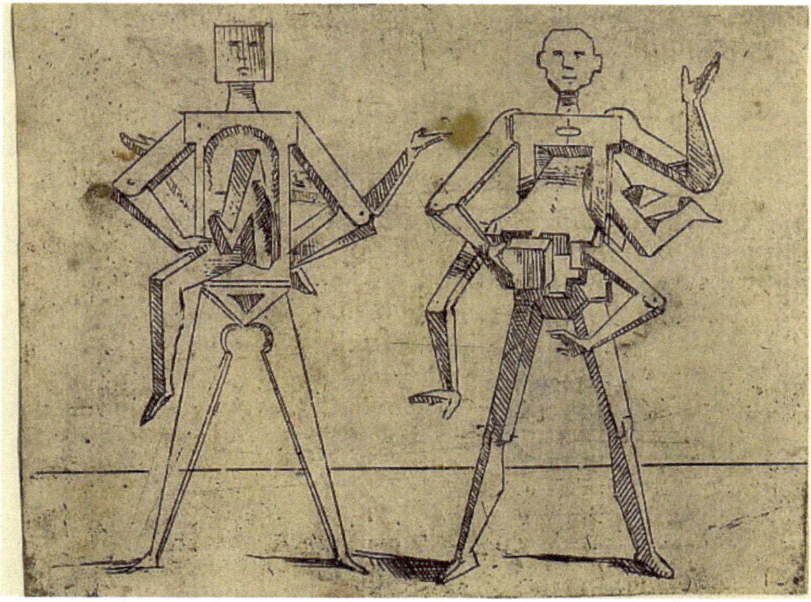

Fig. 8.9 Two abstract sketches of humans with surreal elements, including multiple torsos, heads, and limbs. The two figures are highly geometric and stylized, bizarre representations of human-like beings

causis. Here Liceti, a physician who worked at the universities of both Pisa and Padua, shifted the study of monsters in the direction of natural philosophy by drawing on the medical and anatomical knowledge of his day.[25] Rather than viewing deformed bodies as manifestations of divine wrath, he reframed their existence as a naturally occurring alternatives to the normal body, something to be studied to advance our understanding. And it is in this field of study—anatomy—that we find some of the richest context for understanding Bracelli's images, context which I believe is central to understanding the *Bizzarie*.

Anatomical Influence

Bracelli was no stranger to anatomical work. At the Florentine Accademia del Disegno, where he was a member, statutes assigned responsibility for securing a corpse to the Consoli who would see to its dissection in the nearby hospital of Santa Maria Nuova.[26] All artists were required to attend these anatomies and encouraged to perform private dissections in the winter months.[27] This emphasis on an artist's familiarity with the body's substructures was influenced by developments in Italian medical schools. At the University of Padua, human and animal corpses—dogs, pigs, and monkeys—were displayed on an elevated table at the center of a purpose-built amphitheater. A balustrade surrounded the table to secure space between the lector and his assistants and the audience members, who frequently tried to get close enough to see the fine details in the body. Two hundred and fifty students convened over fourteen days to watch the professor perform a carefully scripted process of post-mortem dissection. From as high up as thirteen meters, students leaned over the bannister to watch the undoing. These would-be physicians, who hoped to preserve body and health among their future patients, had to first bear witness to the dissembling. And as the days passed, the bodies they studied broke down. Organs were removed, arteries and veins collapsed, and muscles were absolved of their strength. Except for the bones, the well-ordered systems of life became slowly less recognizable. Bracelli's presence at local anatomies would have roughly followed the outlines of the academy's procedure, even if performed more rapidly for art students.

Anatomies were more than lectures, however. They were academic shows with close ties to the theatric traditions in Italy, and they represent an important moment in the history of the stage because of their public, performative nature.[28] Though the professor carrying out the dissection was nominally teaching medical students, the audience frequently included local dignitaries, professors from other disciplines, school administrators, and occasionally commoners from the town, all of whom paid a fee in order to attend. The idea of "selling theater" was, this time, a relatively new one and purchase of seats in the anatomical theater coincided with the sale of tickets to dramatic performances.[29] Funds collected from spectators in advance of the anatomy were used to buy supplies for the demonstration. For their part, audience members were hardly silent witnesses but rather vocal participants, shouting questions and—in more raucous performances—challenging the knowledge of the anatomist himself.[30]

The anatomies were also theatrical in their didacticism, with the dissection of the corpse serving as a kind of moral exercise, in much the same way that puppet theater served to present the public with morality plays. The corpse on the table was typically that of a sinner, someone hanged to death, but through dissection the audience could view "the expediency of Creation, visible in the human physique."[31] Through the act of anatomizing and the process of recognizing the perfection of corporeal structures, the body regained a perfection in death that it had lost in life. Bracelli and his peers were not just surrounded by anatomical study; they interpreted it through a vital artistic lens. The dissected body depicted in anatomy texts has been described as "Dead and not dead, oblivious to its own partition, stumbling through the countryside or sprawled over a chair or table."[32] Consider Bracelli's seated figures [Fig. 8.10], which behave in an entirely normal way, working at their desks, with no sense that their bodies have been stripped of humanity. Both sit and work on a surface composed of the same materials that comprises them. The line between nature and artifice has dissolved. It is a visual representation of the materialism that was rapidly emerging to replace older, more teleological frameworks. Historians have often viewed the mechanical philosophy as one of the primary drivers of this materialist turn, and there are certainly links between the two paradigms. But anatomical research did not simply turn humans into machines. As Bracelli's images suggest, after a certain point the dissembled body served to erase distinctions between animate and inanimate. Bracelli does not give us automata; he gives us parts—individual elements—with the potential to combine in a logical, or perhaps functional, manner.

At the very same time Bracelli was etching his bodies, René Descartes was penning L'Homme, positing a theory of the body as complex mechanism. And while the soul tucked neatly into the pineal gland, the rest of the body was an assemblage of parts. Levers, pulleys, and pumps became metaphors for the workings of man. From such a gestalt, the medical community broke the body down into increasingly smaller and distinct components, with studies on individual organs, bones, muscles, and eventually those elements too small to be seen with the naked eye. Such analysis paralleled an artistic shift that called for greater attention to anatomic structures. Italian drawing books, such as Colombina and Esengren's 1623 Discorso Sopra Il Modo Di Disegnare, presented the artist with a fully dissembled body, page after page of anatomical parts, often overlaid with "cages" to assist with proportion [Fig. 8.11]. Torsos pile up at various angles, and an odd number of limbs reach and bend without purpose. Just out of view, one imagines, is the dissection table. Bracelli and his fellow artists at the Academia were certainly familiar with these drawing books, some of which were published by peers in Florence.

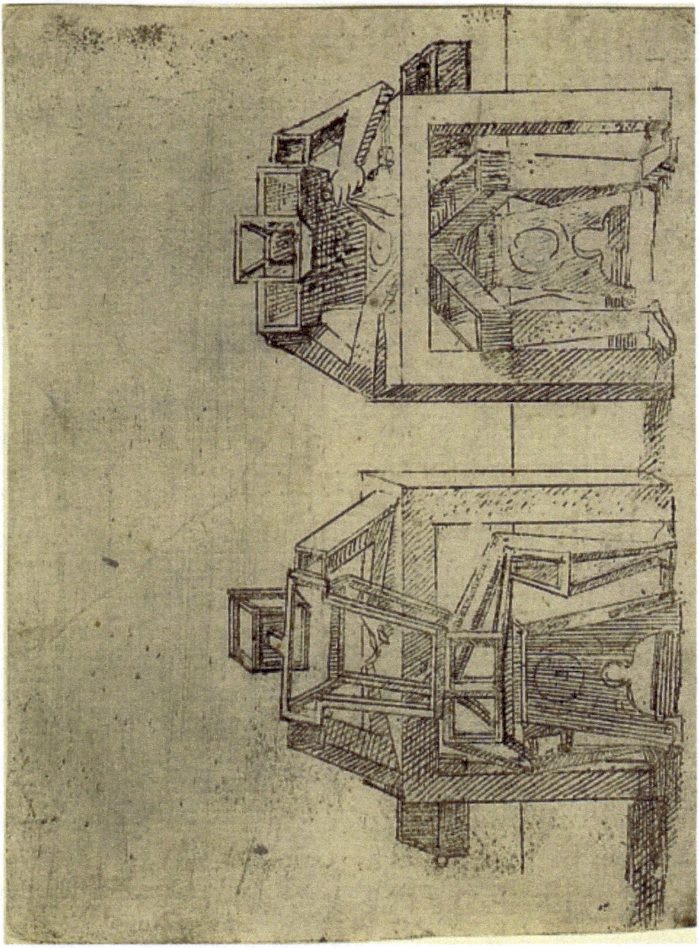

Fig. 8.10 Two abstract sketches of humans, each seated at a drafting table, using a compass to draw. One faces the reader, the other faces away. Both bodies are constructed from minimal geometric shapes, principally cubes and boxes

Fig. 8.11 Italian Drawing Book, Colombina and Esengren

Looking at the images, and considering Bracelli's participation in public anatomies, we can begin to frame the *Bizzarie* not so much as a deconstruction, but as a reassembly project. The parts could be organic or not, but with care they might be attached to create a new being: between male and female, organic and inorganic, natural and monstrous. In a sense, the object of the anatomy became the "living dead man" from Shakespeare's reference, the marionette reanimated in the spirit of pedagogy.

Historian Jonathan Sawday characterizes the culture of dissection as "a network of practices, social structures, and rituals surrounding this production of fragmented bodies…once the body has been partitioned and its interior dimensions laid open to scrutiny, the very categories 'male' and 'female' become fluid, even interchangeable."[33] The links to Bracelli's figures are evident. Consider the engraved copper plates in Fabricius' *Operationes chirurgicae*, first published in Frankfurt in 1592 and subsequently in Venice in 1619 [Fig. 8.12]. In this work he presents the anterior and posterior views of an orthopedic cuirass used to treat orthopedic injuries or physical deformities. The assemblage of parts is reminiscent of early modern automata and effectively inverts the traditional anatomical image: rather than a body stripped of flesh, this exoskeleton of braces contains and supports a void. The Science Museum in London has an articulated manikin made from iron and dating from the sixteenth or seventeenth century [Fig. 8.13]. The catalog note suggests the model was used to teach bone-setting. The similarity to the image in Fabricius' text suggests that either Fabricius or someone familiar with his work had it made. Both the engraved and three dimensional figures are also clearly reminiscent of the marionettes discussed above, with elbows pulled out from the body, as if by strings. It is no surprise, then, that when we turn to Bracelli's images, we see the deeply connected reverberations of this anatomical and artistic tradition.

The forty-six figural plates of Bracelli's *Bizzarie* can be categorized around distinct aesthetic themes. The first group are very organic, having the physiognomy and musculoskeletal attributes found in man. A second category emerges around everyday objects like pots and pans, ropes, or miscellaneous art supplies. Finally, there are the figures that transform the human body into three dimensional geometric shapes—pyramids, cylinders, and cubes. Looking more closely at these groupings, we can begin to see the threads of influence outlined above in the warp and weft of Bracelli's figures. Beginning with the organic group, it is evident that Bracelli pulled on both anatomical and artistic images commonly seen

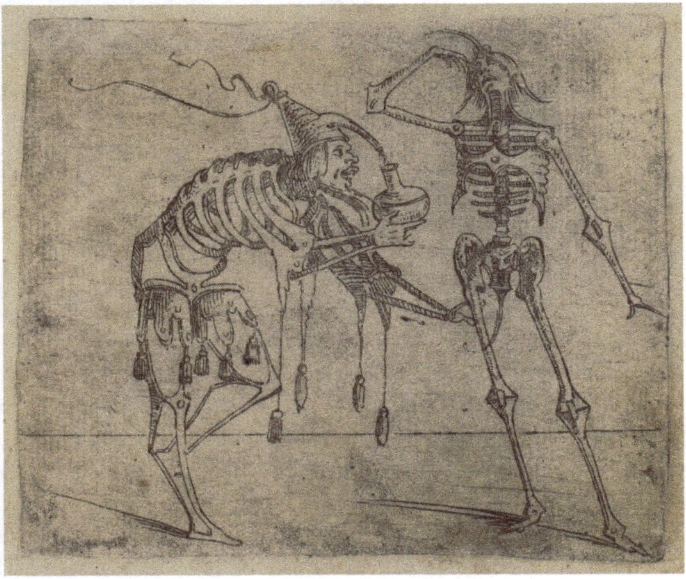

Fig. 8.14 Two skeletal figures engage in conversation. On the left, the figure holds a urine flask and appears to be offering a diagnosis to the figure on the right, who holds their head in seeming dispair. The image is surreal and gestures at doctor/patient interactions

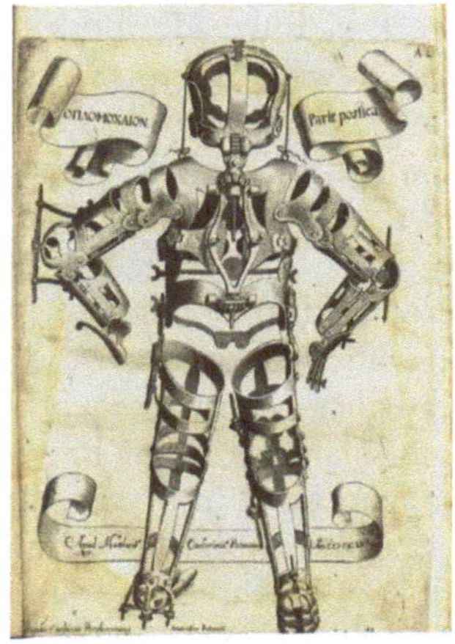

Fig. 8.12 From Fabricius' surgery treatise

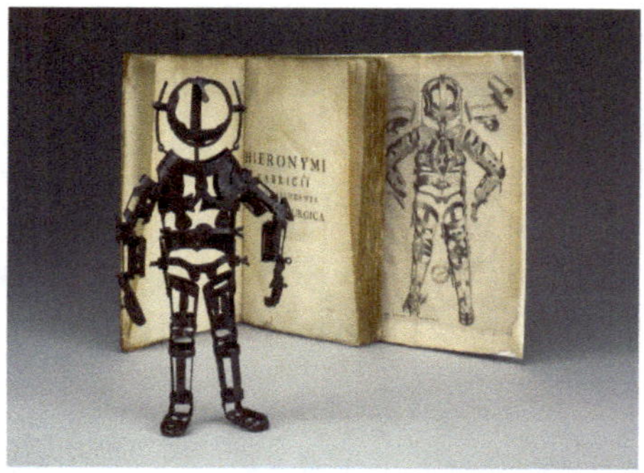

Fig. 8.13 From Fabricius' surgery treatise

at the time. One set of figures becomes the embodiment of earth and air, another fire and water, but both pairs have a human morphology. Likewise, we have a pairing who resemble the mountebanks of the day, the one on the left presumably offering an interpretation of the urine flask—a common diagnostic tool—to the man next to him. [Fig. 8.14] These figures look like corpses in their final days of anatomization, bodies reduced to skeletons. The few rags that hang from their frames resemble the flayed skin of the écorchés. Similar etchings present bodies with minimal musculature or, in the case of Bracelli's Adam and Eve, people whose forms are integrated with natural objects, such as trees. Adam's torso looks human, but it is also a tree trunk, his arms and phallus emerging as tree limbs. Eve is even more human in appearance, with a clearly feminine face, torso, and right arm, while the rest of her dissolves into foliage. All of the images contain a playfulness and naturalism common to the period.[34]

The ludic element of Bracelli's etchings becomes even more pronounced in the nineteen etchings where bodies are composed of everyday objects. Here, like Arcimboldo before him, Bracelli looks to the assembly of human parts—knees, hands, bellies—and replaces them with inorganic substitutes. One pair [Fig below, left] holds swords and flags, a drum, and spears, but while the upper torsos appear normally embodied, their lower halves are composed of wooden stilts. Likewise, on the right we see a pair with legs made from scissors or a shovel, while an arm takes the shape of a spatula, a compass, or a pick. Gender is erased completely, as are human expressions. The reduction here is more than just anatomical; it is teleological. That is, the purpose of human limbs is curbed radically as they are replaced by more mundane objects that accomplish simple tasks. Legs here move the body forward, but knees and hips are erased. Arms that might serve innumerable purposes are reconstituted as simple levers, platforms, or hooks. When talking about works from this period art historian Giancarlo Maiorino observed that, "the artfully natural became the spectacularly artificial."[35] This sentiment captures Bracelli perfectly. It also calls to mind the work of Guilio Casseri, another student of Fabricius. In his work on the laryngeal complex and vocalization, Casseri described the movement of the vocal cords in decidedly mechanical terms, directly comparing their operation to the workings of a surgical instrument called a glossicomium: "The construction of this machine is as follows: its top and bottom sections contain a plank, and on the machine several little pulleys are built up..."[36] Explication of an anatomical or physiological function through comparison with a mechanical process was increasingly important heuristic tool in the period.[37] The mechanical gestalt is evident throughout *Bracelli*'s etchings, often to a grotesque degree. Some figures in the *Bizzarie* are so deformed and anatomized as to blur the lines between the living and the artificial. They become manifestations of the grotesque which Maiorino maintains "is not a way station in which the formless awaits to be 'formed,' but an ambivalent mechanism out to mix the text of nature with that of culture."[38] It is precisely this ambivalence that Bracelli embodies in his forms. (Figs. 8.15 and 8.16).

Fig. 8.15 Two figures, both helmeted, both extravagantly dressed for battle. One is armed with spears and sword, and the other a drum and spears. Neither figures has human legs, but rather surreal lower bodies comprised of stilt-like objects. Their arms appear human

Fig. 8.16 Two abstract figures composed of everyday objects face one another. On the left the figure's torso is a rectangular canvas, and their arms and legs are tools. On the right the figure's legs are a pair of scissors

Examining the third group of Bracelli's *Bizzarie* figures, we see the full impact of period's myriad anatomical developments and mechanical visions of the body. Sixteen of the plates in this collection take on distinctly geometric forms, in a kind of Neoplatonic performance. One pair is comprised of empty boxes, hollow cubes for heads and ribs of descending, hollow compartments. A nod to the naturalism of Platonic thought comes in the birds, a starling and a pigeon, which perch on the hand and head of the figures as if on statues. They are more alive than the figures on which they rest. We can look at this etching as a manifestation of humanized architecture—the body reduced to mere structure. It simultaneously dehumanizes the bodies in the same way that anatomizing them did. Bracelli puts viewers in a liminal space, highlighting the tension a humanist might feel when confronted with a machine-like body. There is also a striking etching [Fig. 8.17] of two figures fighting over a goat. The tools of a cobbler—knives, thread, and awls—are scattered throughout the scene, as are the forms of shoes. And bodies are not just stripped of human anatomy, but the skeletal structure that remains—which may represent the shoemaker's frames—resembles an inorganic spine, a series of vertebrae assembled to make not just the torso but limbs as well. The goat between them, similarly reduced to a frame that is loosely covered in skin, is likely the source of the cobblers' leather. Two final examples display Bracelli's brilliant incorporation of the theatric traditions into the anatomized body. In the first etching [Fig. 8.18] abstracted bodies contort like puppets-gone-awry, as if Callot's *Caprici* had been put on the anatomy table, flayed and dissected, then set loose. They struggle to hold themselves upright, like Vesalius's muscle-men who collapsed as—page by page—their corporeal structures were removed. The poses are dramatic, rehearsed for the stage, but the geometry of their joints and limbs is a stark reminder of the degree to which the human body was becoming a collection of mere objects. The second etching works similarly [Fig. 8.19] as androgynous figures—borderline automata—bend at the waist to look back at the viewer through their legs, their buttocks, head, hands, and feet just empty Platonic solids. With both of these etchings, Bracelli taps into a wonderment of the early seventeenth century, when the organic body and its internal mysteries were increasingly unpacked and understood. They are bodies anatomized, reassembled, and animated in the manner of several Renaissance theater traditions.

Sixty years after Bracelli's *Bizzarie* was published the Italian anatomist Giovanni Alfonso Borelli would consider the entire body in a wholly mechanized fashion. Natural philosophers would engage with the mechanical metaphor differently throughout the seventeenth century, but to couch the anatomical process in strictly mechanical terms misses a dynamic component. Increasingly, the bodies anatomized in the theaters of Bologna and Padua came to look like, and be imagined as, more complex versions of the puppets that entertained the public in the city squares of Florence and Livorno. Bracelli's *Bizzarie di varie figure* is one of the best examples we have of an artist negotiating the converging trends in science, theater, and art. He certainly owes a debt to Callot, Arcimboldo, and Buontalenti, but his images should not be considered derivative. He stands shoulder to shoulder with these great and more

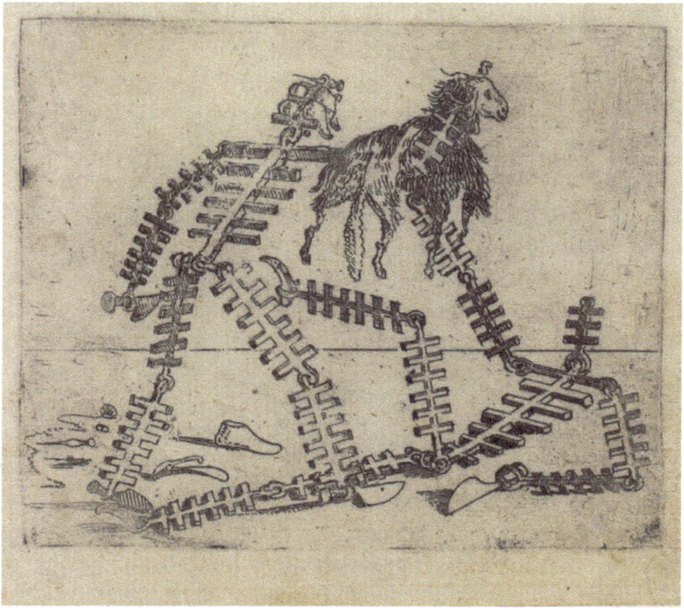

Fig. 8.17 Two highly abstracted figures composed of cobbler's tools, one standing and one on the ground. In between them is a goat, also partially made of a cobbler's frame, and both figures are vying to grab the goat

Fig. 8.18 Two abstract figures, one standing in an arched posture, the other with two hands on the floor, bent over. Each is comprised of geometric forms

Fig. 8.19 Two fibures, both comprised of connected and nested boxes, bending over at the waist and looking back through their legs

recognized artists, and reaches far beyond their aesthetic boundaries to make a statement—at once ludic and alarming—about the human body, mortality, and the self.

Notes

1. Ruth E. Fine and Lessing J. Rosenwald, *Tribute to a Collector* (Washington D.C.: National Gallery of Art, 1982), 23.
2. Background information can be found in several sources, but the best is Sue Welsh Reed, "Commentary" for Bizarre di Varie Figure (1624), from The Lessing J. Rosenwald Collection, Library of Congress.
3. Erika Mary Boeckeler, *Playful Letters: A Study in Early Modern Alphabetics* (Iowa City: University of Iowa Press, 2017), chapter 2.
4. Kenneth Clark, "The '*Bizzarie*' of Giovanbatista Bracelli," *The Print Collector's Quarterly* (1929): 319.
5. Tristan Tzara, "Propos sur Bracelli" in *Bracelli Bizzarie*, facsimile edition of the 1624 (Paris: Alan Brieux, 1963).
6. Arthur R. Blumenthal, *Guilio Parigi's Stage Designs: Florence and the Early Baroque Spectacle* (New York and London: Garland Publishing, Inc., 1986), plate 92. Image from https://collection.mcnayart.org/objects/9246.
7. Remigio Cantagallina, *Battaglia navale rapp. in Arno per le nozze del sermo principe di Tosc.* (Florence, 1608). Image courtesy of the Redwood Library and Athanaeum. As part of the wedding festivites, Cosimo II and Silvio Piccolomini reenacted the battle between Christians and Turks in a four-hour, choreographed play,

directed by Parigi. See Sarah Blick and Laura D. Gelfand, eds., *Push Me, Pull You: Imaginative and Emotional Interaction in Late Medieval and Renaissance Art*, vol. 1 (Leiden: Brill, 2011), 415, note 72.

8. Thomas DaCosta Kaufmann, *Arcimboldo: Visual Jokes, Natural History, and Still-Life Painting* (Chicago: University of Chicago Press, 2010).
9. Ibid., 193.
10. Aby Warburg, *The Renewal of Pagan Antiquity*, trans. David Britt (Los Angeles: Getty Research Institute, 1999), 354.
11. Vaughan Hart, *Art and Magic in the Court of the Stuarts* (New York: Routledge, 1994), 8. See also Corey Tazzara, *The Free Port of Livorno & the Transformation of the Mediterranean World 1574–1790* (Oxford: Oxford University Press, 2017), chapter 1.
12. Martha D. Pollak, *Turin 1564–1680: Urban Design, Military Culture, and the Creation of the Absolutist Capital* (Chicago: University of Chicago Press, 1991), 16.
13. The Bracelli image is from Plate #20 in his *Bizzarie* series. This version comes from the copy in the Metropolitan Museum of Art. (https://www.metmuseum.org/art/collection/search/817371, Accessed 10/14/19).
14. Ben Johnson, *Bartholomew Fair*, Act V, Scene 5. https://ebooks.adelaide.edu.au/j/jonson/ben/bartholomew-fair/act5.html (Accessed October 23, 2019).
15. Edwin T. Bechtel, *Jacques Callot* (New York: George Braziller, 1955), 18.
16. Robert Henke, "The Italian Mountebank and the Commedia dell'Arte," *Theater Survey* 32, no. 2 (1997): 1–29.
17. Sue Reed Walsh and Richard W. Wallace, *Italian Etchers of the Renaissance and Baroque* (Boston: Museum of Fine Arts, 1989), 230.
18. Marjorie Garber, "Out of Joint," in *The Body in Parts: Fantasies of Corporeality in Early Modern Europe*, ed. David Hillman and Carla Mazzio (New York: Routledge, 1997), 39. See also George Speaight, *The History of the English Puppet Theater* (London: G. G. Harrap, 1955), 36.
19. Charles Magnin, *Histoire des marionnettes en Europe depuis l'antiquité jusqua nos jours* (Paris: Michel Lévy frères, 1862), 27–28.

20. John McCormick, Alfonso Cipolla, and Allesandro Napoli, *The Italian Puppet Theater: A History* (Jefferson, North Carolina: McFarland & Company, Inc., 2010), 10.
21. Elena Lazzarini, "Wonderful Creatures: Early Modern Perceptions of Deformed Bodies," *Oxford Art Journal* 34, no. 3 (2011), 419.
22. Benedetto Varchi, *La Prima Parte delle Lezzioni di M. Benedetto Varchi nella quale si tratta della Natura, della Generazione del corpo humano, e de' Mostri* (Florence, 1560).
23. Elena Lazzarini, op. cit., 422.
24. Benedetto Varchi, op. cit., 104–105. Google Books: accessed 10/12/19. See also Zakiya Hanafi, *The Monster in the Machine: Magic, Medicine, and the Marvelous in the Time of the Scientific Revolution* (Durham: Duke University Press, 2000), 19.
25. Lazzarini, op. cit., 428.
26. See Ted Reynolds, "The 'Accademai del Disegno' in Florence, its Formation and Early Years." Ph.D. diss (Columbia University 1974), 140.
27. Domenico Laurenza, *Art and Anatomy in Renaissance Italy*, 33.
28. William S. Heckscher, *Rembrandt's Anatomy of Dr. Nicolaas Tulp: an iconological study* (New York: New York University Press, 1958), 28.
29. Rafael Mandressi, "Of the Eye and of the Hand: Performance in Early Modern Anatomy," *The Dramatic Review* 59, no. 3 (2015): 66.
30. Cynthia Klestinec, *Theaters of Anatomy: Students, Teachers, and Traditions of Dissection in Renaissance Venice* (Baltimore: Johns Hopkins University Press, 2011), 137.
31. Thomas Rütten, "Early Modern Medicine," in *The Oxford Handbook of the History of Medicine*, ed. Mark Jackson (Oxford: Oxford University Press, 2011), 67.
32. Ibid., 119.
33. Ibid., 3.
34. See David R. Smith, ed., *Parody and Festivity in Early Modern Art: Essays on Comedy as Social Vision* (New York: Routledge, 2012), chapter 2.
35. Giancarlo Maiorino, *The Portrait of Eccentricity: Arcimboldo and the Mannerist Grotesque* (University Park, PA: The Pennsylvania State University Press, 1991), 18.
36. Quoted in Cynthia Klestinec, *Theaters of Anatomy*, 163.

37. See Domenico Bertoloni Meli, *Mechanism, Experiment, Disease: Marcello Malpighi and Seventeenth Century Anatomy* (Baltimore: Johns Hopkins University Press, 2011).
38. Ibid., 125.

CHAPTER 9

The Epistemic Category of Vegetative: An Historical Tracking Indicator for Understanding the Evolution of Western Medicine Toward Transhumanism

Sarah Carvallo

"ovum subventaneum, quatenus vivit ut planta conceptus perfectus" Harvey, *De generatione*, 1651, 982.
the constituted egg to the extent that it lives as a plant is a perfect design (My translation)

This paper aims at proposing a historical rereading of the paradigm of vegetative life or vegetative states in order to understand the scientific and historical roots of transhumanism, insofar as transhumanism denies their value to certain forms of human existence such as lives of people in a vegetative state. My philosophical point consists in showing that

S. Carvallo (✉)
Department of Philosophy, University Claude Bernard Lyon 1, Lyon, France
e-mail: sarah.carvallo@ens-lyon.fr

J. Muller (ed.), *Automata, Cyborgs, and Mutants*, Palgrave Studies in Medieval and Early Modern Medicine,
https://doi.org/10.1007/978-3-031-93201-4_9

early modern humanist medicine, then experimental medicine and finally biomedicine effectively nourish the breeding ground of transhumanism. My historical thesis is that, until Bichat, plant life embodies a perfect life, or pure life. The medical category of the vegetative state has served precisely to characterize the essence of life till the nineteenth century. Since Antiquity indeed, vegetal life was a pure life i.e., the quintessence of life, an absolute life. In this sense, we were not only just like plants, but more critically we were plants ourselves. In this perspective, vegetative life was not a pathology, but a life in the full sense of the term, which concentrated the absolute value of life. Consequently, a reflection on the nature of man implied a reflection on the nature of plants. Then, with Bichat, vegetative life becomes a diminished life, an infra-human life that has no value by itself, which can or must be corrected, even suppressed. The value and meaning of the reference to plants changed; with Bichat now, pure life is not worthy, the consciousness becomes the primary indicator of our humanity. In 1800, this modification is contemporary with a change of nomenclature in medicine vocabulary: vegetative life does no longer mean a vital function, but designates the vegetative nervous system to which are attached the functions of respiration, heartbeat, wakefulness and sleep, in distinction to the animal nervous system, on which consciousness depends. This distinction induces a difference in nature between human and vegetative life. During the twentieth century, vegetative states are reduced to a devalued infra-life, meanwhile intelligence and consciousness indicate the specificity of human life. This reversal of value is now correlated with the progress in resurrection technics after World War II, which generate new kinds of patients, whose vegetative functions are artificially kept working. To be in a vegetative state today means, for many people, no longer enjoying a truly human life, to be a vegetable (*legume*, in French), which is quite depreciative.

How shall we understand the epistemic and axiological evolution of the category of vegetative life, which crystallizes a high value from the Antiquity to the nineteenth century, and then characterizes infra-human lives? My argument is that the historical and epistemic trajectory of the category of vegetal life or vegetative state is an indicator that helps understand a main trend of western medicine, which gradually established a difference in nature between plants and men by privileging the criterion of consciousness rather than that of pure life. Actually, this historical and epistemic trajectory reveals a genealogy of transhumanism in modern Western medicine: transhumanism is not at odds with our medicine, but deeply intertwined with it. Our culture and our biomedicine constitute a favorable ground for transhumanism. It is a thesis defended by both the followers and the critics of transhumanism.[1] Among the protagonists, Nick Bostrom underlines the double continuity between humanism and transhumanism *and* between medicine and transhumanism.[2] Max More and his wife, Natasha Vita-More, consider transhumanism as the future of medicine.[3] In this perspective, our humanistic and transhumanist culture is also the fruit of a long-term history of science. And the roots of the transhumanist crisis (to parody Lynn White's article on the roots of the ecological crisis in 1967)[4] are deeper than biomedicine (term quite recent which emerged after WWII).[5] In fact, transhumanism only brings this logic to its completion: it is not an accident, but the very meaning of the evolution of medicine from the moment it concentrates all the value of human life on consciousness. Nevertheless, this genealogy does not imply any determinism: the evolution of medicine has not necessarily predetermined the constitution of transhumanist theses, but in retrospect there is a coherence between the historical evolution of scientific epistemology, which tends to eliminate the vegetative dimension to think the human, and the emergence of transhumanism.

For a long time, plants have been a blind spot of the history of science and philosophy. Today new interest in botany and in the conceptual links between human and vegetative life appears. In Antiquity, Brooke Holmes, Claudia Zatta, Robert Vinkesteijn, Alessandro Buccheri, Arnaud Macé, or Leon Wash helped to fill the gap, and in early modernity, Guido Giglioni, Fabrizio Baldassari, and Fabrizio Bigotti showed the importance of plants and vegetative life as constituent of modern medicine and recapture its epistemological and ontological transformations. There are mixed traces of a global story, which underline the continuity of the vegetative reference from Aristotle to Darwin.[6] Building on those works, this paper focuses on the meaning of medical categories directly referring to plants during the early modern period and identifies three stages (XVth, XVIIth, XVIIIth c.) and three concepts (nutrition, circulation, regulation) which help understand both the permanence of the fundamental value ascribed to plants during the early modern period in order to think about human life and the transformations of the content attributed to vegetative life as to the fundamental vital phenomenon. Those three moments correspond to three figures or conceptions of life inscribed in three medical epistemologies. First, for Fernel and Vesalius (1530–1543), vegetative life designates the nutrition, as well as for Galen, and is based on an analogy between humans and plants. Second, Harvey's circular theory of generation (1651) makes it possible to shift the concept of vegetative from nutrition to circulation; but Harvey does not consider an analogy between the movements of blood and sap. It is Malpighi and Perrault (1671–1680) who apply the circulatory scheme to sap on the same model as blood as no longer being within the paradigm of generalized analogies, but rather within the context of universal laws of nature. Third, with Stahl and Bordeu (1707–1767), the general function of vegetative life turns into a regulation closely associated with sensitivity and secretion, and it is up to the clinic to decide and understand the modalities of vegetative life.

Fernel and Vesalius (1530–1543): The Nutritive Paradigm

During the years 1530–1550, two new university teachings appeared in the field of life studies in Padua. Founded in 1222, the *Studium Patavinum* instituted in 1528 a scientific council responsible for establishing the regular teaching provided by professors already recognized and remunerated by the students. The archives record the realization of a public dissection during a surgery performed in 1546 by Vesalius, recruited on the professorial chair of surgery.[7] In 1533, the first chair of study of the simple was created; in 1545, the first botanical garden and in 1561, the first chair of practical botany. Anatomy and botany shared at that time three trends: first, they both used a method based on the direct observation of things; second, they referred to the ancient treatises, including Galen and Dioscorides,[8] in order to confront ancient texts to their own observations about the bodies or the plants: third, they needed new institutional places—the anatomical theater and the botanical garden, conceived as a theater of the earth.

This epistemic similarity between anatomy and botany is based on the idea that plants and humans share the same form of life in regard to nutrition (including generation and growth): observing human bodies or plants participate to the very same knowledge of life at its first degree. This common view is particularly noticeable in the physiology of Jean Fernel (1497–1558) which embodies a galenism reborn in Paris and was well known in Padua.[9] In book V of his treatise on physiology (1542) dedicated to the faculties of the soul, Fernel describes the vital faculty as a form of life common to men, animals, insects, plants, and roots, acting both at the global level of the individual and in each part of the body.[10]

Against Aristotle's thesis, Fernel chooses the Platonic option which distinguished the natural faculty from other vital or animal faculties. The natural or nutritive faculty finds its seat in the liver and constitutes a specific dimension of life. Common to plants, animals and humans, the nutritional function is thus distinguished from other animal or vital faculties, which correspond to movement and sensation on the one hand, and, on the other hand, to the heat that finds its seat in the heart.[11]

Form, nature, faculty, power, virtue, strength: vegetative life constitutes an innate property (*nutrition*), a hidden power (*res abdita*) because irreducible to a purely material explanation. His theory takes up the galenic description of the natural faculty,[12] since Galen established the nutrition as a general natural faculty, which is therefore characterized by its autonomy vis-à-vis the other vital and animal faculties: innate, it recognizes the qualities to assimilate or expel, which effectively applies to plants that draw from the ground the elements they need. Galen calls this nutritional faculty "vegetative soul" or nature and refers to Plato who explained this immediate sensation common to organs and plants, in *Timaeus* 77b-78d.[13]

From an anatomic point of view, the nutritional function corresponds to the venous system of the human body which is like the branching network of plants.

Vesalius was directly taught by Fernel in Paris and seems to transcribe Fernel's physiological theory in his own anatomy. In Chapter 19 of Book V of *De humani corporis fabrica* devoted to the anatomical method, Vesalius takes up the tripartition of faculties according to the four faculties studied by Galen to explain the anatomy of liver and veins,[14] while contesting some of Galen's anatomical conclusions, but also emphasizing the points of agreement.[15] With regards to the theory of the vegetative faculty, Vesalius takes up Galen's thesis while correcting its anatomical description. The venous and arterial system of man is like a plant in the human body, according to a double morphological and processual identity, because, more fundamentally, there is a unity of life between man and plant at the level of the natural faculty, a term that comes from Galen. This functioning and morphology are not based on a simple metaphor, but on a single natural faculty, which operates in both cases in the same way. The analogy with the plant works as a heuristic key, not only from a functional point of view but also from an anatomical point of view. Anatomical shapes make it possible to understand natural faculties that remain invisible. When describing the venous system and the liver, Vesalius presents indeed a theory of nutrition as a plant life form.[16]

In Book III devoted to veins, Vesalius takes up the theory of nutrition according to the four faculties studied by Galen, who elaborated a theory of the powers (*dunameis*[17]) of the soul. Galen calls "vegetative soul" or nature this nutritional faculty that he conceives as a power (*dunamis*). The galenic theory of the natural faculty[18] assumes that each part of the human body attracts the food it needs, retains it, assimilates it by altering it and rejects the surplus. In his treatise *On natural faculties*, Galen explicitly interprets the *Timaeus*, where Plato develops his thesis of the celestial plant man, which leads him in particular to defend the idea of an unconscious sensation that does not pass through the nerves and is therefore independent of the rational soul, which controls voluntary movement and nervous sensation. From an anatomical point of view, "the veins of the whole animal (are) like branches of a trunk. To the arteries which, similar to the hair of plants, leave the heart to branch out into the lung, correspond the veins of the stomach, spleen and mesentery…"[19] From a physiological point of view, their diversity and multiplicity make it possible to bring the specific food to each part. As R. Vinkesteijn and F. Baldassari have pointed out, Galen pushes the analogy between man and plant to its limit in the form of a quasi-identity: at the embryonic stage, or through certain functions such as nutrition, the human being is a plant.[20] Like Galen, Vesalius takes the analogy to the extreme to expose an identity of the nutritional faculty, a form of life shared between man, animal, and plant. At this semantic level, however, he adds a visual representation absent from the treatises of Galen and Fernel: not only do Vesalius' plates show botanical elements recurrently around the dissected body, but especially the plates devoted to the venous system present a morphological identity with branches, roots and vines.[21] This identity does not suppose a force in the Galilean sense of the term, but a faculty in the galenic sense: the nutritional faculty expresses the identity of a common regime of life shared between man and plant, a particular form of life distinct from rational or motor and sensory faculties.[22]

The first plate represents the ramifications of the portal vein which present the hair of the plants described by Galen [23] (Figs. 9.1, 9.2, and 9.3).

The second plate represents the blood vessels as a whole: it shows the articulation of arterial and venous trunks (large veins from the liver or large arteries such as carotids), twigs and surgeons.[24]

262 ANDREAE VESALII BRVXELLENSIS
VENAE PORTAE ORTVS, IPSIVSQVE
propaginum ſeries. Caput V.

VENAE PORTAE AB VNIVERSIS QVI-
BVS COMMITTITVR PARTIBVS LIBERAE, INTEGRA
delineatio, in ea proportione expreſſa, ad quam ſecundùm præſentem fi-
guram aliquis iecur, bilis ueſiculam, uentriculum, lienem, omen
tum, meſenterium & inteſtina ex illorum magnitudi
ne, ac inſuper in ſuo ſitu depingeret.

NVDAE VENAE PORTAE DELINEATIONIS
characterum Index.

A, A etc. QVINQVE his characteribus portæ uenæ propagines indicantur, per iecoris corpus diffuſæ, & hic ueluti iecoris formam ſecundùm cauam ipſius ſedem exprimentes.
1, 2, 3, 4, 5 Numerus iſte quinq; portæ uenæ ramos (niſi ſubinde etiā pauciores ſint) notat, ex quibus ipſius caudex quodāmodo proximè cōſtituitur, aut in quos is primùm in iecoris ſubſtantiam digeritur.
Venæ

Fig. 9.1 Vesalius, The ramifications of the portal vein

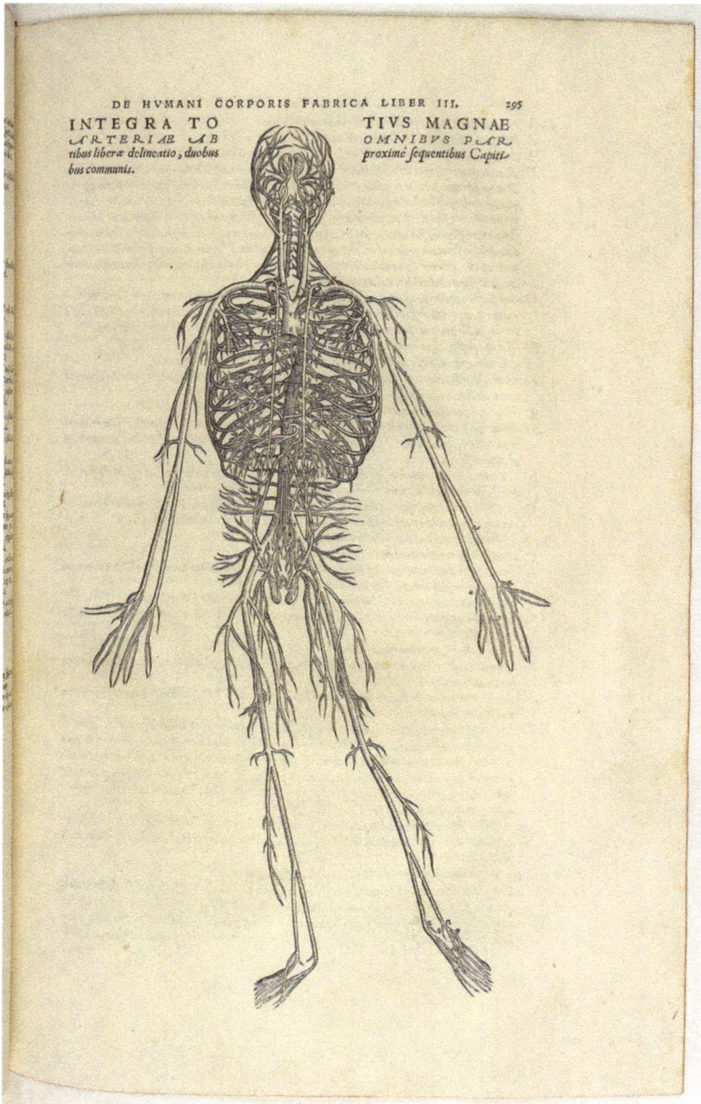

Fig. 9.2 Vesalius, The blood system

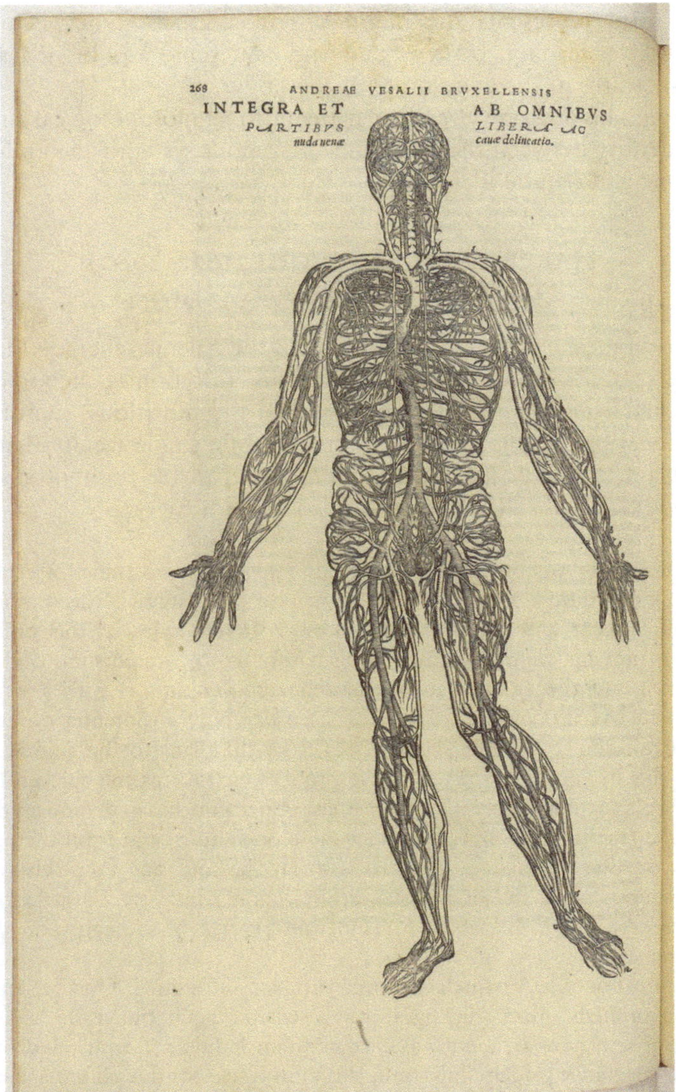

Fig. 9.3 Vesalius, The blood System

The third represents the blood system alternating between trunks (large arteries leading to the organs and cava vena) and branches (small arteries leading to the extremities of the body).[25]

Through his anatomy, Vesalius' plates affirm a morphological and functional identity between the plant and the venous system which refers to a functional and natural identity.

Harvey and Malpighi (1628–1680): The Circulatory Paradigm

Harvey is a moment of transition between the galenic scheme of faculties and the regime of the laws of nature. This transition is anchored in an Aristotelian context[26]: Aristotle considered the nutritious, sensorimotor and rational operations as the development of a single faculty that stops, in the plant, at the vegetative stage, unfolds at the sensorimotor stage in animals and reaches the rational stage in humans.[27]

> This point (as Aristotle calls it) is already moving like an animal. Certainly, the animal differs from the non-animal (a non animali) according to the movement and the sensation (sensu). That is why, as this point is already moving, we rightly say that his nature has recovered souls (induisse animas); and the egg previously impregnated (imbutum) with a vegetative soul has also received motor and sensory power (potentia motiva et sensitiva); and it has passed (transisse) from the plant to the animal; and from this moment it begins a chicken soul, which forms the chicken from the egg and then informs it. The philosopher concludes demonstratively that the faculties are implanted from these operations: and from them, the cause and the principle of life, that is to say the soul, and then this passes into action (since the operations are in action.) Harvey, *De generatione animalium*, 1651, Exercise 16, Third observation of the egg. 1651, 52. My translation.
>
> Punctum hoc (ait Aristoteles) movet jam sese, ut animal. Quippe, animal a non animali, motu distinguitur, et sensu. itaque punctum hoc jam primum sese moveat, merito animas naturam induisse dicimus; and ovum anima vegetative pridem imbutum, jam motiva, et sensitiva potentia insuper donari; and a planta in animal transisse; eodemque tempore animam pulli ingredi, quae ex ovo pullum format, eumque postea informat. Quippe ex operationibus, inesse facultates; et ex his, vitae caussam et principium (animam scilicet) idque actu (cum operationbus actu sint) Philosophus

demonstrative concludit. Harvey, *De generatione animalium*, Ex. 16. Tertia ovi inspectio, 52.

For Harvey, the vegetative soul has no part. Without it, the sensitive soul cannot develop. And it cannot exist in action without associated organs, since it constitutes the act of the living body. Therefore, the sensory and motor organs result from the operations of the vegetative soul. Because it appears first chronologically, it also persists in the sequence that derives from the order of the generation. The temporal order of appearance of the organs manifests the passage from the vegetative stage to the animal and then human stages.

Thus we check what appears obviously in the egg and which Aristotle stated about the generation of perfect animals: that is, that all the parts do not appear at the same time, but according to an order, one after the other; and the first that exists is the genital particle from which all the other parts then come according to its own vigor (virtute) – as if from a principle. It is the same in the seeds of plants (for example, beans or acorns), we first observe the bud (gemma), a protruding tip, which is the principle of the whole future tree. Harvey, *De generatione animalium*, 1651, Ex. 50, On the order followed by generation, 149.

In other words, the regular temporal order reveals an ontological causality and an epistemic rule (*regula*).

We will later say how anatomical observations manifest this universal rule of Aristotle concerning the priority of the parts. But already, we will see how he rightly deduces, according to this rule, the causes of priority (to put it that way) in the generational process.

Quomodo autem, ex observationibus anatomicis, universalis haec Aristotelis regula, de partium prioritate, vera appareat; posterius dicemus. Interea temporis, quo pacto, secundum hanc regulam, ipsemet caussas prioritatis (ut ita loquar) in generatione, sufficienter deducat, jam videbimus. Harvey, *De generatione animalium*, 1651, 177.

According to my lexical research in the treatise on generation, Harvey uses the terms norm or rule (*norma* or *regula*), never law (*lex*). The rule expresses the development of the force that generates action: natural philosophy seeks the final, material, efficient, formal causes that manifest themselves through the living. In the preface, Harvey explains the method to follow:

> It is necessary to investigate the phenomena from the causes, in this case, material and efficient (...) by observing the perfect animals and going up from the point of arrival to the starting point [*a meta ad carceres regredi*] (through the different stages through which they became adults), it seemed to me to gather the traces (of the process); as one cannot go further up the chain, we must admit that we have finally arrived at the principles themselves; and we understand that the plastic force appears from the raw material, and that it proceeds from the main efficient cause and according to its modalities from the principles. Harvey, *De generatione animalium,* 1651, Forword. B2. Not paginated.
> visum est mihi, ad animalia perfecta retrospiciendo (quibus nempe gradibus orta et adulta fuerint) tanquam a meta ad carceres, vestigia recolligere; ut tandem, ulterius non detur regressus, nos ad ipsa principia pervenisse constet; simul appareat, ex qua materia prima, a quo efficiente principali et quomodo ex iisdem procedat vis plastica: tum etiam quos processus in hoc opere natura faciat. Harvey, *De generatione animalium,* 1651, B2 n.p.

Generational circulation expresses the form of the rule which keeps the species identical to itself through the birth and death of the individuals; the *vis plastica* embodies the efficient cause that tends toward the deployment of the forms or souls inscribed in it as potentialities; the material cause consists in the food of the egg and the final cause is the soul which operates through the body. In retrospect, the observer understands that this plastic force was already present at work in the plant or egg (*primordium vegetale).* This retrospective method is based on analogies that connect all orders of living beings together, which are explained in turn by the actualization of material and efficient causes.[28] Starting from the development of the most perfect animals, such as chickens, makes it possible to better see the invisible processes occurring in the lower living beings such as plants.

The ontology of the cause conceived as a plastic force goes hand in hand with the epistemology of a natural philosophy based on the search for causes, and not the description of laws. But through plants, animals, and humans, the generation manifests the same circular structure, in the fact that circulation expresses the fundamental phenomenon of life. On the scale of vegetative life, circulation characterizes the phenomenon of generation through the *vegetal primordium* that circulates through individuals to preserve the species identical to itself: individuals are born and die, generations follow one another, the species remains the same. Animal life appears with the *saliens punctum*, the first manifestation of the blood pulsation that gradually constitutes the heart: this point of origin appears on the fourth day in the development of the chicken's egg. It should be emphasized in this regard that there is no circulation of sap for Harvey because there is no *saliens punctum* in plants: in plants, circulation concerns only the generation, since what circulates is the *vegetal primordium*. It is up to Malpighi to uncover the circulation of sap, but also to renounce the conception of circulation according to the theory of the four causes, henceforth, describing it according to physico-mathematical laws governed by elastic phenomena alone. Nevertheless, for Harvey, vegetative or animal circulation, generational or blood, circulation have a single source in the principle of life, knowing that this principle of life is transformed under an animal figure with the *saliens punctum*. The *saliens punctum* is therefore a new form or figure of the *vegetal primordium*, when it reaches the animal stage. Now at the other end of the chain, generational circulation surpasses the only field of organisms and is part of a meteorological phenomenon: in Book IV of *Meteorology*,[29] Aristotle describes the cycles of transformation of the generative element in all cosmic cyclic movements. As water circulates and transforms its appearance, the primordium of the generation passes from the state of seed to that of individual while being formally preserved. This cosmic anchoring justifies a series of analogies between generational, blood, and meteorological circulations[30]:

> From this circuit, it is clear that this principle of revolution comes from the hen to go in the bud, and, again, from the egg to the hen: it is communicated in an eternal way. That is why, according to Aristotle, it is analogous to the stellar element. It makes possible the fact that the parents conceive and the fact that the seeds or eggs are fertile. Like Proteus, it continues to persist both in parents and in eggs under different figures. Similar to the intelligence (*mens*) or the spirit (*spiritus*) which constantly agitates in this mass the sun that circulates from east to west through the different terrestrial regions according to a perpetual motion, so in the gallinaceae, this little god or divine principle acts either as a plastic force, or as a nutrient force, or as a growth force. It performs a conservative as well as vegetative function, and it remains the same through the generations.[31]

Consequently, the observation of these analogies makes it possible to identify the vegetal primordium with a stellar element, which makes it possible to constitute the three meteorological, generational and blood circulations according to a single causal regime in three forms, which allows a new explanation of the innate heat:

> And I say the same about innate heat and blood: they are not of fire, nor do they derive their origin from fire; but a body, which participates in the divine, and for this reason they do not act according to an elementary faculty, but there is something in the seed which makes it fruitful in itself, and which excels at making the animal by the forces of the elements (excellere in aliqua re); it is a spirit, and nature in this spirit corresponds to the stellar element. Thus, in the same way that in blood there is a spirit or a force that acts above the elementary forces (it is particularly remarkable (conspicua) in the phenomena of nutrition and conservation singular parts of animals), and its nature, i.e. the soul that is in this spirit and in the blood of animals corresponds to the stellar element. Then, it is manifest that the heat that is in the blood of animals during their life is not a fire and does not originate from fire; and our observations show it to satiety. Harvey, *De generatione animalium*, 1651, 247.

This stellar element is characterized as a force in the Aristotelian sense of a latent power that is actualized.[32] The egg vegetates (*vegetare*) and constitutes the heart of the process: its production from the vegetative seeds of the parents, its vegetative life (common to all forms of life, for example until the fourth day for the chicken), then its development in different forms (plants, animals, men): for animals, it is more particularly embodied in the blood circulation. All living things begin as plants. The natural force generates the egg according to the same process as plants[33] and then develops according to its own intrinsic nature in order to realize its specific type. Examining the chicken egg thus makes it possible to know the origin of life regardless of the future developed form (plant, animal or human). Indeed, all living organisms are analogous to eggs at first, whether it is the plant seed, or the embryo of viviparous.[34] Moreover, the development of the egg does not depend directly on the food that surrounds it—the bowels of the mother or the earth—but on an internal principle, the vegetative soul: for the embryo, the ramifications of the umbilical veins function like the roots of plants in the earth. This internal principle is an internal heat, a manufacturing nature (*innatus calor, aut natura opifex*), a faculty of nutrition and growth. In other words, gestation, nutrition, and growth are identified very precisely with vegetation, which expresses the activity of the plastic force (*vis plastica*). This becomes experimentally visible only at the appearance of the *saliens punctum when* the animal stands out from the plant. To the extent that it characterizes the life of the seed, egg, or fetus, it is able to give its own form to each type of living organism and manifests itself in a particularly visible way during the process of differentiation that distinguishes the animal from the plant with the appearance of the *saliens punctum.* The egg resembles fungi, flies, and other parasites that feed on the moods of the tree without forming part of it. This autonomy characterizes the perfection of vegetative life which has, by itself, its own principle of life. To use Aristotle's aphorism, "the egg as it is a vegetative production is perfect by itself, but as it is an animal production still in power, it is imperfect."[35] More generally, the egg process manifests the very same activity of each organ that develops intrinsically like the egg developing because of its own innate faculty.

Now the understanding of the principle of life as a plastic force makes it possible to reject the galenic theory of the natural faculty and to confer onto the egg an exemplary value that is based not on a law of nature but on a norm (*norma*) which expresses the actualization of the plastic force still latent in the egg:

> ...that is why, as in our survey of eggs, I will propose to look for a first genus, which is worth as an example for all the others; thus, as we will know perfectly, it will enlighten all the others by analogy... Harvey, *De generatione animalium*1651, 214.
> ...quare ut in ovis fecimus) unum aliquod genus primo, tanquam caetorum omnium exemplar proponam; quod, utpote notissimum, reliquis lucem afferat, et norma sit, ad quam alia omnia per analogiam quandam referantur. Harvey, *De generatione animalium* 1651, 214

Marcello Malpighi (1628–1694) retains the pattern of circulation, but makes a transition using the microscope and reconfiguring the reflection on nature in a model that is no longer based on analogy, but on the laws of nature. In the treatise *Anatome Plantarum* (1675), Malpighi reduces Harvey's circulatory conception to a mechanical model that he applies to the circulation of sap identical to that of blood in animals. According to the Italian anatomist, circulation must make it possible to bring plants and the whole living process into the regime of the laws of nature, which structures all the science modeled on mechanics. The introduction of the concept of force (theorized by Malpighi and Leibniz in terms of elasticity) into the field of physiology to bring to light the laws of life does not abolish the metaphysical anchoring of medicine, when it thinks of life. By his theoretical gesture, Malpighi does not bring the living into mechanics by reducing life to matter. No more than his master Giovanni Alfonso Borelli (1608–1679) who converted him to the mechanism,[36] does Malpighi claim to evacuate the soul of the living: it remains the principle of life and movement, but anatomy and physiology no longer deal with it directly. From now on, geometry and mechanical reasons are necessary to think about the movement of animals; medicine must perform the same gesture that Galileo performed in the order of the movement of inert bodies. As a result, he leaves the field of Harvey's characteristic ontological and heuristic analogy to identify the operations of universal laws that make it possible to reduce the different phenomena to an identical fundamental process. Thus, the understanding by Malpighi of the category of the vegetative in terms of circulation expresses a new scheme of thought, that he discovers in physics through the idea of the laws of nature and that he applies to medicine. Malpighi endorses circulation as a fundamental phenomenon of life, the signature of life in matter. As in perfect animals, new foods enter the veins under the species of the chyle thanks to the perennial movement of the blood that turns (*commixtus perenni circulationies motu rotatus*) and replaces the old blood, likewise (*sicut in animalibus perfectis... ita in plantis*) in the plants a raw juice flows and takes the place of the older juice accumulated in the utricles. From the functional point of view, it is therefore the same circulation operation that, in animals or plants, performs the same function to evacuate thickened humors as by fermenting them to bring new foods and digest them.[37] The circular movement of blood or of sap also explains the morphology of vessels in animal or of the regular straight lined wood grain of the tree trunk.[38] The circular rings in the tree are also similar to the rings or spirals of the trachea. They induce movements of constriction and expansion to circulate food.[39] The circular veins of tree trunks can also be brought closer to the structure of insect lungs.[40] They all correspond to a peristaltic movement in animals and herbs, which perhaps makes it possible to understand is the *mimosa pudica* as an intermediate species between plants and animals.[41] Below the obvious diversity of the living, there is the same law:

> ... nature does not always retain the same rule (*norma*) for propagating plants, but they are all suitable nevertheless in that they seem to retain at least an analogism. That is why in the fruits there are several trunks; in grasses there are multiple stems, but one can consider only one, or at least take a replacement for them all.[42]

Actually Malpighi accomplished Harvey's discovery of circulation when, in 1661, he observed under a microscope the anastomoses of blood capillaries that Harvey had not seen and which made it possible to establish the material communication between the venous and arterial systems. However, anastomosis is a common and clearly visible phenomenon in plants: wood fibers bind together and form cylinders and then expand, and from there divide into fibers, arteries, or branches "by a network of reciprocal anastomoses, as in blood vessels." By accomplishing the Harveyan discovery, Malpighi also irreversibly transforms it.[43] Now circulation conceptualizes the uniqueness and unity of the phenomenon of life between plants and men, and no longer nutrition; the plant makes it possible to understand "the operations of nature, when they are enveloped in the latent state."[44] In his *Dedication to the Royal Society* in November 1671, Malpighi underlines the identity of life processes which are all submitted to circulation, following two arguments. First, on a physiological point of view:

> As in perfect animals, new foods enter the veins under the species of the chyle, thanks to the perennial movement of the blood that turns (commixtus perenni circulationies motu rotatus), and replace the old blood, likewise (sicut in animalibus perfectis... ita in plantis) in the plants a raw juice flows and takes the place of the older juice accumulated in the utricles. From the functional point of view, **it is therefore the same circulation operation that, in animals or plants, performs the same function to evacuate thickened humors as by fermenting them to bring new foods and digest them**. Malpighi, *Anatome plantarum*, 13. I emphasize.

And second, on an anatomical point of view:

> Thus, from a morphological point of view this time, **we can compare the circular movement of the blood to that of the sap**. Thus, when we observe tree trunks, we see how **we can compare the growth of wood with that of the body by the concretion of a sequence of circular movements**. Malpighi, *Anatome plantarum,* 20. I emphasize.

The laws of constriction and dilation make it possible to explain the production of eggs in plants or chickens, or the circulation of sap and blood, or the dissemination of food. In other words, circulation results from a composition between expansion and constriction that characterize certain properties of elastic movement.[45] From cosmic phenomena to man to plants, forms and processes turn out to be identical. From now on, the concept of circulation makes it possible to unify a set of vital phenomena in a single category that would be the signature of the elastic force in matter. The scheme of circulation works at the four levels of the cosmos, plants, animals, and men, because it obeys to the very same laws.[46] If the search for analogies remains a structuring theme of medicine for Malpighi, its heuristic meaning changes for two reasons: on the one hand, analogy becomes the expression of the identity of the invariant laws of nature through the variety of phenomena; on the other hand, it makes it possible to thematize life as a force according to a notion imported from the physical and mathematical model, distinct from galenic faculty, or from Harvey's divine principle.[47]

This new conceptualization of the unity of nature through the living and inert kingdoms justifies a new experimental practice[48];anatomy and botany follow the same logic; the same experiences manifest the same operations. Now the key lies in the idea of the laws of nature,[49] which makes it possible to design and carry out experiments. Even more radically, experiments in physics, anatomy and botany are all based on the same causalities. Because the processes are identical, the shapes are similar: so when he discovers under a microscope the glomeruli of the kidneys by injecting ink mixed with wine spirit into the emulgent arteries, Malpighi compares them to apples hanging from a tree. Making structures visible presupposes having thought of the great operating rule of nature ("*hanc naturae operandi regulam*"), which ultimately consists in the conjunction of the laws of movement and pressure to move, assemble, separate. The experimental method only follows the path of nature (via) assuming that.

> ... the rule of nature is simple, one and can be easily observed in living bodies forming mixtures (...) nature follows the same path (*eandem viam*) to arouse heat in plants and animals, when it warms the organs through nutrition by arousing a movement through the fistulas that occurs by the pressure in the viscera and muscles in animals or the vibration of the rays of the Vessels...[50]

The concept of law also makes it possible to establish an identity of scale between the large and the small: by developing the microscopic anatomy, Malpighi guides his instrument in the search for functioning analogous to the visible toward the invisible to find the usual method of nature (*consuetam naturae methodum*) in all parts of the body both large and small (*vel maxima, vel minima*).[51] In the order of the very small, living bodies are still arrangements of small machines (*machinulae*) that can be thought of by analogy or seen under a microscope.[52] Finally, it makes it possible to establish the physical causes of physiological phenomena.

Jane Ruby showed how rules and laws fall within the same conceptual field[53]; I showed how, before the theme of the laws of nature, norms refer to the rules of the *fabrica*, that is, the functioning of what was created by God.[54] Malpighi uses the terms *lex, regula or norma*. The work of terminological and conceptual analysis of the laws of nature proves a transformation of the scientific project and goes further than Harvey did. By accomplishing the Harveyan discovery of blood circulation through the observation of capillaries, Malpighi also transforms it irreversibly.[55] In 1727,[56] Stephen Hales extended these experiments by establishing a plant static that sought to calculate the pressure of sap and blood. He highlights in plants the phenomena of absorption and transpiration, which Santorio had studied in his human statics (*De statica medicina*, 1614). Perhaps taking up Boyle's experiments with pressure, Hales established numerical correspondence tables to compare the different pressures in horses, dogs, plants, hydraulic devices, and fountains. By injecting mercury, he observed the pressure variations of the sap along the tree trunk.[57] As Sophie Roux has shown in physics, a strong equivalence structures the relationships between nature, force and law.[58] The laws regulate the relations between phenomena and are expressed in the form of intrinsic force "without which they could not conform to the general laws of nature or movement."[59] Thus, Malpighi explicitly associates force and law: it is the same force of nature (*foecunda naturae vis in gubernandis viventium ordinibus*) that operates "according to an almost identical method, or at least analogous" in all living things (*eadem fere, vel saltem analoga methodo*).[60] In perfect animals, insects or plants, we first see a small egg, such as a bud or an eye (*ovum, gemma, oculus*), then branches or vessels (*rummy*) that develop in a circular way until they appear visibly.

Malpighi thus instituted a dynamic of life, at the very moment when Leibniz conceived a dynamic of bodies, which gave an internal principle to mechanics by adding force to Cartesian movement. The concept of force or life changes and finds a mechanical relevance that results in movements that are first invisible, then visible, but they are always the same movements governed by the same laws. As François Duchesneau points out,[61] Leibniz and Malpighi each develop their own theory in parallel.[62] When he tries to generalize his dynamics to living beings, Leibniz takes up Malpighi's observations and gives them a philosophical meaning. Force refers to an active principle that explains the organization of life, which Leibniz also calls plastic nature and relates to the machines of nature: vegetation becomes the vital force which operates following laws of nature. It was not until the 1690 s that Leibniz synthesized his ideas in physics with the analyses of doctors such as Malpighi, Leeuwenhoek or Swammerdam, in the *New System of Nature and Communication of Substances*, in 1695 and the *De ipsa natura* in 1698.[63] By now clearly distinguishing between the living and simple aggregates, Leibniz seeks to explain the organization of the living by laws and internal forces. As early as 1687, he found in Malpighi a principle of unity of the living that makes it possible to think about the common forces acting in plants and man:

> I wish I could explain the differences or degrees of other immaterial expressions that are thoughtless, in order to distinguish simple bodily substances, the living, and animals as much as one can distinguish them, but I have not meditated enough on this, nor examined nature enough to be able to judge the forms, by comparing their organs and operations. Mr. Malpighi, based on very considerable analogies, has a strong tendency to believe that plants can be understood under the same genus with animals, and are imperfect animals.[64]

So Leibniz went beyond the ether hypothesis of elasticity in the 1670 s—even if he still kept it in the *Tentamen anagogicum* in 1688—and introduced the notion of a dynamic force as early as 1689–1690, in the treatise *Dynamica de potential*[65]; he participated in the debate on the living forces in the years 1685 and then elaborated his dynamic theory as a counterpoint to the publication of the *Principia Mathematica* of Newton (1687), which he never ceased to contest.[66] He then proposes to think of the vital force as an immanent action *actio in se ipsum*: "any power of immanent and therefore vital action."[67] Every action presupposes an individual subject. In physiology, the subject of blood circulation is the heart, an organ capable of putting itself into action by itself, an action that diffuses in the form of circulation (*Replica XXI to Stahl, Point 17*).[68] Within the framework of dynamics, Leibniz will deepen implicit metaphysics of the medicine of his time, as if he wanted to naturalize metaphysics according to the expression of Michel Fichant.

Bordeu: The Regulatory Paradigm

Fifty years later, Montpellier vitalist medicine doctor enacts a new rise in power of the paradigm of vegetative life through the medical notion of regulation, which encompasses now circulation, secretion, excretion, breathing, nutrition, growth, and generally speaking all forms of vital intelligence. Regulation becomes the most fundamental vital activity, which subsumes all the others as particular cases. The concept of regulation refers to Stahl and its theory of tonic movement, which includes three functions (circulation, secretion, excretion) under a single general function.[69] Regulation expresses an intrinsic sensitivity of the living body. Correlated to regulation through the observation of the glandular activity, sensitivity cannot be reduced to a physical notion such as irritability. It becomes the specific marker of life, and it operates through regulation. It fluctuates following degrees and it operates at two levels: First, as an internal force in life located in the glands; second, as a global phenomenon, which traduces the vital organization of the organism, and reveals a continuity between sensitivity and intelligence via the concept of organization and integration.

Théophile de Bordeu (1722–1776) develops the reflection on the identity of the regulatory phenomenon in plants and animals through his attention to glandular secretion, which constitutes a form of sensation, is itself in continuity with intelligence through the ability to recognize specific stimuli and provide an appropriate response. The theory of secretion makes it possible to understand the sensitivity to the local (glands) and global (the body) scales from the tonic action.[70] Therefore, regulation becomes the mark of the difference between the automaton unable to regulate itself because unable to adapt to external stimuli, and the living body which is characterized by sensitivity. Regulation operates through three vital actions. First, it induces the form of the equilibrium that finds its local scale in the mucous tissue and glands. This equilibrium depends no longer on the only blood circulation, but on the regulation between the local scale of tissues (glands or mucous membranes) and the global scale of the organism

> The result of these efforts is a kind of reciprocal balance between action and reaction, which depends both on the cohesion of the parts that make up the balloons, and on those of the productions that bind them to each other. This equilibrium makes a kind of tonic movement that can be considered simply passive: it [the mucous tissue] contains the parts in the reciprocal relations that they must have; and it acquires more or less energy by the action of nerves and vessels, which makes it cause of many phenomena. Théophile de Bordeu, *Recherches sur le tissu muqueux* (1767) in *Œuvres complètes*, Paris, 1818, 751.

Second, regulation explains how the organs are planted in glands and mucous membranes—as trees are planted in the ground—in which they vegetate, i.e., unfold their vegetative life:

> It is in this spongy organ, thus conformed, that the different parts, the viscera, the muscles and the glands are placed: they are, so to speak, planted in this parenchymal substance, in which they vegetate, covering themselves with several layers; they extend and arrange themselves by the force of their germs, or the ends of the nerves that are specific to them: all these parts are, in young subjects, only species of buds that come to vegetate in the cell tissue, as the branches, fruits and leaves of trees extend into the air, or rather as the roots vegetate and bypass in the earth. Théophile de Bordeu, *Recherches sur le tissu muqueux*, in *Œuvres complètes*, Paris, 1818, 752.

Third, regulation operates the transmission of signal between the local and the global, like plants with their environment: there is a general communication network between glands, organs, organisms, and between organisms or with environment, to the point that generation appears now only as a special case of this general communication. In this perspective, generation becomes also a regulatory phenomenon.

> Let us therefore conclude that each organ of the body has, by its emanations resulting from its vital activity, some relationship with the flowers that spread in the air a seminal and living emanation that gives an idea of the seed of animals, and all other exhalations to which their parts are subject. Bordeu, *Analyse médicinale du sang*, 1775 *Œuvres complètes*, Paris, Caille et Ravier, 1818, t.2, 945.

So, medicine assumes not only a role of natural history, but also of natural philosophy: it differs from anatomy in that it seeks pure life, the origin of life, and its primitive functioning. Clinic doesn't calculate laws but understand the vital principle operating through holistic processes. It embodies the true knowledge of life, insofar as, through the physician, it is life itself which is knowing itself.

> ...what anatomists usually demonstrate is, so to speak, only the framework, the carcass or the skeleton. Doctors go further: they follow life to its last reduced, to the monad or the living atom, united to the spiritual soul, nestled in the spongy body, drooling or liquid, and sitting especially or eminently on the nervous gender, which is itself only a kind of singularly spun and organized glue. Bordeu, *Analyse médicinale du sang*, 1775, *Œuvres complètes*, Paris, Caille et Ravier, 1818, t.2, 937.

Bordeu makes compatible the ancient scheme of pure life with the scientific and clinical method of regulation. For Bordeu, to know life is to recapture the intelligence of life in the sense of a subjective genitive, a life that knows and develops even in the absence of any consciousness of the living. In turn, the vitalist doctor aims to know life, to bring this self-knowledge that the living has theoretical and therapeutic knowledge.

Conclusion

By transferring the paradigm of vegetative life from nutrition to circulation, then to regulation, medicine brings vegetative life into the realm of the laws of nature. Even if the content and the epistemic context evolve, the category of the vegetative draws a fundamental continuity in the need to mobilize plants to know and understand human life. There is a pure life common to plants and men that constitutes the essence of life. Actually, the scientific category of plants also implies an axiology that assigns an absolute value to life from its vegetative state.

Conversely, from Bichat onward, there is a reversal of the values according to which a vegetative life is no more a human life, has no intrinsic value and can be manipulated according to the interests of the others. Some few remarks about Bichat and Nietzsche will help understand the epistemological and axiological turn which occurred since the beginning of the nineteenth century Marie-Francois-Xavier Bichat (1771–1802) distinguishes the organic life embodied by plants—or animals or men at the very beginning or the very end of their existence—and the animal life which implies desire, mobility, and even consciousness for human beings. Moreover, he operates a dissociation between life and intelligence: intelligence no longer designates the property of life itself following a subjective genitive lecture (the intelligence of life means that life by itself is intelligent), but the consciousness of an individual. Consequently, Bichat, like Johann Christian Reil (1759–1813) in Germany, considers that a man who sleeps and digests leads a plant existence that has no dignity. Consequently "the crime of destroying one's fellow man is more relative to animal life than to organic life." (Marie-Francois-Xavier Bichat, *Recherches physiologiques sur la vie,* Paris, Flammarion, GF, 1994, 88.) Because the death of plants does not move us "the state of the animal that natural death will annihilate is close to that where it was in the womb of its mother, and even that of the plant, which lives only within, and for which all nature is in silence." (Bichat, *Recherches physiologiques sur la vie,* 1994, 111).

At the end of the nineteenth century, Nietzsche will only radicalize the social conclusions of this devaluation of the vegetative life of some patients. The German language names *fortvegetieren* the kind of life that patients lead when they have lost consciousness and become totally dependent on doctors.

"Der Kranke ist ein Parasit der Gesellschaft. In einem gewissen Zustand ist es unanständig, noch länger zu leben. Das Fortvegetiren in feiger Abhängigkeit von Ärzten und Praktiken, nachdem der Sinn von Leben, das *Recht* zum Leben verloren gegangen ist, sollte bei der Gesellschaft eine tiefe Verachtung nach sich ziehen. Die Ärzte wiederum hätten die Vermittler dieser Verachtung zu sein, – nicht Rezepte, sondern jeden Tag eine neue Dosis *Ekel* vor ihrem Patienten [...]." (Nietzsche, *Götzen-Dämmerung*, Leipzig, 1889, Abs. 36)
The sick person is a parasite of society. In a certain state, he is unable to live even longer. The survival in cowardly dependence (Forvegetiren) on doctors and practices induces that the meaning of life, the right to life has been lost, and should entail a deep contempt in society. The doctors would have to be the mediators of this contempt – not prescriptions, but every day a new dose of disgust in front of their patient [...]. My translation.
"Das feige Fortvegetiren in einer absurden Abhängigkeit von Ärzten und Praktiken sollte in der öffentlichen Meinung Verachtung mit sich bringen. Die Ärzte sollten den Muth haben, ihren Kranken diese Verachtung täglich zu kosten zu geben." Nietzsches Folioheft aus dem Jahre 1888. (*Sämtliche Werke: kritische Studienausgabe,* ed. G. Colli, M. Montinari, Bd. 14, München, de Gruyter, 1988, 430)

> The cowardly fortification (Fortvegetieren) in an absurd dependence on doctors and practices should bring contempt in public opinion. Doctors should have the courage to give their sick people this contempt on a daily basis. My translation.

From then on, life in the vegetative state is no longer a pure life, a perfect life, but becomes an inhuman life that must be suppressed. Because he knows the functional laws of life, the doctor becomes the one who can judge the value of life according to the criterion of the individual's consciousness. Transhumanism is rooted in this change of the doctor's attitude, which is also a conceptual change in the notion of life or vegetative state, which becomes an amputated, diminished life.

Notes

1. John Harris recalls the effectiveness of all the meliorative techniques now commonly practiced by medicine. He argues that one cannot indicate a difference in nature between care and improvement. John Harris, "Enhancement et éthique," *Journal international de bioéthique*, 22/3–4 (2011): 136–151, 144. Carl Eliott, *Better than well: American medicine meets the American dream* (New York: W.W. Norton and Company, 2004).

   "Transhumanism is in fact the uninhibited version of biomedicine's aspirations to *do better than good,* to use the famous slogan of the pharmaceutical company *Pfizer.*" Nicolas Le Dévédec and Johanne Collin, "Le médicament augmenté: l'usage du médicament dans les discours transhumanistes et ses significations sociales," *Journal international de bioéthique et d'éthique des sciences*, Vol. 29/3 (2018): 93–108.
2. Nick Bostrom, "A History of Transhumanist Thought," *Journal of Evolution and Technology*, 14 (1) (2005), http://jetpress.org/volume14/bostrom.html; et *id.*, "A Short History of Transhumanist Thought," *Analysis and Metaphysics*, 5 (1–2) (2006): 63–95. Emmanuel Taïeb, "Transhumanism and perfect health," *Notebooks* [Online], 99–100 (2019–2020) posted January 01, 2020, accessed January 30, 2020: 126, http://journals.openedition.org/quaderni/1571; https://doi.org/10.4000/quaderni.1571.
3. Max More, "Letter to Mother Nature," Robert A. Freitas Jr., "Welcome to the Future of Medicine," in *The Transhumanist*

Reader, ed. Max More and Natasha Vita-More (Chichester: Wiley-Blackwell, 2013).

4. Lynn White, "The historical roots of our ecological crisis," *Science* 155 (1967):1203–1207.
5. Alberto Cambrosio and Peter Keating, "Qu'est-ce que la biomédecine ? Repères socio-historiques," *Med Sci* 19/12 (2003):1280–1287, https://doi.org/10.1051/medsci/200319121280.
6. Mix traces a global story: continuity until today of the vegetative reference (*Life-Concepts from Aristotle to Darwin: On Vegetable Souls*, Palgrave-Macmillan, 2018). C. Zatta (*Interconnectedness: The Living World of the Early Greek Philosophers*, Academia Verlag, 2017) analyzed it for presocratics; B. Holmes (Brooke Holmes, "Pure Life. The Limits of the Vegetal Analogy in the Hippocratics and Galen" in *The Comparable Body-Analogy and Metaphor in Ancient Mesopotamian, Egyptian, and Greco-Roman Medicine*, ed. John Z. Wee (Boston-Leiden: Brill, 2017), 358–386) and A. Buccheri ("Analogies d'analogies botaniques: épiclèses divines et métaphores du développement humain," in *Des dieux et des plantes. Monde végétal et religion en Grèce ancienne*, ed. A. Gartziou-Tatti and A. Zografou (Liège 2019), 69–98) at Hippocrates; R. Vinkesteijn develops it from Galen ("The vegetative soul in Galen," in *Vegetative Powers: The Roots of Life in Ancient, Medieval and Early Modern*, ed. Fabrizio Baldassarri and Andreas Blank (Cham: Springer, 2021), 55–72), Giglioni developed the importance of plants to think about the living ("Plant-animal imagination: Life and perception in early modern discussions of vegetative power," in *Vegetative powers: the roots of life in ancient medieval and early modern natural philosophy*, ed. F. Baldassari and A. Blank (Cham: Springer, 2021), 325–345); Hirai studied it from the focus at Renaissance (*Medical humanism and Natural philosophy. Renaissance Debates on Matter, Life and soul* (Leiden: Brill, 2011)), Fabrizio Baldassarri and Andreas Blank show how Aristotelian, Platonic, Stoic, Platonic and Galenic traditions informed the medical theories of early modernity (eds., *Vegetative powers: The roots of life in ancient, medieval and early modern natural philosophy*, (Cham: Springer, 2021)). The literature review highlights a new interest in the vital functions determining generation, nutrition or sleep that correspond precisely to the functions attributed

to the vegetative soul. Fabrizio Bigotti proposes a long-term approach to recapture the epistemological and ontological transformations of the plant schema in "Vegetable life: applications, implications and transformations of a classical concept 1500–1700," in *Vegetative powers, op. cit*. 383–406.

7. David Soulier, "D'un *Studium* à l'autre: la *peregrinatio academica* des professeurs anatomistes en Europe (années 1540–1550)," *Cahiers de la Méditerranée* 102 (2021), http://journals.openedition.org/cdlm/14030. Jerome J. Bylebyl, "The School of Padua: Humanistic Medicine and mortality in the Sixteenth Century," in *Health, Medicine, and Mortality in the Sixteenth Century*, ed. C. Webster (Cambridge: Cambridge Univ. Press, 1979), 347–355.
8. Chairs of practical medicine are created in Bologna (1534), Padua (1533) or Pisa (1544). They include a teaching of botany and pharmacology for medical students from the commented reading of two fundamental ancient texts, the *De materia medica* of Dioscorides, and the *De simplicium medicamentorum temperamentis et facultatibus* of Galen. In 1533 the first chair for the study of the simples was created. In 1545, is created the first botanical garden; in 1561, the first chair of practical botany. In 1606, the first practical manual for the preparation of herbariums appeared in Padua. On theater as part of an epistemic genre, V. Giacomotto-Charra, "*Quelques réflexions sur la notion de genre épistémique à la Renaissance,*" *Réforme, Humanisme, Renaissance* 1, 86 (2008): 185–197.
9. Nancy G. Siraisi, "Giovanni Argenterio and sixteenth century medical innovation. Between Princely patronage and academic controversy," *Osiris 6, Renaissance medical learning: evolution of a tradition* (1990): 161-180. Vivian Nutton, "André Vésale et l'anatomie parisienne," *Cahiers de l'Association Internationale des Études Françaises* 55 (2003): 239–249.
10. Jean Fernel, *The Physiologia of Jean Fernel* (1567), trans. J.M. Forrester (Philadelphia: American philosophical Society, 2003), 362–376.
11. John Henry, "Introduction." In Jean Fernel,*The Physiologia of Jean Fernel op. cit*, p. 256–279. Fernel published this treatise for the first time under the title *De naturali parte medicinae* (1542) and republished it in its second edition under the title of *Physiologia*

(1567). Hirai Hiro, « Alter Galenus:Jean Fernel et son interprétation platonico-chrétienne de Galien», *Early Science and Medicine*, 2005, 10 (1), 1–35, 25.
12. Galen, *De usu partium* IV 17; V 4; *On natural faculties* I, 1, 2, 10–17; II 6–7, III, 1–15.
13. Plato, *Timaeus* 77b-78d, 1992, 455–456.
14. Vesalius, *De Humani corporis fabrica libri septem* (Bâle: Oporinus, 1543), 257–262. Galen, *De usu partium* IV 17; V 4; *On natural faculties* I, 1, 2, 10–17; II 6–7, III, 1–15. "But I have not undertaken for the moment to expose the false theories of Galen, who is by far the most eminent professor of anatomy, and I would even less like to pass from the outset as ungodly towards the source of all our goods, and too disrespectful of his authority. I am not unaware, in fact, how troubled the doctors (unlike Aristotle's disciples) are when they note today during a single session of Anatomy that Galen deviated more than two hundred times from the true description of the harmony of the parts of the human body, their usefulness and their function.".

    www.biusante.parisdescartes.fr/vesale/?e=1&p1=00006&a1=f&v1=00302_1543x00&c1=2). *The factory of Vesalius*, edited and translated by J. Vons and S. Velut (Paris: Bibliothèque interuniversitaire de santé, 2014–2016), http://www3.biusante.parisdescartes.fr/vesale/debut.htm, consulted the 7th of october 2022.
15. Vesalius, *De Humani corporis fabrica,* 1543, p. 289, and more generally: "Rather, I await the recognition of those who seek the truth, because I have remembered the teachings of Galen and I study for myself the true making of man, even if my discourse sometimes deviates a lot [from his]." Vesalius, *De Humani corporis fabrica,* I, p. 83, *La fabrique de Vésale, op.cit.*
16. Vesalius, *De Humani corporis fabrica*, 548.
17. R. Vinkesteijn, "The vegetative soul in Galen", op. cit., 55–72, in particular 56–61.
18. Galen, *De usu partium* IV 17; V 4; *On natural faculties* I, 1, 2, 10-17; II 6–7, III, 1–15. On Galen reader of the Timaeus, Aileen R. Das, "Galen and the 'Medical' *Timaeus,*" in *Galen and the Arabic Reception of Plato's* Timaeus (Cambridge: Cambrige University Press, 2020), 30–68.
19. Galen *De usu partium*, XVI, xiv, Daremberg, t.2, 197.

20. F. Baldassari, "In the Beginning Was the Plant: The Plant-Animal Continuity in the Early Modern Medical Reception of Galen," in *Galen and the Early Moderns*, ed. M. Favaretti Camposampiero and E. Scribano (*International Archives of the History of Ideas* 236, 2022), 55–81, 58; R. Vinkesteijn, "The vegetative soul in Galen," *op. cit.*, 55–72.
21. Vesalius, *De Humani corporis fabrica*, 1543, 510. Fernel, 2003, I, c.7. Owsei Temkin, "The Classical Roots of Glisson's *Doctrine* of irritation and Seventeenth-Century Anatomy,"*Bull Hist Med* 38 (1964): 297–328, 309. Vesalius, *De Humani corporis fabrica*, 548.
22. For Galen it is actually the same power that we find both in plants and in the human liver (...). Since this power is present in plants itself, it is easier to analysis. R. Vinkesteijn, "The vegetative soul in Galen," op. cit., 61 and it infers a similarity in substance from the similarity of the powers that can be observed, in so far as the substance manifests itself through its powers.
23. Vesalius, *De Humani corporis fabrica*, 257; 262. Galien, *Sur les facultés naturelles* III 15, Daremberg, t.2, 318.
24. Vesalius, *ibid*, 295.
25. *Ibid.*
26. As Baldassari reminds, Harvey follows the vegetal analogy of his peer physicians. (...) he has accommodated a Galenic tenet within a new physiological system. Baldassari, *"In the Beginning Was the Plant," op. cit*, 76.
27. For Aristotle, man temporarily lives as a plant as an embryo dependent on his mother for food and then he becomes an animal when he acquires sensation and becomes potentially locomotor. But for Galen the reason for the identity between embryo and plant does not correspond to dependence on the mother, or the lack of certain functions as for Aristotle, but to the identity of powers (Vinkesteijn, 2021, 64). For Galen, the vegetative soul contained in seeds is nothing but power, the nature that builds plants and animals.
28. Sarah Carvallo, "La circulation sanguine comme pierre de touche. Harvey, Riolan, Descartes," *Lato* Sensu 3/1 (2016): 85–92.
29. Aristotle, *Meteorologica* (trans: Lee HDP), (Cambridge, MA: Harvard University Press, 2014).
30. Martin, Craig, "Francisco Valles and the Renaissance Reinterpretation of Aristotle's Meteorologica IV as a Medical Text," *Early Science and Medicine* 7/1 (2002): 1–30.

31. Harvey, *De generatione animalium* 1651, exercitatio 28, 109. I translate.
32. *Ibid.*, 91; 98; 116; 122; 126.
33. *Ibid.*, 110.
34. *Ibid.*, 26. Harvey takes up Aristotle's thesis On the generation of animals III 2,753b.
35. Harvey, 1651, 79.
36. Giovanni Alfonso Borelli, *De motu animalium*, 2 vols., Cork, C. Boutesteyn et al., 1685, Dédicace; Domenico Bertoloni Meli, "The Collaboration between Anatomists and Mathematicians in the Mid-Seventeenth Century with a Study of Images as Experiments and Galileo's Role in Steno's *Myology*," *Early Science and Medicine* 13 (2008): 665–709, 677–684. Ugo Baldini, "Giovanni Alfonso Borelli e la rivoluzione scientifica," *Physis* 16 (1974): 97–128.
37. Malpighi, *Anatome plantarum, Omnia opera* (London: Apud Robertum Littlebury, 1686), 13.
38. *Ibid.*, 20.
39. ... veluti circularia involucra sese invicem amplexantur, et continent, et ab utriculorum ordinibus a peripheria versus centrum horizontaliter productis pervaduntur. Malpighi, *De viscerum structura exercitatio anatomica, Praefatio, Opera omnia*, 1686, 3.
40. Interdum tamen utriculi diversis angulis pollentes continuatum canalem efformant; quandoque ovales sacculi lenes, et lucidi, altero etiam fine impervii, quales in pulmonibus insectorum observare est, deprehenduntur, et frequenter plures vesicularum, sese invicem coangustantium, ordines eodem ligneo quasi tubo contenti spirals fistulas efficiunt. Malpighi, *De viscerum structura exercitatio anatomica, Praefatio, Opera omnia*, 1686, 3.
41. ...ex quo fortasse deduci postest ratio motus in mimosa planta observatus. An hujusmodi spirales fistulae sint traches, pro solo aere vel saltem halitu devehendo a Natura institute, examinabo; et an simul aere succum exportent, et an aerem ab imo telluris, an ab exteriori ambiente hauriant, inquiram. Malpighi, *Anatome plantarum,* 1686, 4. Descartes is also interested in this sensitive plant, Letter from Descartes to Mersenne, August 23, 1638, AT II 329. On the importance of this plant in the seventeenth century, Guido Giglioni, "Touch Me Not: Sense and Sensibility in Early Modern Botany," *Early Science and Medicine* 23/5–6 (2018): 420–443. This plant continues to be the subject of scientific studies on

plant intelligence, M. Gagliano, M. Renton, M. Depczynski and S. Mancuso, "Experience teaches plants to learn faster and forget slower in environments where it matters," *Oecologia*, 175 (2014): 63–72.

42. Licet perpetuam Natura No Servet In compangendis vegetantibus normam, quaedam tamen conveniunt, Vel saltem analogismo parum distare videntur; quare in fructibus plures trunci, in herbis ipsis multiplies caules, interdum unus, Vel saltem vicarius, deprehenditur. Malpighi, *Anatome plantarum*, 1686, 5.
43. *Ibid.*, 7.
44. ...involuta and latentia Sunt haec Naturae opificia, *ibid*.
45. Malpighi, *Dissertatio epistolica de formatione pulli, 1672.*
46. Malpighi is not the only one to translate physiology into laws: in France, Jean Pecquet (*Experimenta nova anatomica*, 1651) studies the movement of the chyle in a purely mechanical way by rejecting the explanation by attraction in the galenic sense of natural faculties, according to which each part of the human body attracts the food it needs, retains it, assimilates it by altering it and rejects the surplus. With Mariotte's help, Perrault observed the circulation of sap and the peristaltic movement of plants and animals; he presents his experiences to the Royal Academy of Sciences. "Especially we should examine whether they do not other not this in common with animals, that there is Had in them a main part that Gave The soul and movement to all others..." Perrault, *History of The Academy*, 1773, vol. 1,19. On the change induced by this experimental approach at the Academy of Sciences, S. Mazauric, *Fontenelle et l'invention de l'histoire des sciences à l'aube des Lumières* (Paris: Fayard, 2007), 7–23.
47. Sarah Carvallo, *L'homme parfait. L'anthropologie médicale de Harvey, Riolan et Perrault (1628–1688),* (Paris: Classiques Garnier, 2017); D. B. Meli, *"The Collaboration Between Anatomists and Mathematicians...,"* *op. cit*, 670–677.
48. Malpighi applies indeed he method of ligation developed by Harvey, Aselli and Steno, to uncover the circulation of blood in veins and arteries, or chyle in milky pathways, to the circulation of sap in the tree: in any case, swelling occurs upstream of the ligation. Everything is carried out according to the same laws. Malpighi, *De hepate, Opera omnia*, 69. Domenico Bertoloni Meli, "Early modern experimentation on live animals," *Mechanism,*

Experiment, Disease: Marcello Malpighi and Seventeenth-Century Anatomy, (Baltimore: Johns Hopkins University Press, 2013), 114–117, 253.
49. Malpighi uses the term *regula, norma* or *Lex.*
50. simplex, una et facilis videtur in conflandis viventium corporibus naturae norma… Malpighi, *De viscerum structura exercitatio anatomica, Praefatio, Opera omnia*, 1686, n.p.
51. Malpighi, *De viscerum structura exercitatio anatomica, Opera omnia*, 1686, 66.
52. François Duchesneau, *Leibniz, le vivant et l'organisme*, (Paris: Vrin, 2010), 28.
53. Jane Ruby, "The Origins of Scientific Law," *Journal of the History of Ideas*, heft. XLVII, n° 3 (1986): 341–359, in particular 347–350.
54. S. Carvallo, op. cit., 2019. Recall that the *Norma* Latin translates into French as the rule, which designates the tool of the architect, God being himself conceived as a great architect, *opifex.*
55. Malpighi, *Anatome plantarum, Omnia opera* (London: Apud Robertum Littlebury, 1686), 7.
56. Stephen Hales, *Vegetable staticks*, London, 1727.
57. B. Meli, *op. cit*, 2006, 187–189.
58. Sophie Roux, "The Laws of Nature in the Classical Age: The Terminological Question," *Revue de Synthèse* Springer Verlag/ Lavoisier (2001): 531–576. halshs-00806468.
59. Roux *ibid.*, 543.
60. Malpighi, From anatome plantarum, *Opera omnia*, 1686, 29, The nuance between identity and analogy lies in the presence of the soul in man. Following Luigi Belloni, François Duchesneau calls microstructuralism the infinite regression of the machine in the machine that supposes the operativity of laws everywhere and at all scales of life, F. Duchesneau, *Leibniz, Le vivant, l'organisme*, 2010; L. Belloni, "Malpighi and the founding of anatomical microscopy," in *Reason, experiment and mysticism* (New York, 1975), 95–110, in particular 108.
61. Raphaële Andrault, "Leibniz and the knowledge of the living," in *Leibniz. Lectures et Commentaires*, ed. Mogens Laerke, Christian Leduc and David Rabouin (Paris: Vrin, 2017), 171–190.

62. Leibniz, *Directiones ad rem medicam* (1671), IV, xii, 10, A VI, VI, 453–454. Justine Smith, *Divine machinery: Leibniz and the Sciences of Life* (Princeton: Princeton University Press, 2011), 33–37.
63. Leibniz, *Système nouveau de la nature et de la communication des substances*, 1695, rééd. in *Système nouveau de la nature et de la communication des substances et autres textes*: 1690-1703, éd. Christiane Frémont, 1994, 65–90. Catherine Wilson, "*From Ipsa Natura*. Sources of Leibniz's Doctrines of Force, Activity and Natural Law," *Studia Leibnitiana*, vol. XIX, 2 (1987): 148–172; R. Palaia, "Naturbegriff und Kraftbegriff im Briefwechsel zwischen Leibniz und Sturm," *Leibniz 's Auseinandersetzung mit Vorgängern und Zeitgenossen*, *Studia Leibnitiana*, Supplementa 27 (1990): 157–172.
64. Leibniz, Letter to Arnauld, Oct. 1687, A II, 2, 253.
65. Leibniz, 1689–1690, *Dynamica de potentia et legibus naturae corporeae, rééd. in Mathematische Schriften,* vol. VI, ed. Carl I. Gerhardt (Hildesheim/New York, Olms, 1971), 281–514.
66. One issue in the controversy between Newton and Leibniz concerns the meaning of the explanation, mathematical according to Newton, causal according to Leibniz. The epistemological issue is redoubled by an ontological question: do forces be exercised at a distance or do they act as an internal principle of the body? Leibniz defends the intrinsic reality of force and its causal nature. For him, the scientific explanation must go back to a principle inherent in the body: it therefore requires a mathematically formulated physical perspective and a metaphysical analysis theorizing the body as a substance: "[…] the notion of force to which I have devoted a special science, Dynamics, sheds a lot of light on the intelligence of the true notion of substance." Michel Fichant, "De la puissance à l'*action*: la singularité stylistique de la dynamique," *Revue de Métaphysique et de Morale* no. 1, La métaphysique de Leibniz (1995): 49–81; Pauline Phemister, *Leibniz and the natural world. Activity, passivity and corporeal substances in Leibniz's philosophy* (Dodrecht: Springer, 2005).
67. Leibniz, *Opuscules philosophiques choisis,* ed. P. Schrecker (Paris: Vrin, 1978), 102. and Just before: "For my part, if, as I believe, I have understood the notion of action well, I believe that it implies and justifies the unanimously accepted philosophical principle, that all action is the action of an individual subject. And I find this

principle so true that its reciprocal is also true, that is, not only is everything that acts an individual substance, but also that every individual substance acts without interruption, and I do not even except the body, because one never notices absolute rest." Leibniz, 1698, here 1978, 101.

68. Leibniz, *Stahl-Leibniz. Controverse sur la vie, l'organisme et le mixte*, 2004, 136–137.
69. Sarah Carvallo, "Les effets rétrospectifs de Stahl sur Descartes. Cartésianisme et anti-cartésianisme en Europe entre 1650 et 1875," in *Chemins du Cartésianisme. Cartésianisme et anti-cartésianisme en Europe (1650-1750)*, ed. Raffaele Carbone and Antonella del Prete (Paris: Classique Garnier, 2017), 211–240; S. Carvallo, "Histoire du tonus: ruptures ou continuité ?" in *Du nouveau dans les sciences*, ed. S. Carvallo and S. Roux (Paris: Vrin, 2006), 71–120.
70. "Each gland, each organ exerts — through its department, or zone of influence — an action on the glands and organs that surround it. This immediate action is coupled with a remote action, by means of the humors secreted and excreted by the gland. But every element and part of the body is dependent on the activities and moods of other organs. When these reciprocal influences are balanced, it is the state of health, when any cause comes to break this harmony it is the beginning of the pathological process." Dominique Boury, "Théophile de Bordeu: source et personnage du *Rêves de D'Alembert*," *Recherches sur Diderot et sur l'Encyclopédie* [Online], 34 (2003): 20, http://journals.openedition.org/rde/154.

PART III

The Quest for Human Perfection in Transhumanism

CHAPTER 10

Between Thinking Tools and Extended Minds: Is There Any Room for Natural Organs of Knowledge and Action?

Guido Giglioni

Introduction: Trees, Wood and Matter

> The tree outside the window taps very gently on the pane …I want to think quietly, calmly, spaciously, never to be interrupted.
> Virginia Woolf, *The Mark on the Wall*, 1917

A tree is not a piece of wood. This seemingly anodyne thought came to my mind a few summers ago while two woodcutters hired by the city in which I live—Civitanova Marche, Italy—were felling three old maritime pines in the small park near my apartment block. When I asked the woodcutters what they were doing, they explained to me that the trees were threatening the life of little children. The trees' fault was to have been

G. Giglioni (✉)
University of Macerata, Macerata, Italy
e-mail: guido.giglioni64@gmail.com

J. Muller (ed.), *Automata, Cyborgs, and Mutants*, Palgrave Studies in Medieval and Early Modern Medicine,
https://doi.org/10.1007/978-3-031-93201-4_10

shaped in extravagant poses by the wind during the many years of their lives. Reasons of health and safety prevailed over majestic works of natural art sculpted by the wind for almost a century.[1] Trees of those shapes, apparently, can kill children: *pinus pinaster* ... and *homicida*, one should add to the Linnean definition of their botanical variety. Far from killing children, trees are, in fact, routinely killed by the children's parents, who seem to glibly ignore that the trees and the children of the future belong to each other in a tacit pact of mutual respect. That summer day, a reasonable conversation with the woodcutters was impossible, and not simply because they were taking orders "from above." When I said that I would be mourning the death of the three *pinastri* whom I got to know in the past four years, the two woodcutters burst out in a guffaw while the trees were swiftly being cut into pieces and shredded into sawdust. In this chapter, I would like to resume the point I was trying to make that summer when I had the thought that a tree is not a piece of wood. Trees are not timber nor lumber. I would argue that, while not entirely selves (for their apprehension of reality lacks the focus of self-consciousness), trees are natural organs of knowledge and action. As such, they are an integral part of the vital economy of the universe.[2]

By natural organ of knowledge and action, I mean a living device—a set of motions—which, despite lacking conscious volition, is nonetheless able to contribute to the global production of information and energy occurring at every instant in our universe. Unlike human organs of knowledge and action, which are dependent on the self-sentient abilities of the human mind and whose applications cover a wide range of instantiations, from logical procedures to robots, natural organs enjoy a limited but essential amount of autonomy in the environment in which they operate. The idea of natural organs of knowledge and action occurred to me while working on certain features of Aristotle's philosophy. We owe to Aristotle the notions of *organic* life and *organic* thinking, where the adjective "organic" has a specific technical meaning: life and minds need organs—that is, tools—to function. Just as animate bodies are made up of parts shaped in suitable ways (organs), so minds capable of reasoning rely on logical tools (organs). As Aristotle's metaphysics shows very well, however, tools are puzzling and disquieting entities from a philosophical point of view. Since ancient times, they are closely associated with ideas of freedom and oppression, for tools may work as agents of both liberation and enslavement. As we will see in this chapter, Aristotle's tools cover the whole gamut of knowledge and action, from reason (λόγος) to lack of reason (ἄλογος φύσις), from human deliberation (προαίρεσις) to natural spontaneity (αὐτόματον). Although tools have constantly suffered from negative publicity among philosophers, for they remind them of issues of instrumental action and hierarchical segregation, in fact, tools can act as natural levelers every time allegedly disinterested actions are there to conceal deeper and opaque interests and every time false hierarchies need to be redrawn according to new, more urgent priorities.

One of the aims of this chapter is to use the notion of natural organs of knowledge and action to address contemporary issues concerning animation and automation. I will sketch out the hypothesis that natural beings, all of them, may function as receptors and transmitters—perhaps even makers—of knowledge. To do so, I have decided to borrow from Aristotle such key categories as natural productivity (φύσις), method of making (τέχνη), seemingly accidental and seemingly unconscious agency (αὐτόματον). As these categories still resonate with heuristic and epistemological potentials, they can contribute to contemporary debates on the nature of thinking and its relationship with the life of the cosmos. Furthermore, Aristotle's metaphysics of tools is broad and ambitious to the point of being sometimes incautious and foolhardy. In its vast scope, it includes man-made entities (from syllogisms to bedsteads) as well as humans themselves (slaves and women). This level of complexity and puzzlement acts as a unique argumentative catalyst for the mind of unbiased inquirers.

Continuing with my Aristotelian borrowings, in the second part of this chapter I will focus on Pietro Pomponazzi's revolutionary redefinition of the Aristotelian notion of imagination and, more specifically, on his reading of a specific passage from *De intellectu* commonly attributed to Alexander of Aphrodisias, the influential Aristotelian commentator active between the second and the third century AD. In *De immortalitate anima* (1516), by redefining the conceptual knot that ties together mental representations, anatomical organs and objects of knowledge, the philosopher Pomponazzi (1462–1525), active in important centres of Renaissance philosophy such as Padua, Ferrara and Bologna, reconsidered the relationship between human thinking, the vital processes in one's body and the larger context of nature's causality. In his seminal account of the imagination as body of representations (*corpus ut obiectum*), that is, as an organized system of perceptual states capable of interfacing one's mental space with the vast body of nature, the *obiectum* becomes the principal tool through which reality is representable and knowable to the mind.

In the end, by adopting Pomponazzi's articulate redefinition of thinking, body and imagination, I will compare the natural organs of knowledge and action to imaginations (*obiecta*) without substantial selves (*subiecta*), that is, representational states of matter which do not require the involvement of human consciousness (a self-sentient self) to be activated as transformative powers. I will maintain that, while in Aristotle's cosmology such entities as plants and seeds can be viewed as examples of natural organs of knowledge and actions, for they act as conveyors of active energy issuing from nature, Pomponazzi shifts the focus on animal representations of reality as catalysts of material transformations in the surrounding world. In the treatise *De incantationibus*, published posthumously in 1556, he describes the cognitive patterns conveyed by actual imaginations as representational forces (*species spirituales*) capable of producing real things (*ideae rerum fiendarum*) so as to explain seemingly supernatural phenomena in natural terms.[3] Through his notion of *imaginationes* as reflective frames of one's body, Pomponazzi expanded the meaning of natural organ of knowledge and action to include all kinds of changes in nature. I will suggest that this evolution paved the way for positions advocating the understanding of matter as one active substance. Together with vegetative powers, active seeds and spontaneous processes of growth, the notion of imaginations as states of heightened perceptual energy working as natural receptors of information unencumbered by mental awareness provided a glimpse into the unconscious life of matter as a self-organizing power in which all individual bodies participate in building the perceptual network underlying the universal body of nature.[4]

In the final section of this chapter, I will hybridize my spuriously—let's say, unauthorized—Aristotelian notion of natural organs of knowledge and action with the concept of extended mind, theorized in the late 1990s by the philosophers Andy Clark and David Chalmers. I will suggest that the natural organs of knowledge and action, with their emphasis on the active autonomy of self-organizing matter, can expand the ontological scope covered by the theory of the extended mind, while defusing some of the most problematically anthropocentric assumptions that still underlie that theory. The thinking tools of the ever-expanding space of the human mind, a rather recent outcome of the evolutionary life of consciousness, need to be attuned to the original tools of active matter.

Thinking and Tooling: Re-engaging with Aristotle's Metaphysics in the Anthropocene

> Wood is a pleasant thing to think about. It comes from a tree; and trees grow, and we don't know how they grow.
> Virginia Woolf, *The Mark on the Wall*, 1917

Where does the mind stop and the rest of the world begins?[5] This question, fascinating as well as challenging, was asked in 1998 by Clark and Chalmers and it has recently gained a new meaning. In the past few years, humankind has found itself in the unprecedented situation of having to extend the scope of its mental powers through digital tools. This has brought back to the fore ideas that had been worked out during the Renaissance, at a time of disruptive change and technological innovation. What I would call Renaissance thinking tools represented an attempt to show in concrete terms—through books and diagrams, *tabulae* and *machinae*—that thinking is embodied, spatially framed, object-oriented and socially constructed—or, to put it even more simply, that the mind has a body, lives in a world, handles objects and communicate with other minds through languages and systems of coded information.[6] By now these statements may sound all too familiar after decades of critical theory, social constructivism, new rhetoric and new historicism have rightly alerted us to the many ways in which context and materiality condition our understanding of reality. In fact, as I am going to argue in this chapter, the Renaissance idea of thinking tool may still have an effective impact on the rather fusty discipline of history of philosophy, for, more than any other discipline, history of philosophy seems to remain under the spell of such totemic constructs as consciousness, pure reason, self-transparency, and anthropocentric dualisms.

Epistemologically speaking, this chapter draws inspiration from current discussions in two specific field studies. The first is the already mentioned research program investigating the extended mind, developed by Clark and Chambers since the late 1990s.[7] The other research field is the non-binary understanding of the relationship between nature and culture that has been advanced in the past years by the anthropologists Eduardo Viveiros de Castro, Philippe Descola and Eduardo Kohn.[8] I argue that views regarding the possibility of subjecting the things of nature and the things of culture to a non-binary consideration can already be encountered in rather sophisticated approaches in Renaissance philosophy and science, although these views were rapidly overcome by the greater appeal that a different perspective, based on the metaphysics of consciousness and the mechanization of nature, exercised on the minds of early modern learned Europeans.

I began toying with the idea of thinking tools while working on Francis Bacon's philosophy of the mind. Bacon (1561–1626) envisaged the human mind as a hardly controllable source of formidable power that needed to be both assisted and restrained by a whole array of thinking tools starting with a method—an *organum*—that could give sound directions on how to conduct one's own thoughts. Bacon believed that the mind required to be constantly held in check by a comprehensive set of instruments and implements, beginning with one's own *organic* body (i.e., a corporeal entity made up of anatomical organs), the hands and other bodily parts, the senses (both external and internal), the imagination, with its ability to handle inner and outer images, diagrams and *tabulae*, and then language—possibly more than one language—and the instruments made by means of one's body, including the pen, the tablet, the volumen, the codex, the book, the library (here I am limiting my examples to reading and writing technology), and so on through the ever expanding world of the human *artes*. Within this universe of knowledge, the encyclopedic program outlined in Bacon's *Advancement of Learning* (1605) stands as the mirror-like image of his model of instrumented and tutored mind—its map as it were. Bacon's view of the extended mind proceeds through an expanding range of organs, both anatomical and artificial.[9]

The question then becomes whether a thinking tool is something accidental to the activity of harvesting knowledge or whether it is an integral part of the cognitive process. This question, which has recently been scrutinized by Clark with many insightful results,[10] sparked off a heated debate during the Renaissance, with endless discussions among schoolmen, humanists and science practitioners arguing whether a method was to be considered—to take up the words of the Paduan philosopher Jacopo Zabarella (1533–1589)—as a form of knowledge production in itself (one of the five *habitus intellectuales* such as *scientia*, *prudentia*, *ars*, *sapientia* and *intellectus*) or as a set of formal criteria needed to discipline the act of thinking (*habitus instrumentalis*).[11] In general, we now tend to accept the view that tools, be they physical or mental, are an essential component of human thinking.[12] They are ways in which the mind projects its inquiring power onto the external reality, for thinking is no introspective and solipsistic affair. Indeed, Bacon maintained that to be left alone in the company of one's own thoughts represented one of the utmost danger for humans, all the more so because thoughts, in their natural and unsupervised states, are slaves to appetite. For this reason, thinking should constantly reach out into the world and demand a collaboration with other thinking beings.[13] As a result, the idea of a close interdependence of thinking and tooling sparked numerous debates during the early modern period. It concerned the setting up of disciplinary methodologies (as in medicine, rhetoric and mathematics) and, more generally, philosophical discussions aimed at devising a technology of thinking tools capable of improving the functioning of the human mind (Bacon, Descartes and Spinoza, to mention the most representative cases).[14] It also prompted the question whether any means of natural production of knowledge could be envisioned outside the sphere of the human mind. I call this subset of tools natural organs of knowledge and actions.

The world of premodern and early modern theories of reproduction is an area where we can start investigating this particular kind of tools. It may sound odd today, but Renaissance diatribes about generation and the animation of the seed are an ideal place where to look for discussions regarding interrelated issues of methodology and technology. A central issue concerned the question of whether a seed was to be looked at as an instrument (an *organum*) of the formative faculty of the body not to be materially left in the thing made (as argued by Aristotle and Galen) or whether it could be viewed as a self-acting tool. The reason that at the time theories of generation kept intersecting the fields of epistemology and technology depended on the lingering presence of the Aristotelian framework. As we will see, in Aristotle's philosophy, the ὄργανον considered as both an anatomical part and a vital tool was a key element to understand any kind of technological transfer of knowledge. Aristotle associated the functioning of life with the use of organs. He defined an anatomical organ as a part of the body that the soul can use to fulfil a specific end. Undifferentiated (inorganic) matter, being devoid of shape, could not be handled by the soul and therefore was deemed to be inanimate. By contrast, an organic body was animate, that is, alive, when it was predisposed to use tools, from the transfer of fluids and nutrients in plants to the modulation of locomotive patterns in animals, from the exquisite accomplishments of the human body to the menial services of human slaves. It is therefore safe to say that Aristotle's views on knowledge production and technological innovation presupposed the existence of a vital link between natural and human actions, a feature that in contemporary terms is known as biomimicry. Bacon may have been influenced by the Aristotelian notion of τέχνη when in the *Novum organum* (1620) he introduced the concept of hyperinduction as a way of connecting the progress of knowledge (induction) with the possibility of "superinducing" change onto nature through coordinated and controlled sequences of natural motions. As nature is the grand arena of all vital endeavors (*tentamenta*) of matter, so the human mind can make apt and wise use of these natural attempts at transforming reality.[15]

For all these reasons, when we examine the notion of natural organs of knowledge and action in the Renaissance, we cannot help being reminded of Aristotle's definition of τέχνη in various parts of his oeuvre. In Book 12 of the *Metaphysics*, for instance, human craft is defined as "a generative principle in something other than the thing moved," while nature is "a generative principle in the thing itself" (3, 1070 a).[16] Technology is the art of channeling and transferring natural sets of motions into man-made objects or into natural objects treated as tools (the slave, once again). In Book 2 of the *Physics*, Aristotle argues that physicians can heal themselves only by accident, that is, not because they are essentially and spontaneously self-healers, but only because they apply to their own bodies the τέχνη of healing.[17] Likewise, Bacon states that the mind can think in a suitable and safe way when it applies to itself the tool of acquired learning (*doctrina*).[18] In both cases, with respect to the Aristotelian physician and the Baconian thinker, the production of knowledge occurs through the mediation of a tool that acts as a conveyor of information and energy.

A traditional historiographic view, still widely embedded in textbooks of history of philosophy and science, tends to regard the philosophical *novatores* of the early modern period as anti-Aristotelian. And yet on this point Bacon remains very close to Aristotle, in the same way in which he is close to Aristotle in the domain of moral philosophy when he endorses the concept of virtue as practice: "Cultured people not only use their mind and exercise virtues, but continuously correct themselves and make progress in virtuous behavior."[19] In the *Eudemian Ethics*, Aristotle establishes a striking connection between the notion of ἀρετή (virtue) as a means of self-fulfillment and the concept of instrument (ὄργανον) as a mental prop (1248a). The passage is worth reporting in full because of the way in which motion, mind and life are knitted together within the broader context of cosmological order:

> this is what we are investigating—what is the starting-point of motion in the soul? The answer then is clear: as in the universe, so there, everything is moved by God; for in a manner the divine element in us (τὸ ἐν ἡμῖν θεῖον) is the cause of all our motions. And the starting-point of reason is not reason but something superior to reason. What, then, could be superior even to knowledge and to intellect, except God? Not virtue (ἀρετὴ), for virtue is an instrument of the mind (ἡ γὰρ ἀρετὴ τοῦ νοῦ ὄργανον).[20]

In this excerpt, ἀρετή means moral excellence. Whether virtue is the result of character or practice, ethical striving depends on the intellect and is one of the tools that the intellect uses to reach full actualization. Like the principles of medical science in the previous example involving a self-treating physician, virtues are organs of moral improvement applied to specific circumstances by the mind. In the *Physics*, Aristotle says that it is through the τέχνη of sculpture that Polyclitus becomes a sculptor, while Polyclitus *qua* Polyclitus is an accidental sculptor. In this example, sculpture as a craft is actualized into a set of motions that lead the accidental sculptor to the making of the object (195ab). Along similar lines, in the *Generation of Animals*, Aristotle identifies the process of animal reproduction with the transmission of a sequence of motions conveyed by the seed. As with the art of sculpting (ἀνδριαντοποιΐα), the seed is a tool that transfers the capacity for life from the male parent to the offspring and it is there actualized as a pattern of carefully choreographed sequential motions. In the same work, Aristotle explains that a specific form is imprinted on a specific object by the mind of a carpenter, which moves the hands which in turn move the tools (ὄργανα) which finally move the material. Like semen in reproduction, "in the tools lies in a certain sense the motion (κίνησις) of the τέχνη" (730b). What is more, in this example the form that animates an object is the εἶδος that has been transmitted from the mind. Imagination, as we are going to see in the next section, has a key role in the making of natural organs of knowledge and action.

As mentioned in the introduction, Aristotle's metaphysics of tools includes the highly problematic case of slaves, whom he defined as the animate and detachable tools of a master (*Politics*, I, 5, 1254b). They, too, are natural organs of knowledge and actions, but the amount of information and activity released by them is obtained in a morally and politically questionable manner.[21] Aristotle's equation in his *Nicomachean Ethics* (1161b)—ὁ γὰρ δοῦλος ἔμψυχον ὄργανον, τὸ δ' ὄργανον ἄψυχος δοῦλος: slaves are animate tools, tools are inanimate slaves—brings brutally to the fore the disreputable implications of instrumental agency. Aristotle's theory of the natural slave assumes that certain humans are not completely human because they are constitutively unable to be free from the urge of the appetitive life. As animated tools, slaves lack the ability to control their nature. Like children and women, but in different respects, slaves are unable to restrain themselves because they lack the authoritative power of deliberation. Indeed, Aristotle's metaphysics of tools is premised on the primacy of the metaphysics of disinterested thinking: any activity relying on tools, including the use of tools for thinking, is slavish. In this sense, natural slaves are natural users of bodies. Aristotle's metaphysics of tools reflects precise ontological hierarchies: the soul is to the body as humans are to animals, as man is to woman, as the adult is to the child, as the free is to the slave. The principle underlying this set of hierarchies is the difference between ruling and being ruled. As there are in nature beings that are unable to rule themselves, so these need to be ruled by those beings that are capable of self-control. Aristotle's comprehensive and compact system of hierarchical order ends in an embarrassing declaration of anthropocentrism: "man alone of the animal has reason (λόγος)" (*Politics*, 1253a9).

This convergence of reproduction, technology and instrumental agency is where the Aristotelian intertwinement of thinking and tooling comes very close to the contested category of spontaneous agency (τό αὐτόματον). The semantic area covered by the notion of spontaneity in Aristotle's natural philosophy is not easy to circumscribe, for it often intersects different cultural concepts. The adverb αὐτόματως, which describes the action of being moved by one's own impulse, can often be rendered in English with a simple "by itself." Significantly, this type of action covers three meanings: the domain of chance occurrences, the process of seedless generation and the mechanical life of the marionettes. While Aristotle situates his discussion of the αὐτόματον firmly in the domain of nature, the spheres of the intellect and the gods are also parts of the picture. For instance, in a passage in the *Generation of Animals* (715b), Aristotle uses the verb αὐτοματίζω ("to act of oneself," "to act spontaneously," "to happen by chance") to explain phenomena like putrescence and plant parasitism:

> some [plants] are formed out of seed (ἐκ σπέρματος), others as it might be by some spontaneous activity of nature (ὥσπερ αὐτοματιζούσης τῆς φύσεως) – they are formed when either the soil or certain plants become putrescent, since some of them do not take shape independently on their own (οὐ συνίσταται καθ' αὐτὰ χωρίς), but grow upon other trees, as for instance the mistletoe does.[22]

In this passage, Aristotle speaks of φύσις αὐτοματιζούσα as nature acting "spontaneously." This means that αὐτόματον is that which is not able to be put together as an organized being by itself (συνίσταμαι) through the set of motions encoded in a seed (σπέρμα). Nevertheless, it is still capable of acting as resulting from the active power of nature. But then the question becomes: What is the foundation of this active power of nature? Aristotle's φύσις αὐτοματιζούσα is close to his renowned but most elusive definition of nature as δαιμονία mentioned in the short treatise on divinatory dreams: "Broadly speaking, since some nonhuman animals also dream, dreams are not sent by gods, nor do they occur for this reason. They are demonic, though, for nature is demonic, but not divine."[23] We might say that natural organs of knowledge and action are "demonic" in that they function in teleologically surprising ways without being directed by human or divine consciousness.

Piecing together the concepts of godhead, demonic presence and natural action within the framework of Aristotelian philosophy is not an easy task from an interpretative point of view, as commentators have known since ancient times. For Aristotle, life and knowledge are not sent by gods or demons. They are imparted to nature by the intellect (νοῦς), which in Aristotle's cosmology is a divine, but impersonal force. As stated in the excerpt above from the *Eudemian Ethics*, god is first and foremost a principle of motion. Aristotle rejects notions of direct divine guidance, personal guardianship by demons and deities or possession by gods. In a universe in which changes occur through sequential and coordinated transfers of motion, the system of causal relationships is structured as a sophisticated chain of interdependent means and ends. The organization is teleological, but the process is unconscious: the actualization of a potential resource is instrumental to the realization of a higher goal. Tooling is therefore essential in the grand mechanism behind the activity of the cosmos. Only the unmovable mover enjoys disinterested contemplation in its purest form; all the rest is tooled to serve the actualization of the universe. This is what allows unconscious nature (ἄλογος φύσις) to be remarkably tuned to divine movement.[24] Philip van der Eijk rightly characterizes the Aristotelian model of δαιμονία φύσις as a "psycho-physiological mechanism" that is predicated on the specific natural constitution of the receiving powers.[25] From a cosmological point of view, Aristotle's notions of αὐτοματιζούσα, ἄλογος and δαιμονία nature, together with his overarching metaphysics of the material intellect, helped conceptualize matter as actively predisposed to the action coming from the intellect. Aristotelian matter, as we will see in the next section, is an inherently receptive being.

Relying on all these cases examined by Aristotle, we can describe a natural organ of knowledge and action as an adaptable device activated by a set of natural motions liable to be transferred from subject to subject. Τέχνη itself is one of such organs. These organs are characterized by varying degrees of functionality (working as means to an end), autonomy (being capable of initiating or conveying motions) and spontaneity (being impervious to conscious control). Three possible layers of meaning, therefore, constitute a natural organ of knowledge and action: the vital, the material and the representational. As mentioned above, plants, seeds and human slaves are concrete instances of natural organs of knowledge in Aristotle's universe. They testify to a domain of nature that does not attain the level of actualization that is the mark of true cognition and true action. Plants are more vital than stones and slaves are more animate than machines. However, neither plants (the intelligence of insentient nature) nor slaves (the work of imperfect humanity) affect in any way the ideological primacy of anthropocentric reason. Can we radicalize the meaning of natural organs and—against Aristotle—go so far as to envision the possibility of fully autonomous and fully self-organizing tools of nature, so much so that thinking tools (where "thinking" is a gerund) turn into thinking objects (where "thinking" is a participle) or self-active tools?[26] When the tools for thinking take on a life of their own, they become ensouled beings, capable of perceiving and acting autonomously. This is how trees and slaves turn into free agents (which in Aristotle's philosophy is not allowed), while their role as spontaneous αὐτόματα can be replaced by machines. Pushed to such limits, as we are going to see in the next section, matter itself becomes the most active natural organ of knowledge. This case is extreme, though, and presupposes a materialist ontology that is foreign to the spirit of Aristotle's philosophy. Indeed, it is a true reversal of the founding hierarchies of his metaphysics. As a result, the Aristotelian paradigm breaks down and its ever-lingering naturalism mutates into vital materialism. An example of this outcome, in which matter is viewed as one self-moving and self-acting organ, is Francis Glisson's theory of active substance. Interestingly enough, as a Harveian anatomist who followed the metaphysics of Francisco Suarez, Glisson (1599?–1677) was both a Baconian and an Aristotelian.[27]

The Pomponazzian Turn: The Imagined Body as a Representational Tool

Everything's moving, falling, slipping, vanishing ... There is a vast upheaval of matter.
Virginia Woolf, *The Mark on the Wall*, 1917

Natural organs of knowledge and actions—plants and slaves, among them—surreptitiously emerge out of Aristotle's metaphysics of tools as beings that demand ontological recognition and manifest a role that is more active than anticipated in the life of the universe. They disrupt the hierarchical divisions imposed on the cosmos by a metaphysics that privileges conscious control instead of perception from below and they herald the conceptual plausibility of self-organizing matter. As historians, we should scrutinize the reasons that led a seventeenth-century anatomist such a Glisson to elaborate a theory of active matter moving from Aristotelian premises. One of the directions that this kind of investigation may take is by looking for possible sources and precedents. Among the precedents, sixteenth-century readings of Aristotle's theory of the soul are among the most valuable repositories of information. During that time, the return of Alexander of Aphrodisias's interpretation of Aristotle's material intellect prompted a materialist drift in peripatetic exegesis.[28] The Italian philosopher Pietro Pomponazzi, the notorious author of a mortalist but not materialist interpretation of the human soul in Aristotle's oeuvre, made an original use of Alexander's theory of the intellect in his work as a reader of Aristotle.[29]

The reason why, in Pomponazzi's opinion, Aristotle had regarded the human soul as a perishable substance depended on the fact that the soul relied on the body via the imagination to perform its functions. For Pomponazzi, however, this reliance was of a representational, not physical nature: "the human intellect cannot be entirely released from matter, for that intellect is always moved by a representation (*obiectum*) even with respect to the knowledge that it can have of itself and when it grasps the universal in the singular."[30] In the brief history of thinking tools and natural organs of knowledge and action which I am outlining in this chapter, Pomponazzi plays a crucial role as he dramatically reoriented the Aristotelian theory of the intellect by turning what originally was a cosmological entity into a mental function. For Pomponazzi, the imagination is the representational power that through the *obiectum* helps the mind get out of the loop of introspective self-reflection. The principal conceptual pivot in Pomponazzi's demonstration of the mortality of the soul consists in distinguishing between the body as a *subiectum*, that is, as corporeal substance, from the body as an *obiectum*, that is, as a cognitive function. In this discussion, we can also replace the term "body" with that of instrument, *organum*. The *organum* Pomponazzi is thinking about when he speaks of organs and organic functions is no anatomical part of the body, important as this may be (whether it is the heart or the brain). Rather, this *organum* is a cognitive tool, that is, the imagination or the imagined body, i.e., the *corpus ut obiectum*, as represented to the mind by the imagination. In this case, too, Pomponazzi heralded a remarkable shift in the meaning of bodily *organum*. He considered the human soul to be *materialis* because he believed that the intellective part of it—the mind—needed the body as imagined reality: not as a physical organ, but as a thinkable object, a sort of representational prism relentlessly multiplying images out of itself.[31] Human thinking, Pomponazzi argued, requires objects to be thought through and about. As a result, this type of intellect is characterized by spatial extension and temporal succession. For humans, bodies are the organs of their minds insofar as by body one understands a mental object, a universe of imaginable things. With Pomponazzi, the mind becomes stretchable because it incorporates the body as its thinking tool.

The most important consequence one may draw from the emphasis that Pomponazzi played on the *obiectum* (the mental image) as *organum* is that our mind becomes what it makes of the reality that surrounds it. We are not our allegedly disembodied and immortal selves, but the result of what the imagination represents of the world. As this type of mind depends on objects of representation, there cannot be any human thinking without objects and these objects are its thinking tools. The body that is therefore needed by the human mind as its tool allows thinking to process all sorts of objects (images, food and the world outside). This body is also needed for the mind to operate properly in the world it constantly represents to itself. Through the imagined body and the imagination, reality becomes representable to the mind and can be acted upon by the latter. Within this Aristotelian framework reoriented by Pomponazzi's mortalism, thinking is active receptivity.

Between the thirteenth and the seventeenth centuries, another (competing) reading of the Aristotelian theory of the mind, the Averroist position, transferred the question of human knowledge to a most rarefied plane of universal objectivity by divesting the intelligible content of any link it could still have with individual human bodies and individual representational faculties.[32] In other words, for Averroes and the Latin Averroists the intellect was radically *inorganicus*, i.e., not depending on anatomical organs. Pomponazzi addressed and solved the vexed issue of the *intellectus organicus* by saying that the organ in question was the imagination, and not the physical body.[33] The organ of the mind was a body of representations. Understood as the Aristotelian *locus specierum*, the intellect was still everywhere in the human body.[34] The modality of this presence (*esse in toto corpore*), however, was radically changed. The space in question was mental as well as physical: mental, in that the imagination acted as the primary power in the cognitive construction of reality; physical, in that all sublunary reality was made of the same principal elementary forces.

To shed light on this momentous shift, Pomponazzi referred to a specific passage in *De intellectu* attributed to Alexander of Aphrodisias. In this treatise, the body as a whole, resulting from the appropriate blending of material and elemental forces, was presented as the ultimate instrument of the intellect. The following is how Pomponazzi sets out the discussion in *De immortalitate animae*:

> Rather correctly, Alexander of Aphrodisias stated that the body as a whole was the organ of the intellect (*instrumentum intellectus*), for the intellect comprehends all faculties and not a specific [material] part. The reason is that, if the latter were the case, the intellect would not be able to know all things. The same would be true of any of the sensitive faculties. Although the body as a whole is stated to be an organ of the intellect as if it were a substance, nevertheless it is not truly by being a substance that the body as a whole is to be understood, for the process of knowing through the intellect is not received in that corporeal manner.[35]

In this passage, what I found particularly revealing is that, in the middle of a crucial argument aiming at demonstrating the material nature of the human mind, Pomponazzi feels the need to refer to a specific locus in *De intellectu* attributed to Alexander of Aphrodisias. To understand the specific context of this discussion, dealing with the complicated issue of the relationship between active and potential intellects, I think it is important to read Alexander's passage from a relatively recent English translation of the original Greek text:

> Right from the first deposit of sperm, the intellect in actuality is what pervades everything and exists in actuality, as also in the case of any other body at all. Whenever it is also active through our potentiality, then this is called our intellect and it is we who think, as if one were to imagine a craftsman at one time active in his craft without instruments, but at another with instruments, too, where his activity with respect to the craft is directed towards matter. In the same manner, also the divine intellect is perpetually acting (that is why it also exists in actuality) and yet so instrumentally, whenever such an instrument comes into existence from the combination of bodies and their temperate blend. This intellect at that point engages in an act that indeed is of a material kind and this is our intellect.[36]

The general premise of the discussion is that the intellect is a universal force that pervades every single aspect of nature, including matter and bodies, actualizing all the various δυνάμεις of the cosmos. When this force actualizes the δύναμις characterizing the human species, humans become intellects of their own. The human mind, as is suggested by the author of *De intellectu*, is like a craftsman who needs instruments to implement their craft. As we have seen, in the excerpt from *De immortalitate animae* quoted above Pomponazzi says that—"rather correctly"—Alexander stated that the body as a whole should be viewed as the organ of the intellect. In doing so, Pomponazzi interprets Alexander's "material intellect" (νοῦς ὑλικός) as a corporeal faculty resulting from the blending of elemental powers (δυνάμεις). For Pomponazzi, Alexander's potential intellect is a material form, a bodily *instrumentum*, which he equates to his representational *instrumentum*, the body of representations or the *obiectum* as imagined body.

In Girolamo Bagolino's translation of *De intellectu*, published in 1516, the same year of the publication of Pomponazzi's treatise on the mortality of the soul, the passage we have quoted and commented upon can be found in Chapter 4, which Bagolino entitled "Quo pacto intellectus agens recedere dicatur et accedere ad nos," that is, "How the active intellect is said to move away from us and to come close to us."[37] Bagolino's Latin translation recognizes the difficulty faced by the author of *De intellectu* in postulating the immortality of the active intellect while also explaining that this intellect, despite its coming from the outside, wasn't changing places or moving spatially. On the other hand, the material intellect, thus called because it acts as matter for the active intellect, was said to be "engendered out of a mixed body or something similar to a mixture so that it can function as an organ (*instrumentum*) for this intellect, which is present to this mixture, since it is in every body" ("Quum igitur ex mixto corpore genitus fuerit aut aliquid eiusmodi ex mixtione ut instrumenti locum possit huic intellectui exhibere, qui huic mixtioni assistit, cum sit in omni corpore"). Here is the rest of Bagolino's Latin version, which I have translated into English:

> This body as organ is called the potential intellect, that is, a certain power that is capable of receiving the intellect in actuality. Therefore, when it apprehends this organ, then the active intellect acts through an organ, dealing with matter and through matter. This is when we are said to understand, for the intellect is made up of both the faculty that is the instrument of the divine intellect (which Aristotle calls potential intellect) and the activity of the agent intellect. When one of the two is missing, it is impossible for us to understand.[38]

The material intellect is an instrumental cognitive power in that it works as the tool of the active intellect. The human mind thinks by receiving the concepts elaborated by the active intellect using the potential intellect as its cognitive material. Just as a craftsman sometimes uses tools, sometimes not and yet the craft is acting all the same, so the active intellect uses the potential "material" intellect as a tool when it operates in nature. Within nature, the intellective functions of the human mind are part of the universal process of actualization that is relentlessly performed by the active intellect within the sphere of sublunar change.

De intellectu, commonly attributed to Alexander of Aphrodisias, had been known during the Middle Ages since the end of the twelfth century through a Latin translation by Gerard of Cremona (c.1114–1187), who in turn had translated the Arabic version by Isḥāq ibn Ḥunain (d. 910). Alain De Libera has studied the many permutations that this text underwent between the original Greek, its Arabic translation and Thomas Aquinas's appropriation.[39] Pomponazzi was probably well acquainted with Bagolino's version, but it seems likely that in *De immortalitate animae* he was referring to Gerard's translation. One way or another, he extracted a powerful argument from the text in question. Throughout the discussion, when the functions of the active and the material intellects are examined from the point of view of the human mind, they appear to belong to different systems within the universe of human knowledge. The material intellect provides the conditions for the mental object to be thought through, as knowledge cannot simply be reduced to a self-reflective act of the intellect devoid of any content. Every time the human mind exercises its thinking function, an object is also involved, and this object is, generally speaking, matter as the domain of the universally thinkable. By contrast, when the discussion is seen from the point of view of nature, thinkable matter is both the representable *obiectum* and the very possibility that this matter be identified with the body of nature. Alexander qualified the receptive intellect as ὑλικός ('material') to clarify in analogical terms the speculative core of the original analogy introduced by Aristotle in the *De anima* III, 4–5 (429a–430a). Just as matter's activity in nature corresponds to its suitability for being acted upon, so in the mental universe the *obiectum* is the representational platform which makes the intelligible products of the intellect thinkable. As a receptor, matter is the material, that is, the universally pliable medium needed for any activity to be fulfilled. Cognitively speaking, the matter of a concept is its disposition to be thought. Thinkability is therefore the counterpart of materiality. Matter is active insofar as it is receptive, that is, insofar as it has the ability to be affected and changed by an object. In his own treatise on the soul, *De anima*, Alexander had closely connected the notion of matter with that of receptivity: πᾶν γὰρ τὸ δεκτικόν τινος ὕλη ἐκείνου, that is, "everything that is capable of receiving something is the matter of that something."[40] In order for matter to be the ultimate receptor of all things it cannot have a nature of its own. Matter is the absolutely other that can become everything. Receptivity means a complete disposition to assimilated.

In this way, through the history of the medieval and Renaissance reception of Alexander's theory of the mind, the metaphor of the material intellect comes full circle and, in the end, appears to recover the gist of the literal meaning. Wood is the material used to make things. Wood (ξύλον) is therefore matter (ὕλη). Matter, in turn, is the universal stuff of which all things are made up (ὑποκείμενον).[41] De Libera has captured this aspect of the discussion when he stresses "the paradoxical ontological status of Alexander's material intellect." This intellect is "neither thing, nor being [in actuality], but power or receptivity."[42] Regardless of whether the metaphysical framework underlying *De intellectu* is of a materialistic cast or not, De Libera points out how the Middle Ages attributed "the materialist theory of the material intellect" to Alexander, and this resulted in "an Arcimboldian theory which Averroes criticized and the Middle Ages largely appropriated."[43] And so it happened that the Alexandrian notion of material intellect became a metaphysical haven through which philosophers and physicians could smuggle concepts of a highly sensitive nature regarding matter, the imagination and the changeable fit between knowledge and reality. In this story, Pomponazzi had a central role. Against the background of Averroist and Alexandrian exegetic traditions, he scaled down the cosmological scope of the intellect while dramatically enhancing the cognitive focus of the imagined body, the *obiectum* of the extendable mind.

I argue that the growing attention paid to the possibility of natural organs of knowledge and action can be seen as an integral part of this evolution. The most disquieting meaning of natural organs is the one according to which matter in itself is capable of knowledge production and knowledge transfer, for it has within itself the technological resources to change and modify itself. This is when the Aristotelian metaphor of matter as the pliant platform of all thinkable reality within the human mind turns itself into a self-active organ and then into living matter. In this way, the boundaries of naturalness and artificiality collapse entirely. Today natural organs of knowledge and action are the target of distinctive research programmes with different goals in view: the clone and the robot, the virus and the algorithm, the biological and the computational approaches to the making of artificial intelligence—in other words, ever more complex syntheses of nature and artifice. These programmes are likely to converge into something in the middle paving the way for all sorts of hybrids of neural and cyber thinking.[44] It is here that, ironically, Aristotle comes back to life with his view of τέχνη as biomimicry. In the next and final section, I will refer to the philosophical paradigm of the extended mind to further substantiate my claim that there can be ways of engaging in the contemporary debate by drawing on early modern discussions regarding thinking tools and natural organs of knowledge and action.

Extended Minds and Natural Organs: Makers, Receptors, Projectors?

> Even so, life isn't done with; there are a million patient, watchful lives still for a tree, all over the world, in bedrooms, in ships, on the pavement, lining rooms where men and women sit after tea, smoking cigarettes. It is full of peaceful thoughts, happy thoughts, this tree.
> Virginia Woolf, *The Mark on the Wall*, 1917

During the Middle Ages and the early modern period, one of the ways in which philosophers and physicians addressed the issues of the ontological status of the mind and its extendable nature was by asking whether, in order to perform its mental operations, the intellect needed anatomical organs (*intellectus organicus*) or not (*intellectus inorganicus*). At the end of this survey of mine, which has been recklessly short for limits of space, we reach the point in which the mind is confronted with the ultimate polar tension created by the interplay of literal and metaphorical meanings, the most radical of which are active matter and extended mind. The problem with the literal meaning of "organ" is that the reality of living and thinking matter is brutal and perhaps too crude. The problem with its metaphorical sense is that the human mind may stretch too thinly until it disappears into thin air.

As hinted at the beginning of this chapter while expanding on the notion of Renaissance thinking tools, the early modern emphasis on elaborating the correct method of knowledge inaugurated a powerful philosophical movement concerned with a new technology of thinking. From Bacon's *machina intellectus* to Spinoza's *automa spirituale*, thinking tools were regarded as essential components in one's search after the truth.[45] The seventeenth-century dictionary of philosophy by Rudolph Goclenius records this momentous conceptual shift with its usual clarity and scholarly accuracy. Organs, be they parts of the body or bodies themselves, are tools of the mind. The entry "Organicum" consists of three parts. In the first, Goclenius establishes a series of analogies between bodies, tools and matter:

> That which is organic is the same as that which is instrumental, or that which is provided with an organ or uses an organ. Such are the organic parts, which Zabarella somewhere calls matter and Aristotle in *De anima* calls bodies.[46]

In the second part, through Averroes who interprets the important locus of *De intellectu* (which we have commented upon in the previous section), Goclenius connects minds with tools and advances the hypothesis that, in a way, the intellect may be considered to act 'organically':

> All natural bodies, that is, all bodily organic parts, are tools of the soul. The intellect is said to be inorganic, as it does not have a specific organ which it uses. The intellect, however, can be called organic because all the organs of the organic faculties serve and refer to it. In a remote sense, therefore, that is, through the mediation of the lower powers, the intellect uses all the organs. In his commentary on Book 3 of Aristotle's *De anima*, Averroes states that the human body taken as a whole is the tool of the intellect, for the intellect rules over all the parts and faculties of a human being and over the human body as a whole. Therefore, the intellect uses organs, not by itself, but through the lower powers.[47]

In the third section of the entry, finally, Goclenius reiterates the importance of thinking tools for the practice of philosophy: "Logic is called organic knowledge because it is the tool of philosophy."[48] That is to say, the human mind thinks 'organically'.

Goclenius's entry is valuable on several counts. It confirms the view that instrumentality is an integral part of the enterprise of knowledge at a time in which, especially through the technology of the printing press, a whole host of new organs were seen as necessary for the handling and dissemination of knowledge. This type of instrumentality can also be called materiality, for thinking needs materials for implementing its activity. Even the Averroist intellect, the most abstract and incorporeal of all the models of intelligible life that had been worked out within the Aristotelian tradition, uses human bodies to harvest its representations of reality. Logic, finally, being the specific discipline that is committed with the task of managing human λόγος, is "organic" because thinking is organized hierarchically according to divisions that involve nature, the intellect and human science.

The notoriously disquieting image of the Averroist universal intellect using human beings as automata capable of releasing representations of reality (*obiecta*) pops up in Goclenius's entry and offers us a glimpse of how contested the divide between natural spontaneity and mechanical control still was at the beginning of the seventeenth century. As already mentioned, in Book 2 of *Physics*, while discussing the interplay of life and technology, Aristotle had distinguished between natural and man-made things. A recurring example was the difference between a tree (δένδρον), wood (ξύλον) and a bedstead (κλίνη). While wood is the matter resulting from the vital coalescence of the four elementary δυνάμεις (heat, cold, moisture and dryness), the tree is a natural being and the bedstead is the product of human τέχνη. In this case, too, the material of which the bedstead is made up continues to have the natural principles of change within itself, so much so that Aristotle did not rule out the possibility that a buried bedstead may revert to being a tree. This outcome can be used as evidence that in fact an artificial object remains natural throughout. Aristotle mentioned the fifth-century BC philosopher Antiphon in the course of his argument:

> Antiphon took it as an indication of this that if a man buried a bedstead and the sap in it took force and threw out a shoot it would be tree and not bedstead that comes up, since the artificial arrangement of the material of the craftsman is merely an incident that has occurred to it, whereas its essential and natural quality is to be found in that which persists continuously throughout such experiences.[49]

Apparently, Antiphon had explained that any correct use of names depended on their denoting actual natural kinds.[50] Today, the progress of robotics allows us to think of the possibility of animate beds being produced in the future, further eroding the boundaries dividing natural from artificial beings. Meanwhile, the category of sentience has expanded to the point of including natural beings previously deemed to be unable to perceive. For Aristotle, a tree was alive but not self-moving nor sentient. Today, by contrast, trees are seen as moving and intelligent beings, capable of communicating and establishing thriving communities in the form of forests and subterraneous networks of mycorrhizal conglomerations. The three old pines I mentioned at the beginning of this chapter had been striking the most elaborate poses for a century, year after year, to react to the blow of the winds and to find the most perfect balance on earth with the strong grip of their roots. According to the anthropologist Eduardo Kohn, taken as forming forests, trees are thinking beings.[51]

One step forward, and the dualism of natural and human-made productions can be integrated into a broader view in which knowledge belongs to nature first and to human beings only secondarily. While human thinking tools, as we noted with Pomponazzi, shift the focus from consciousness to the *obiectum*, that is, the imagined world, the natural organs of knowledge and action re-establish the correct proportion between human minds and the surrounding world. To introduce a few concluding remarks, I suggest that, in their most radical meaning, the natural organs of knowledge and action re-assign to matter the power to organize itself and produce its own instruments of knowledge, including the human minds.

Let's resume the question posed by Clark and Chalmers in the late 1990s: Where does the mind stop and the rest of the world begins? In the article mentioned at the beginning of this chapter, the two authors suggested that our cognitive processes should be seen as unremittingly extending into the environment. How should we describe this extension into the environment? The question is all the more urgent because, more than two decades after the appearance of the article, the issue of the relationship between the human mind and the material environment has become way more delicate and precarious. Is it appropriate to promote a model of thinking in which the world is progressively being colonized by the human mind? Can nature be seen as the world of the imagination that humans construct by applying their minds to the interface of the imagined body, the *obiectum*? Clark and Chalmers presupposed the coupling of "biological organism and external resources."[52] Indeed, this coupling promised a refreshing expansion of what it is like for humans to be thinking beings:

> if the view is taken seriously, certain forms of social activity might be reconceived as less akin to communication and action, and as more akin to thought. In any case, once the hegemony of skin and skull is usurped, we may be able to see ourselves more truly as creatures of the world.[53]

To be sure, Pomponazzi would have agreed with Clark and Chalmers that making the mind dependent on the body as *subiectum* (skin and skull) rather than on the body as *obiectum* (extendable mind) is tantamount to sanctioning the hegemony of physical brain and its neural appendages. This kind of hegemony also means superiority of blood and soil, of uncompromising literalism and monolithic identities, a whole reservoir of questionable assumptions that is no longer affordable for the mind to entertain. It's not simply that they are not plausible; they are immensely costly. On the contrary, to understand cognition as being "continuous with processes in the environment" expands the range of the mind without yielding to egotistic and self-deluded demands. Here language takes on a key role, by becoming "a central means by which cognitive processes are extended into the world:"[54]

> Without language, we might be much more akin to discrete Cartesian "inner" minds, in which high-level cognition relies largely on internal resources. But the advent of language has allowed us to spread this burden into the world. Language, thus construed, is not a mirror of our inner states, but a complement to them. It serves as a tool whose role is to extend cognition in ways that on-board devices cannot. Indeed, it may be that the intellectual explosion in recent evolutionary time is due as much to this linguistically-enabled extension of cognition as to any independent development in our inner cognitive resources.[55]

In this case, too, I would like to involve Pomponazzi in this discussion, and to the list of resources underpinning this intellectual explosion I would add the extraordinary effects that the imagination and its fictive powers produce in the human mind, especially when they are combined with linguistic assets. This vast expansion of the representational domain of the *obiectum*, however, demands that we renegotiate the boundaries that separate the human mind from the external world. Compared to the 1990s, the burden that we were supposed to spread into the world by thinking through languages and imaginations, far from providing stretchable and inspiring visions of things, revivifying flows of information and establishing symbiotic integrations of neural and computational systems, has taken the form of unrecyclable wreckage, polluting emissions and a drastic decline of biodiversity. It is now evident that, in fact, codes and images can have a seriously destructive impact on our lives. This means that mind-expansion should not be seen as yet another form of anthropo-expansion, simply more sophisticated. This kind of expansion should be saluted as a magnification of representability in its cosmic sense and not as the next ideological ruse (think of sustainable growth) to violently encroach on all that is not a human mind. Thinking can only be expansive insofar as it is communicative and not exploitative. Any expansion of thinking is first of all an expansion of life. As such, it's nature's tentative perception transcending the constraints of individuated bodies and egoic selves.

It may look clumsy and quaint to combine in one discussion extended minds, early modern thinking tools and the Aristotelian notion of τέχνη, but the recovery of essential elements of Aristotle's philosophy adds to the contemporary debate in significant ways. Aristotle made the epoch-making decision to connect the spheres of life and thinking to the domain of tools (ὄργανα) and tool use. As I have argued in this chapter, Aristotelian tooling had remarkably broad implications. A tool could be a logical apparatus (a method to connect more easily to the life of λόγος), but it also could be a slave (a human being considered to be constitutively unable to use λόγος to control his or her appetitive life). Aristotle argued that life required divisions and shapes to emerge out of homogeneous matter. Likewise, for minds to be operative, they needed distinctions and schemata to punctuate the otherwise fluid process of thinking and desire. But the mixing of shapes with schemata was the fateful act that originally engendered the ambivalent relation of life and machine. Because of the Aristotelian belief in a fundamental proximity of reality and definition, the organs of life (the shaped body) and the organs of the mind (the schematic body of knowledge) can turn into apparatuses of cognitive exploitation and power control. In different degrees and different ways, in Aristotle's view plants, children, women and slaves are not fully rational and are unable to control their impulses because they are ill defined. Here is where the natural organs of knowledge and action, issuing as it were from the back door of Aristotle's metaphysics, can reveal their unexpected revolutionary potential: they give us the speculative resources to think of the possibility of self-active tools. The fact, however, is that self-active tools, to avoid being contradictory entities, can only work within the context of a universe in which every part of it is alive and sentient and all parts function together as organs of the whole. In other words, self-active tools presuppose active matter.

As I am adding the final touches to this chapter, the city gardeners of Civitanova are raising in the middle of the main piazza a towering Christmas tree previously felled somewhere by human beings—a *pinus sylvestris*. Cutting trees of this tallness and magnificence in 2024 is both fatuous and cruel. While in the summer the *pini pinastrae* that have been shaped by their century-old reactions to the sea winds are blithely cut down because they might kill children, in the winter it is the *pini sylvestres* that are cut down because they have to entertain children. We are dealing with an infantilized humanity at a time when childhood should in fact help us recover the true sense of an imagined future. One might certainly claim that cutting a tree for Christmas will not accelerate climate change. And yet to change our mind and mental habits about the life of trees is now essential if we want to preserve the rest of the world.

Notes

1. On winds as organs of natural and cultural transformations, see Lyall Watson, *Heaven's Breath: A Natural History of the Wind* (New York: New York Review Books 2019 [1984]); Tim Robinson, *Listening to the Wind* (Minneapolis: Milkweed Editions, 2019 [2006]). Francis Bacon devoted one of his natural histories to the study of the winds and their power to change things. There he noted that, "in locis proximis iuxta mare, flectunt fere se arbores et incurvant, quasi aversantes auras marinas. Neque tamen malicia est; sed venti maritimi, ob humiditatem et densitatem, sunt tanquam ponderosiores." See Bacon, *Historia ventorum* (1622), in *Works*, ed. James Spedding, Robert Leslie Ellis and Douglas Denon Heath, 14 vols. (London: Longman, 1857–1874), II, 19–78 (33).
2. On the sentient life of trees, the literature is now immense. Here I only refer to a few significant studies in both the science and philosophy of plants: Matthew Hall, *Plants as Persons: A Philosophical Botany* (Albany: State University of New York Press, 2011); Daniel Chamovitz, *What a Plant Knows: A Field Guide to the Senses* (New York: Scientific American / Farrar, Straus and Giroux, 2012); Emanuele Coccia, *The Life of Plants: A Metaphysics of Mixture*, tr. Dylan J. Montanari (Cambridge: Polity, 2018 [2016]); Stefano Mancuso, *Botanica* (Sansepolcro: Aboca 2021 [2017]); Paco Calvo, *Planta Sapiens: Unmasking Plant Intelligence* (London: The Bridge Street Press, 2022). Mancuso, who is one of the leading lights in the field of plant neurobiology, has recently published a novel on the lives of trees: *La tribù degli alberi* (Turin: Einaudi, 2022). The two woodcutters of Civitanova Marche are probably laughing about it.
3. Pietro Pomponazzi, *De incantationibus*, ed. Vittoria Perrone Compagni (Florence: Olschki, 2011), 24–25. See, in particular, 24: "Quare nihil inconvenit si idea, quae est in mente nostra, quae est species, producat aliquando suum ideatum secundum esse reale mediantibus instrumentis corruptibilibus, quae sunt spiritus et sanguis, ubi passum fuerit dispositum." *Imaginationes*, in this view, are organs of a second order, whose correct functioning is assured by organs of the first order, such as blood and vital transmitters (*spiritus*).

4. The idea of imaginations as states of heightened perceptual energy also has remote Aristotelian origins. See *De motu animaliuum*, 7, 701b. I have discussed the possibility of materialistic outcomes issuing from the tradition of Aristotelian naturalism, especially in early modern medicine and biology, in Guido Giglioni, "Scaliger Bacon Harvey Glisson: A Trajectory in the Early Modern History of Vegetative Life," in *Mechanism, Life and Mind in Modern Natural Philosophy*, ed. Charles T. Wolfe, Paolo Pecere and Antonio Clericuzio (Cham: Springer, 2022), 11–25.
5. Andy Clark and David Chalmers, "The Extended Mind", *Analysis*, 58 (1998): 7–19. Now in Andy Clark, *Supersizing the Mind: Embodiment, Action, and Cognitive Extension* (Oxford: Oxford University Press, 2011), 220–232.
6. Among the sixteenth-century authors who theorized the centrality of thinking tools within the universe of mental expansion is Charles de Bovelles. See Guido Giglioni, "La natura impara a leggere e scrivere: Mente, mano e mondo nel *Liber de sensu di* Charles de Bovelles," *Bruniana et Campanelliana*, 24 (2018): 447–460. On Renaissance concepts of thinking tools, see Dominique Couzinet and Guido Giglioni, "How to Collect and Store Information in the Early Modern Period: From Learning to Knowledge," *Bruniana et Campanelliana*, 26 (2020): 469–474. The topic of early modern thinking tools has also been the subject of a recent colloquium in Paris entitled "Old and New Organons: Technologies of Thinking between the Sixteenth and the Seventeenth Centuries" (27–28 June 2022), organized by Couzinet and Giglioni. The organizers are working on a volume which will collect the proceedings of the two-day conference.
7. After the groundbreaking article published in 1998, Clark and Chalmers have been developing the idea of the extended mind in various directions. See Clark, *Supersizing the Mind*; id., *Surfing Uncertainty: Prediction, Action and the Embodied Mind* (Oxford: Oxford University Press, 2016); David J. Chalmers, *Reality + : Virtual Worlds and the Problems of Philosophy* (London: Allen Lane, 2022). Thinking outside the brain is the theme of Annie Murphy Paul's *The Extended Mind* (Boston and New York: Houghton Mifflin Harcourt, 2021), which provides a synthesis of the principal features of the theory.

8. See Eduardo Viveiros de Castro, *Cosmological Perspectivism in Amazonia and Elsewhere* (Manchester: HAU Network of Ethnographic Theory, 2012); id, *Cannibal Metaphysics*, ed. and tr. Peter Skafish (Minneapolis: University of Minnesota Press 2014 [2009]); Philippe Descola, *Beyond Nature and Culture*, tr. Janet Lloyd (Chicago: The University of Chicago Press, 2013 [2005]); Eduardo Kohn, *How Forests Think: Toward an Anthropology beyond the Human* (Berkeley and Los Angeles: University of California Press, 2013).
9. See Guido Giglioni, "Lire, penser, vivre: Les interactions entre *inductio* et *instauratio* dans le *Novum Organum* de Francis Bacon," *Archives de Philosophie*, 84 (2021): 17–31.
10. Especially in *Surfing Uncertainty*, Andy Clark has shown the extent to which the expansion of mental powers is today intertwined with the developments of artificial intelligence, robotics and neuroscience.
11. See Jacopo Zabarella, *Opera logica* (Venice: Paolo Meietti, 1586 [1578]), 4–20.
12. See Daniel C. Dennett, *Intuition Pumps and Other Tools for Thinking* (London: Allen Lane, 2013).
13. Here I only mention one place in Bacon's oeuvre out of many in which the *vis insidiatrix* of the mind and its *mala complexio* are described as the source of all sorts of human failures. See Francis Bacon, *Novum organum*, in *The* Inastauratio Magna *Part II*: Novum Organum *and Associated Texts*, ed. Graham Rees with Maria Wakely (Oxford: Clarendon Press, 2004), 34.
14. See Guido Giglioni, "Medicine of the Mind in Early Modern Philosophy," in *The Routledge Handbook of the Stoic Tradition*, ed. John Sellars (London: Routledge, 2016), 189–203; id., "Estranged Minds and Their Corpuscles: Bacon, Helmont and Locke on Dysfunctional Thinking," *Studi lockiani*, 4 (2023): 153–182.
15. Bacon, *Novum organum*, 200–208. See also Giglioni, "Lire, penser, vivre," op. cit., 28. On Bacon's notion of hyperinduction (*superinductio*), see Guido Giglioni, *La saggezza dell'induzione: Come Francis Bacon rilegge e riscrive il libro della natura* (Pistoia: Metilene, 2015), 103–107.
16. Aristotle, *Metaphysics*, ed. and tr. Hugh Tredennick, 2 vols. (London and Cambridge, MA: William Heinemann and Harvard

University Press, 1935), II, 129 (with slight changes to the translation).

17. Aristotle, *Physics*, ed. and tr. Philip H. Wicksteed and Francis M. Cornford, 2 vols. (London and Cambridge, MA: William Heinemann and Harvard University Press, 1957 [1929]), I, 109 (192b).
18. Francis Bacon, *Advancement of Learning*, ed. Michael Kiernan (Oxford: Clarendon Press, 2000), 50.
19. Francis Bacon, *De augmentis scientiarum*, in *Works*, I, 481: "Literatus contra non tantum utitur animo virtutesque exercet, sed continuo emendat se et in virtutem proficit" (unless otherwise indicated, all translations from Greek and Latin are mine).
20. Aristotle, *Eudemian Ethics*, ed. and tr. H. Rackham (London and Cambridge, MA: William Heinemann and Harvard University Press, 1935), 467 (with slight changes to the English translation).
21. In his excellent article on Aristotle's view on slaves and women, Holt N. Parker shows how logically defective Aristotle's argument of natural slavery is. See Parker, "Aristotle's Unanswered Questions: Women and Slaves in Politics 1252a1260b," *EuGeStA*, 2 (2012): 71–122.
22. Aristotle, *Generation of Animals*, ed. and tr. A. L. Peck (London and Cambridge, MA: William Heinemann and Harvard University Press, 1942), 9: "Ἔστι δὲ καὶ ἐπὶ τῶν φυτῶν τὸν αὐτὸν τρόπον· τὰ μὲν γὰρ ἐκ σπέρματος γίγνεται τὰ δ' ὥσπερ αὐτοματιζούσης τῆς φύσεως. γίγνεται γὰρ ἢ τῆς γῆς σηπομένης ἢ μορίων τινῶν ἐν τοῖς φυτοῖς· ἔνια γὰρ αὐτὰ μὲν οὐ συνίσταται καθ' αὑτὰ χωρίς, ἐν ἑτέροις δ' ἐγγίγνεται δένδρεσιν οἷον ὁ ἰξός."
23. Aristotle, *On Divination during Sleep*, 463b 12–15: "Ὅλως δὲ ἐπεὶ καὶ τῶν ἄλλων ζῴων ὀνειρώττει τινά, θεόπεμπτα μὲν οὐκ ἂν εἴη τὰ ἐνύπνια, οὐδὲ γέγονε τούτου χάριν, δαιμόνια μέντοι· ἡ γὰρ φύσις δαιμονία, ἀλλ' οὐ θεία."
24. See Aristotle, *Magna moralia*, 1207a: "Ἔστιν οὖν ἡ εὐτυχία ἄλογος φύσις."
25. See Philip van der Eijk, "Divine Movement and Human Nature in *Eudemian Ethics* 8, 2," *Hermes*, 117 (1989): 24–42 (31).
26. I would like to thank Dominique Couzinet, Michele Merlicco and Leonardo Graciotti for several inspiring conversations during which we have been exchanging ideas on such issues as living matter,

thinking tools, thinking objects and natural organs of knowledge and action.

27. On the kind of vital materialism grounded in radical Aristotelianism which is characteristic of Glisson's philosophy, see Guido Giglioni, "Sentient Nature and the Great Paradox of Early Modern Philosophy: How William Harvey and Francis Glisson Reinterpreted Aristotelian ΦΥΣΙΣ," in *Natureza, causalidade e formas de corporeidade*, ed. Adelino Cardoso, Marta Mendonça and Manuel Silvério Marques (Ribeirão: Húmus, 2016), 9–28; id., "Francis Glisson and the IrritableLife ofNature," in *Encyclopedia of Early Modern Philosophy and the Sciences*, ed. Dana Jalobeanu and Charles T. Wolfe (Cham: Springer, 2020) < https://doi.org/10.1007/978-3-319-20791-9_620-1 > ..
28. On the return of Alexander of Aphrodisias in the Renaissance, see F. Edward Cranz, "Alexander Aphrodisiensis," in *Catalogus Translationum et Commentariorum*, ed. P. O. Kristeller, 1 (1960), 77–135; Eckhard Kessler, "Alexander of Aphrodisias and His Doctrine of the Soul: 1400 Years of Lasting Significance," *Early Science and Medicine*, 16 (2011): 1–93.
29. For a recent general account, see Leen Spruit, "The Pomponazzi Affair: The Controversy over the Immortality of the Soul," in *The Routledge Companion to Sixteenth Century Philosophy*, ed. Henrik Lagerlund and Benjamin Hill (London: Routledge, 2017), 225–246.
30. Pietro Pomponazzi, *Tractatus de immortalitate animae*, ed. and tr. Thierry Gontier (Paris: Les Belles Lettres, 2012), 119.
31. Ibid., 31–33, 89.
32. See Guido Giglioni, "Accidental Intellects: Pomponazzi on Human Imagination, Body and Mortality," in *Tradizione e illuminismo in Uriel da Costa: Fonti, temi, questioni dell'*Exame das tradições phariseas, ed. Omero Proietti and Giovanni Licata (Macerata: eum, 2016), 121–150 (147–148).
33. Pomponazzi, *Tractatus de immortalitate animae*, op. cit., 121: "ipsum [intellectum] indigere eo [organo] ut obiecto, non ut subiecto."
34. Ibid.: "quoniam intellectus humanus est in materia quasi per quandam concomitantiam, et ipsum intelligere quodam modo est

in materia, sed satis accidentaliter, quoniam intellectui, qua intellectus est, accidit esse in materia; non tamen in aliqua parte corporis ponitur ipsum intelligere, sed in toto cathegorematice sumpto."

35. Ibid., 121–123: "Non inconvenienter igitur Alexander posuit totum corpus esse instrumentum intellectus, quoniam intellectus omnes vires comprehendit, et non aliquam partem determinatam; quoniam sic non omnia cognosceret, sicut neque aliqua virtutum sensitivarum. Quamquam autem sic totum corpus ponatur instrumentum intellectus quasi ut subiectum, non tamen est vere ut subiectum, quoniam intelligere non recipitur eo modo corporali."
36. [Alexander of Aphrodisias], *De intellectu*, in *Two Greek Aristotelian Commentators on the Intellect*, ed. Frederic M. Schroeder and Robert B. Todd (Toronto: Pontifical Institute of Mediaeval Studies, 1990), 56.
37. Alexander of Aphrodisias, *Libellus de intellectu*, ed. and tr. Girolamo Bagolino (Verona: Andrea B. C., 1516), sig. F^{rv}: "Sed cum vellet intellectum immortalem ostendere et evitare dubitationes quas afferunt de intellectu extrinsecus invecto, qui cogitur locum permutare, cum non possit ex quo incorporeus esse nec in loco esse nec transgredi et modo ad hunc modo ad illum accedere, ex propria sententia dixit haec de intellectu. In quocunque ipsum esse quod dicitur mortale corpus. Dixit quoque intellectum esse in ipsa materia ut substantiam in substantia. Et actu semper suis actionibus fungi. Quum igitur ex mixto corpore genitus fuerit aut aliquid eiusmodi ex mixtione ut instrumenti locum possit huic intellectui exhibere, qui huic mixtioni assistit, cum sit in omni corpore."
38. Ibid.: "Corpus vero et hoc ipsum est, hoc inquam instrumentum dicitur intellectus potestatis, potestas scilicet quaedam idonea ad suscipiendum intellectum in actu. Quum itaque hoc instrumentum apprehenderit, tunc ut per organum et ut circa materiam et per materiam operatus est. Et tunc dicimur nos intelligere. Nam intellectus compositus est et ex potestate quae est instrumentum intellectus divini quem intellectum potestatis Aristoteles vocat et illius operatione. Quorum cum alterum desit, impossibile est nos intelligere."
39. On the philological and exegetical complexities of *De intellectu* attributed to Alexander of Aphrodisias (a "*blob* théorique"), see Alain De Libera, *L'unité de l'intellect: Commentaire du* De unitate

intellectus contra averroistas *de Thomas d'Aquin* (Paris: Vrin, 2004), 174–175.

40. Alexander of Aphrodisias, *De anima*, in *De anima liber cum mantissa* ed. Ivo Bruns (Berlin: Reimer, 1887), 81. On materiality as receptivity in Alexander's theory of the intellect, see Miira Tuominen, "Receptive Reason: Alexander of Aphrodisias on Material Intellect," *Phronesis*, 55 (2010): 170–190.
41. See Aristotle, *Physics*, 133 (195a).
42. De Libera, *L'unité de l'intellect*, op. cit., 170.
43. Ibid., 175.
44. Recently, Thierry Hoquet has explored this emerging universe—"au-delà et en deçà" of the human boundary—in his *Les Presque humains: Mutants, cyborgs, robots, zombies … et nous* (Paris: Seuil, 2021).
45. See Giglioni, *Medicine of the Mind in Early Modern Philosophy*, 197–199.
46. Rudolph Goclenius, *Lexicon philosophicum quo tanquam clave philosophiae fores aperiuntur* (Frankfurt am Main: Widow of Matthias Becker, 1613), 281: "Est idem quod instrumentarium seu instrumentale, vel habens organon, utens organo, ut partes organicae, quae Zabarellae alicubi veniunt nomine materiae, et Aristoteli nomine corporum ipsorum 2. de anima text. 37." Here Goclenius may be thinking of Iacopo Zabarella, *In tres Aristotelis libros De anima commentarii* (Venice: Francesco Bolzetta, 1605), 86^{v}b: "Sic igitur natura tanquam primarium agens facit materiam, vel saltem dispositiones materiales propter receptionem formae, ut in viventibus facit organicas partes propter animam, tanquam propter finem, qui movit quidem naturam, sed est causa finalis ipsius naturae, id est partium organicarum."
47. Goclenius, *Lexicon philosophicum*, 281–282: "Omnia naturalia corpora animae instrumenta sunt, id est, omnes partes corporeae organicae. Intellectus dicitur inorganicus, qui non habet proprium organon, quo utatur. Potest tamen intellectus dici organicus, quia ei inserviunt et ministrant omnia organa potentiarum organicarum. Remote igitur, id est, per potentias inferiores, utitur organis omnibus. Averroes commento 5. tertio libro ait quod totum hominis corpus instrumentum sit ipsius intellectus, cum ipse hominis omnibus partibus et facultatibus, et hominis toti corpori imperet, et ita instrumentis utatur, non per seipsum,

sed per potentias inferiores." See also Averroes, *Commentarium magnum in Aristotelis De anima libros*, ed. F. Stuart Crawford (Cambridge, MA: The Mediaeval Academy of America, 1953), 393–398. The thesis that "all natural bodies as it were are tools of the soul" (*omnia naturalia corpora sunt quasi instrumenta animae*) is discussed by Thomas Aquinas in his *Sentencia libri De anima*, Liber II, Lectio 7, n. 14: < https://www.corpusthomisticum.org/can2.html > .

48. Goclenius, *Lexicon philosophicum*, 282: "Dialectica dicitur doctrina organica, quia est instrumentum philosophiae." The entry "Instrumentum" (p. 246) in the same dictionary emphasizes the meanings of receptivity and divine transmission. See also Rudolph Goclenius' entry "Organicum" in *Lexicon philosophicum Graecum, opus sane omnibus philosophiae alumnis valde necessarium cum perspicientia philosophysici sermonis plurimum etiam ad cognitionem rerum utile* (Marburg: Rudolf Hutwelcker and Petrus Musculus, 1615), 155: "Organicum est quod habet organon seu partes instrumentarias, vel quod fit organo ut causa ministeriali."
49. Aristotle, *Physics*, 111 (193a).
50. On Antiphon's position, see J. S. Morrison, "The Truth of Antiphon," *Phronesis*, 8 (1963): 35–49 (esp. 43–45); Martin Ostwald, *Nomos and Phusis in Antiphon's Περὶ Ἀληθείας*, in *Cabinet of the Muses: Essays on Classical and Comparative Literature in Honor of Thomas G. Rosenmeyer*, ed. Mark Griffith and Donald J. Mastronarde (Atlanta: Scholars Press, 1990), 293–306.
51. Kohn, *How Forests Think.*
52. Clark and Chalmers, "The Extended Mind," op. cit., 18.
53. Ibid., 18.
54. Ibid., 10–11. In *Supersizing the Mind*, Clark defines language as "a form of mind-transforming cognitive scaffolding" which controls and guides "the shape and contents of our own thinking" (44).
55. Clark and Chalmers, "The Extended Mind," p. 18.

CHAPTER 11

A Transhuman Ontology: Spinoza and the Good Change

Emanuele Costa

Introduction

The primary concern of this essay will be to establish a metaphysical groundwork for the exploration of the notion of "human nature", crucial for the assessment of the transhumanist project as one seeking to "recogniz[e] and anticipat[e] the radical alterations in the nature and possibilities of our lives resulting from various sciences and technologies".[1] The passing reference to the "nature of human lives" expressed in this manifesto will therefore be taken into serious consideration from a metaphysical and ontological point of view, in order to assess the philosophical tenability of an ethical project stemming from the projected transhumanist metaphysics.

In the first section, I will propose that an ideal source for the construction of such transhumanist metaphysics is the work of Baruch Spinoza. His *Ethics* put forward an ontological framework for human nature based

E. Costa (✉)
Vanderbilt University, Nashville, TN, USA
e-mail: emanuele.costa@Vanderbilt.Edu

J. Muller (ed.), *Automata, Cyborgs, and Mutants*, Palgrave Studies in Medieval and Early Modern Medicine,
https://doi.org/10.1007/978-3-031-93201-4_11

on the mereological consideration of the parts and elements that compose the human body, which I aim to reconstruct in its fundamental bare bones. In the second section of the essay, I shall move to demonstrate how this Spinozist understanding of human nature can ground a transhuman approach to change and alteration that is capable of withstanding most anthropocentric criticism on both ethical and metaphysical fronts. Finally, in the third section of the essay, the Spinozist-transhuman framework developed through the piece will be put to test against the criticism leveled by Fukuyama in his 2009 op-ed *Transhumanism*, namely that transhumanism is misguided in its assessment of fundamental human goods.[2] My responses to Fukuyama's criticisms will be grounded mainly on anti-anthropomorphic foundations, expounding the Spinozist suggestion of mereological change as the metaphysical underpinning of human nature.

Spinoza's Metaphysics of Individuals

The core of Spinoza's metaphysics of bodies and individuals is presented in a central and clearly demarcated section of the *Ethics*, which is commonly known as the *Physical Digression*. This section aims at clarifying the nature and structure of the human body. Once Spinoza has established that mind and body are parallel and reciprocally mimetic entities, he can therefore claim that the study of the body can be beneficial to exploring the structure of the mind as well.[3]

Spinoza begins his analysis by establishing that "bodies are distinguished from one another by reason of motion and rest, speed and slowness, and not by reason of substance".[4] By refusing to assign any sort of substantiality to individual bodies, Spinoza outsources the establishment of their numerical identity to relational notions, such as motion and rest, speed and slowness, which—as we will see briefly—are subject to transmission and modification, as long as they remain within certain parameters. In other words, a body (an "Individual", as Spinoza calls it shortly thereafter) is what it is depending on the arrangement and equilibrium of its components. The paradigmatic definition of individual, for Spinoza, is on display "when a number of bodies [...] communicate their motions to each other in a certain fixed manner, we shall say that those bodies are united with one another and that they all together compose one body or Individual, which is distinguished from the others by this union of bodies".[5] The crucial notion, here, is that of a "certain fixed manner", which Spinoza calls *ratio*. The *ratio*, the proportion of movement and harmony that unites the various components of the body, is identified by Spinoza as the true essence of the individual.

By contrast, Spinoza is not committed to a specific number, or nature, or shape of the components as essential for the retainment of the essence of the individual. "If, of a body, or of an Individual, which is composed of a number of bodies, some are removed, and at the same time as many others of the same nature take their place, the body, or the Individual will retain its nature, as before, without any change of its form".[6] The same happens in the case of an enlargement, a restriction, a multiplication, a division, and any other sort of transformation that the body or individual may undergo.[7] The crucial requisite, for Spinoza, is the preservation of the same *ratio*, the same internal equilibrium, the same balance between the parts. For Spinoza, the so-called puzzle of Theseus' Ship would not occur in the first place.[8] As long as the *ratio* of the ship is maintained in the same state, no matter how many of its components are changed, the ship will rightfully be identified as one and the same.

The above determination of identity might already be sufficient to entice a proponent of transhumanism as it grounds the mereological possibility of modifying "parts" on one's body as long as the balance and *ratio* of those parts is preserved, without losing one's identity as a definite human being. However, Spinoza's philosophy can be observed to push even further along these transhumanist routes, insofar as it provides ethical argumentations for the opportunity of such modifications. The third book of the *Ethics*, in fact, is the first that properly explicates its titular topic. It's important to remember that Spinoza's *Ethics* is not a normative ethics in the contemporary sense of the term. Instead, it provides an elucidation of the fundamental components of human action and a justification of their coherence and consistency. In illustrating these fundamental components, Spinoza affirms that "each thing, as far as it can by its own power, strives to persevere in its being", and that "the striving by which each thing strives to persevere in its being is nothing but the actual essence of the thing".[9] Without entering the deep weeds of Spinozian interpretation, I would like to suggest that the "actual essence" mentioned in this passage coincides with the "nature of the Individual" that has been the subject of our study, above. Following this suggestion, we can see that the Spinozist attitude towards the preservation of the individual implies that the balance among its parts and components is projected to be continued for "no finite time, but an indefinite time".[10] For Spinoza, then, the body is intrinsically determined by its own essence—the *ratio* that holds together its core equilibrium, something similar to a gravity well pulling together its parts—to preserve such state and, if possible, to improve it.[11]

To dispel any doubt that Spinoza could mean anything but the preservation of the life of the body with his definition of "striving" as the actual essence of any being, we can look at one of his earlier works, the *Metaphysical Thoughts*, where he defines "life" as "the force through which things persevere in their own being. And because the force is different from the things themselves, we quite properly say that things themselves have life".[12] As it is plain to see, the definition of life offered in this text is quite congruent with the definition of striving (or *conatus*) that appears in the *Ethics*. Thus, one can rightly assume that when Spinoza affirms that the actual essence of the individual is expressed in the preservation of its being through an indefinite[13] time, he is portraying the preservation and flourishing of a being's life through the maintenance of its *ratio* of motion and rest, of the equilibrium among its parts.

On these metaphysical foundations, Spinoza builds his metaethics by claiming that the notion of "good" is merely a perspectival description of "what we certainly know to be useful to us;" by contrast, evil encompasses "what we certainly know prevents us from being masters of some good".[14] As Deleuze insightfully glossed, this allows Spinoza to give a new meaning to the "classical thesis that evil is nothing". Any part of my body may agree with a great number of things, but those agreements and relations between parts "do not necessarily coincide with the preservation of a *particular* relation, which may be dissolved, that is, cease to be realized. In this sense there is no evil (in itself), but there is that which is bad (for me)".[15]

As we saw above, the actual essence of individuals entails their preservation and continuation into being. Spinoza insists on the impossibility of understanding this preservation as a mutation in nature or essence, such as the ascendence into a different *kind* of being. One example clarifies this point:

> when I say that someone passes from a lesser to a greater perfection, and the opposite, I do not understand that he is changed from one essence, or form, to another. For example, a horse is destroyed as much if it is changed into a man as if it is changed into an insect. Rather, we conceive

that his power of acting, insofar as it is understood through his nature, is increased or diminished.[16]

Spinoza's definition of good, when combined with this notion of transition into a greater level of perfection, specifically requires that the preservation of an individual's characteristic *ratio* and equilibrium among its parts be maintained, lest the individual is destroyed by its transition into a different ***kind*** of nature. Thus, whatever transformations an individual may go through in order to preserve its being and increase its power of acting and existing, they cannot so radically modify the internal structure of the individual to the point of transition out of its nature. As I will discuss in the next section, this is perhaps the chief point of contrast between Spinozist and transhumanist ethics, as the latter is adamant in addressing those changes and mutations that are radical in character, which Spinoza deems destructive to the nature of the individual.

Be that as it may, the alignment of intents between the two ethics remains important, insofar as both seek to augment and ensure the preservation and increase of power of the (human) individual, through a recognition of "good" as that which aids in this effort. In the next section, I aim to explore this alignment further. I shall propose the adoption of a Spinozistic conception of "limit" to the modifications that allow the individual to preserve and expand its "striving" while remaining "good" (i.e., in accordance with its nature), without severing the equilibrium that constitutes the nature of each body.

Spinozist Transhumanism

The overall goal that unites the various philosophies known under the umbrella-term "transhumanism" is the consideration of scientific and technological advancements that might transform, improve, and otherwise modify the human condition in the proximate future. Moreover, the authors recognizing themselves as "transhumanists" have at times made explicit that these efforts are so radical that they "call into question the appropriateness of calling "it" [the resulting human condition] "human" anymore.[17] Calling into question and denying such appropriateness, however, are two quite distinct positions, especially within the metaphysical framework that I have developed in the previous section by following some general indications of Spinoza.

As we saw above, there is no guarantee that *any* transformation will allow the individual to retain its identity, as the latter is founded on the maintenance of a certain *ratio* which can be preserved and indeed augmented by *some* transformations but not others. In fact, according to Spinoza, the individual.

> [...] ceases to exist when its parts are determined from without to enter into a different relation, which is not compatible with the former one [...] the degrees of power [of different individuals] necessarily come into conflict in existence inasmuch as the extensive parts that pertain to one degree under a certain relation can be conquered by another under a new relation.[18]

In other words, individuals striving through the universe are seeking to preserve and augment their power by exchanging parts and components with the external environment to their advantage (or, better said, to the advantage of the realization of their essence). Simultaneously, they strive to preserve themselves from being "conquered" and dismantled by other individuals that strive through the universe with a *ratio* of their own to preserve and augment.

The apparent risk for the human individual (which is, after all, our focal case study) is to modify itself while seeking preservation and augmentation to the point of severance, upon which it would cease to retain its characteristic *ratio* and would become something of a different nature, which Spinoza would describe as a destruction and sundering of the previous individual. In other words, "the fluidity of nature, as Spinoza envisages it, is [...] limited. Although the essence or power of an individual can alter in certain ways and to a certain extent, there are also boundaries to the changes it can undergo, and when those boundaries are crossed the individual is destroyed".[19] This would indeed be a considerable obstacle to the construction of a Spinoza-inspired transhumanist account, since—as I hinted at the opening of this section—certain transhumanists consider their account as challenging to the very notion of human nature as commonly understood.

Luckily for the aspiring Spinozist transhumanist, Spinoza himself offers the possibility of an ontological safety to avoid these conclusions. Within the same argumentative arc that establishes the preservatory striving of an individual as its actual essence, Spinoza also introduces the thesis that "No thing can be destroyed except through an external cause". The proof that he offers for this thesis is, in his own words, self-evident, as "the definition of anything affirms, and does not deny, the thing's essence, or it posits the thing's essence, and does not take it away. So, while we attend only to the thing itself, and not to external causes, we shall not be able to find anything in it which can destroy it".[20] While perhaps excessively epistemic in its formulation, the proposition expresses Spinoza's general belief that since the universal law of preservation demands that things are drawn by nature to persevere in their being, it would be impossible for them to contain the causes of their own demise. This does not mean, of course, that individuals are invulnerable from an external point of view—in fact, they can be affected negatively by a great many things. Spinoza's point is that insofar as things are within their power, they will strive to achieve survival and improvement of their condition.

The core of Spinoza's position is radically anti-humanist. Human beings do not enjoy any special status, any apex predator certificate of distinction, any guarantee that they have been created in God's image. Instead, humans are intrinsically bound to the texture of nature. Spinoza declares himself an enemy of those who conceive "man in nature as a dominion within a dominion. For they believe that man disturbs, rather than follows, the order of nature, that he has absolute power over his actions, and that he is determined only by himself".[21] This position situates Spinoza against a supposed "notion of human transcendence, the idea that man is free from determinations that would remain of a merely physical nature".[22] Incidentally, we may signal that Spinoza's anti-humanism (or anti-anthropocentrism) does not collocate him in opposition to transhumanists, who are happy to acknowledge that "there is no fundamental dichotomy between humanity and the rest of the world"[23] and that humans enjoy a phenomenal and not ontological specialty[24] of status.

The argumentation regarding the impossibility of self-destruction follows from this anti-humanism in what may seem a counterintuitive movement. Just like all of nature seeks to preserve its being and, if possible, to improve it, so also human beings will follow this universal striving. To suppose humans as capable of destroying themselves is to recast the actions of the other beings, which always surround them and at times prevail on them, as secondary; this delusion of self-determination is embedded in anthropocentrism, according to Spinoza, and must be dispelled if we are to create a true ethical theory.

Now, how can this anti-humanist stance be of use in the development of a transhuman metaphysical perspective? My proposal is to recognize how the defeat of anthropocentrism is instrumental to the progressive modification of our conception of individual human nature. The de-sacralization of a supposedly superior "human nature", both in general and in individual instances, delivers as a dowry the possibility of transforming such nature, up to the liminal stages of its demise into non-humanness (which Spinoza would equate with destruction). Spinoza's general anti-humanism translates into an ethics of care for the individual humans, who can (and must) improve their individual nature in accordance with the circumstances that surround them, aiming towards an increase in power and perfection.

At the same time, the open-ended perspective on essence and human nature that Spinoza grants us is highly useful for an unbiased understanding of good and evil from a transhuman point of view. In the previous section, I introduced Spinoza's definition of "good" as that which we know can aid us in the attaining of survival and augmentation. This notion of good is supposed to guide the reader of the *Ethics* through their durational lives, which can be described as a "self-producing movement within—and through—a system of encounters".[25] Thus, the spirit of the Spinozian ethics is comprehended as the attempt to describe an optimization of the striving of individual beings, who traverse their durational lives immersed in a multitude of encounters, some of which are "good", some of which are "bad" (in the idiosyncratic sense we described above). The end goal, once again, is to preserve the striving itself, and—where possible—to augment it through the addition of more power to act (more capacity to obtain "good" encounters) and a more extended duration; not infinite, but indefinite. As it is plain to see, these goals align cleanly with those set by transhumanist authors, insofar as they elucidate their philosophy as an attempt to develop and justify the application of technological and scientific means to extend and ameliorate human life.

It is for the reasons explained so far that, in both philosophies, the concept of *change* occupies a crucial role. The justification of change is central to the effort of understanding the transformations, both necessary and useful, to the preservation and thriving of the individual's existence. For transhumanism, change is the goal and just expectation of the age which we inhabit.[26] Human beings dwell within a changing state, and the interpretation of such state is the proper occupation of philosophy. For Spinozists, change is all but the tissue of reality, and the navigation of the various encounters which affect the individual's duration and life is the goal of Spinoza's *Ethics*.

However, for Spinoza, it is important to distinguish change as improvement from change as transmutation. As we have seen above, the latter would in fact result in a destruction of the individual in question, even though it may generate the birth of a new individual. From the point of view of the initial individual, a transmutation would still be a destruction, an act that is metaphysically impossible due to the restraints imposed by Spinoza on the contents of an individual's nature and essence.

For Spinoza, change—when endogenetic and responsive to the essence of the individual—is necessarily an increase in power. The opposite case (i.e., a decrease in our power of acting or a transmutation that would mean the demise of our individual's existence) is veritably understood as an overcoming of the individual's striving at the hand of some external cause.[27] Similarly, for transhumanist philosophers such as Nick Bostrom, "human nature in this broader sense is dynamic, partially human-made, and improvable".[28] This does not immediately call for a *radical* (transmutative) modification of human nature; instead, Bostrom's formulation aligns with Spinoza's viewpoint insofar as it presents an increase in an individual's existential power of acting that contemporarily increases its striving, i.e., its essence.

It is important to add that, for Spinoza, this change or increase is *exponential*. That is, the modifications that an individual essence undergoes increase its capacity for further change and stretch the boundaries of its individuation. As James reminds us, "an individual's essence [...] incorporates the potential to acquire and exercise new powers, which in turn create the conditions for acquiring and exercising yet further powers, and so on".[29] However, given Spinoza's anti-anthropocentric stance, this individual essence's push towards increase and acquisition is not entirely self-reliant, insofar as it depends on the "encounters" or interactions with its environment. A transhumanist interpretation of this theory of essence increase would therefore be able to recast its demands as a regulation (or deregulation) of the kind of external thing that the individual can incorporate in its striving, as beneficial to its exponential increase. Yet, as we have mentioned before, this deregulation comes with a self-imposed boundary of its own. "As well as allowing for change, an individual's nature imposes limits on the powers it can acquire or lose",[30] insofar as the transmutation or metamorphosis into a *new* kind of being would imply the demise of the original being, which Spinoza assesses as metaphysically impossible as an endogenetic process.

The considerations presented in this section enabled us to offer a new metaphysical framework for transhumanism, inspired by the ontology of individuals designed by Spinoza. Our proposed framework sees the essence of individuals as a fundamental force or striving for survival and increment, regulated by a metaphysical impossibility of radical transmutation or metamorphosis and fueled by interactions with the opportunities for new powers that can be found in the interaction with the individual's environment. Moreover, we have observed how—under this paradigm—the essence of individual humans is itself subject to change and modification, increasing and decreasing as its power of striving and acting does. Armed with this array of conceptual considerations, we now move to observe whether this framework can withstand some of the harshest criticism that has been levelled against transhumanism, specifically on the subject of "human goods".

The Resilience of Spinozist Transhumanism

In his 2009 op-ed titled "Transhumanism", Francis Fukuyama summarized his long-standing crusade against the claims and hopes of transhumanist authors for the benefit of the general public. The main argument presented in this text concerns the possibility of individuating and understanding "what constitutes a good human being". This "goodness" allegedly underlies, or justifies, the transhumanist transformation, which shifts away from superfluous and contingent conditions while maintaining a hold of the characteristics that allow a transhuman or posthuman individual to remain, after all, human.[31]

What argument does Fukuyama propose to support his thesis that transhumanism misunderstands the identity and extent of "human good"? While his text is not entirely philosophical in presentation, we can extract from it that his argumentation ties together "human goods" with allegedly necessary "bad" human characteristics. For example, according to Fukuyama,

> If we weren't violent and aggressive, we wouldn't be able to defend ourselves; if we didn't have feelings of exclusivity, we wouldn't be loyal to those close to us; if we never felt jealousy, we would also never feel love. Even our mortality plays a critical function in allowing our species as a whole to survive and adapt (and transhumanists are just about the last group I'd like to see live forever).[32]

While each of these examples is individually analyzable and susceptible to criticism, a more fatal flaw of Fukuyama's objection lies in the assumption that for each "human good" there is a correspondent "human flaw" which is necessary and inextricable for the subsistence of the relative good to which it corresponds. This is, of course, a simplified and reductivist outlook on human nature, one that—we may add—is rather pessimistic in character. Nick Bostrom responded to this criticism along similar lines:

> If there is value in being human, it does not come from being 'normal' or 'natural,' but from having within us the raw material for being humane: compassion, a sense of humor, curiosity, the wish to be a better person. Trying to preserve 'humanness,' rather than cultivating humaneness, would idolize the bad along with the good. One might say that if 'human' is what we are, then 'humane' is what we, as humans, wish we were. Human nature is not a bad place to start that journey, but we can't fulfill that potential if we reject any progress past the starting point.[33]

Moreover, it must be noted that Fukuyama's objection relies on an entirely subjective identification of which bad characteristics correspond to the intrinsic positive features of human nature that Fukuyama would want to—understandably—preserve, such as love, adaptation, loyalty, and survival.

Fukuyama himself should be wary of such throwaway associations, given that his position ultimately concerns a more systematic worry. Namely, his final thesis claims that "modifying any one of our key characteristics inevitably entails modifying a complex, interlinked package of traits, and we will never be able to anticipate the ultimate outcome".[34] Firstly, let me make clear that *if* Fukuyama is correct, and the "package of traits" that is definitory of human nature is indeed complex and interlinked, then there is no telling which specific positive traits are associated with other, equally specific, negative traits. For example, to revamp a couple of the examples provided in the text above, we may be able to eliminate jealousy without simultaneously surrendering love, or we may retain our capacity for self-defense without maintaining aggression and violence as "key characteristics" of what it means to be human. Eliminating specific negative features of our shared human experience does not necessarily entail losing positive ones that are allegedly correspondent, *precisely because*, as Fukuyama affirms, the traits that constitute "human nature" are complex and interlinked in such ways that the outcome of their modification is unpredictable.

Secondly, Fukuyama's claims about human nature depict it as an immutable and unmodifiable solid entity. In another work of his, he defines the concept as "the sum of the behavior and characteristics that are typical of the human species, arising from genetic rather than environmental factors".[35] The word "typical", in Fukuyama's intention, is meant to ward off the immediate objection that may arise from observing that, even within a considerably compact population, individuals "show considerable variance within the same species" in terms of characteristics and behavior.[36]

Drawing on these considerations, Fukuyama acknowledges that the intraspecies variance is particularly relevant

> with cultural animals like human beings: since behaviors can be learned and modified, the variance in behavior is inevitably greater and will reflect the individual's environment to a greater extent than for animals incapable of cultural learning. This means that typicality is a statistical artifact – it refers to something close to the median of a distribution of behavior or characteristics.[37]

However, this characterization immediately contradicts the requirement of "genetic rather than environmental" proposed above. In fact, if behaviors (and characteristics) are "inevitably" modified through cultural interaction, the statistical artificiality of what Fukuyama considers "typicality" loses the normativity that it requires to function as a foundational piece of his notion of human nature. By the lights of this argument, we can infer that the essence of a species must be a more fluid concept, one that—thanks to the criteria developed in the previous sections—we can qualify on the basis of the relative flexibility granted to individuals as they encounter other individuals in the course of their striving through an environment.

So, to sum up: Fukuyama's criticism to transhumanist authors amounts to a disbelief that they have correctly identified human goods. To back up this disbelief, Fukuyama points to a supposed notion of human nature which would ground such goods. However, this "human nature" is so complex and interlinked that any modification to a specific feature belonging to it, even a negative one, would threaten the remainder of it. Particular mention is made of those characteristics and behaviors that Fukuyama, on a non-environmental and exclusively biological basis, reputes crucial for a cogent understanding of human nature, even though these characteristics and behaviors are inevitably statistical in character and subject to modification by the environment.

Following Fukuyama's own considerations, then, we can employ our Spinozist transhumanist ontology to defend a more resilient notion of human nature. In fact, without denying the complexity and interconnectedness of an individual essence it is possible to treat the artificial and statistical character of "nature" as described by Fukuyama as a justification for a disregard of the normativity of such conception of our species' essence. Conversely, within such statistical understanding of typicality our transhuman Spinozist conception of essence as dynamic and individual acquires a more tenable weight; meanwhile, the statistical aspect can be delegated to a unificatory role joining together the different individuals that populate a given species.

What kind of consequences does this understanding bear towards an identification of "human goods", which is after all the crux of Fukuyama's attack on transhumanism? Detaching the notion of "good" from species-specific normativity makes it available for individual natures as a self-determining notion that guides the individual's path through existence. The resulting plethora of human behaviors is still available for classification as an artificially construed and statistically significant depiction of species-*typical* behavior and features, as wished by Fukuyama. However, the significance of this shift is that it allows individuals to follow their singular pathway to survival and increase, free from an externalist normativity that would allegedly force them to adapt to their own essence. Thanks to the ontological framework developed in the previous section, we can claim that the human essence (which is distinct from the essence of human beings in the plural) coincides with the individual's striving to survival and improvement. Furthermore, we can observe that essence itself is dynamic and morphologically diverse, compatible with a variety of modifications that can occur on the basis of the environment and the objects with which the human individual interacts. What the transhumanist facet of our account adds to the Spinozist ontological foundation is a form of non-exclusivity regarding the character of these modifications. As Bostrom claims, "from the transhumanist standpoint, there is no need to behave as if there were a deep moral difference between technological and other means of enhancing human lives".[38] The plural form of "lives" only apparently opposes the allegedly singular form of "nature".

In fact, the former provides a more immediate association with the notion of individual essence that I have been defending through this essay. By contrast, the latter coincides with the pragmatic striving that self-identifies "good" objects to interact with, based on principles of self-preservation. Such principles notably include a preservation of the boundaries of self-identity that Spinoza posed as limits to which transformations are positive and which constitute a destructive metamorphosis.

The "good" resulting from the application of the ontological framework presented in this paper is, for all intents and purposes, singular and individual in character. In many cases, this notion of "good" will be overlapping with a notion of "change" which, as we have seen above, can ground further interactions and modifications, conducing to the sort of dynamism described by James's interpretation of human nature in Spinoza. However, agreeing with Fukuyama that a statistical notion of typicality provides (after all) a passable artificial guideline for species-specific behavior and characteristics, we can infer that a similarly species-specific notion of "human good" can be extracted and analyzed, despite the irreducible individuality of "human nature".

Notes

1. Max More, "Transhumanism: Towards a Futurist Philosophy", *Extropy* no. 6 (1990, revised 1996): 1.
2. See Francis Fukuyama, "Transhumanism", *Foreign Policy*, October 23 (2009), https://foreignpolicy.com/2009/10 23/ transhumanism.
3. Baruch Spinoza, *Ethics*, in *The Collected Works*, trans. Edwin Curley (Princeton (NJ): Princeton University Press 1985), E2p13, I:457–458.
4. Baruch Spinoza, *Ethics*, E2PhysDigL1, I:458.
5. Spinoza, *Ethics*, E2PhysDigD, I:460.
6. Spinoza, *Ethics*, E2PhysDigL4, I:461.
7. See Spinoza, *Ethics*, E2PhysDigLL5-7, I:461.
8. See David Rose (et al.), "The Ship of Theseus Puzzle", in *Oxford Studies in Experimental Philosophy*, ed. Tania Lombrozo, Joshua Knobe, and Shaun Nichols (Oxford: Oxford University Press, 2020), 158–174.
9. Spinoza, *Ethics*, E3pp6-7, I:498–499.
10. Spinoza, *Ethics*, E3p8, I:499.

11. On the notion of "improvement" or "increase" of power as central to the understanding of "striving", see Vallteri Viljanen, *Spinoza's Geometry of Power* (Cambridge: Cambridge University Press, 2011), 109; and Jacob Adler, "Spinoza's Physical Philosophy", *Archiv für Geschichte der Philosophie* 78:3 (1996): 180.
12. Spinoza, *Metaphysical Thoughts*, II.6, §197, in *The Collected Works*, op. cit, I:326; see also Sean Winkler, "The Conatus of the Body in Spinoza's Physics", *Society and Politics* 10:2 (2006): 105.
13. It is important to remark, here, that for Spinoza there is a qualitative distinction between infinite and finite, whereas the distinction between indefinite and finite is a quantitative distinction. In other words, Spinoza is here projecting the survival of the individual through a non-determinate duration, which is still constructed on a series of successive instants of time.
14. Spinoza, *Ethics*, E4dd1-2; 546.
15. Gilles Deleuze, *Spinoza: Practical Philosophy*, trans. Robert Hurley (San Francisco: City Lights Books 1988), 33.
16. Spinoza, *Ethics*, E4pref, 545–546.
17. Charles T. Rubin, "What is the Good of Transhumanism?", in *Medical Enhancement and Posthumanity*, ed. Bert Gordijn and Ruth Chadwick (Dordrecht: Springer, 2008), 137.
18. Deleuze, *Spinoza: Practical Philosophy*, op. cit., 76–77.
19. Susan James, "Spinoza on the Constitution of Animal Species", in *A Companion to Spinoza*, ed. Yitzhak Melamed (London: Wiley-Blackwell, 2021), 366.
20. Spinoza, *Ethics*, E3p4 and demonstration, 498.
21. Spinoza, *Ethics*, E3pref, 491; see also Yitzhak Melamed, "Spinoza's Anti-Humanism: An Outline", in *The Rationalists: Between Tradition and Innovation*, ed. Carlos Fraenkel, Dario Perinetti, and Justin E.H. Smith (Dordrecht: Springer, 2011), 147–166.
22. Warren Montag, *Bodies, Masses, Power: Spinoza and His Contemporaries* (London: Verso, 1999), 37.
23. Nick Bostrom, "The Transhumanist FAQ: A General Introduction (Version 2.1)", accessed July 14th, 2022, http://www.transhumanism.org/resources/FAQv21.pdf, 38.
24. I am here using the term "special" in the same acceptation as Kant's description of the fundamental ideas of reason, which demand a "special" metaphysics: the soul, the world, and God.

25. Philippe Danino, “Spinoza: la philosophie par la rencontre”, *Archives de Philosophie* 81:1 (2018): 91.
26. Transhumanists do, of course, recognize the fundamental notion that “not all change is good” and that important distinctions can be offered in this regard. See e.g., Nick Bostrom, “In Defense of Posthuman Dignity”, in *Bioethics* 19:3 (2005): 205.
27. On this alternative, see Karolina Hübner, “The Trouble with Feelings, or Spinoza on the Identity of Power and Essence”, *Journal of the History of Philosophy* 55 (2017): 40–41.
28. Bostrom, “In Defense of Posthuman Dignity”, op. cit., 213.
29. James, “Spinoza on the Constitution of Animal Species”, op. cit., 367.
30. Ibid., 368.
31. Fukuyama, “Transhumanism”, op cit.
32. Ibid.
33. Bostrom, “The Transhumanist FAQ”, op. cit,. 36.
34. Fukuyama, “Transhumanism”, op. cit.
35. Francis Fukuyama, *Our Posthuman Future: Consequences of the Biotechnology Revolution* (New York: Farrar, Straus and Giroux: 2002), 130.
36. Ibid.
37. Ibid.
38. Bostrom, “In Defense of Posthuman Dignity”, op. cit., 213.

CHAPTER 12

Crossing the Boundaries of Life: Reanimation in Later Medieval Central Europe

Hannah Elmer

Introduction

In the later Middle Ages, the boundaries between life and death could be quite porous, but they were not arbitrary. The dead might return to the community of the living in various ways: people might appear as ghosts, as dream-visions, or more corporeally as revenants, but they could also simply be revivified. In contrast to the ghosts and revenants, the reanimated person rejoined the living in the same soul-body configuration as before. The surrounding community, moreover, was often significantly involved in a person's death. In some cases, surviving members made efforts to bring the deceased back to life, and these efforts might very well be successful (though it is important to emphasize that death itself did not tend to be medicalized).[1] By examining reanimation—the

H. Elmer (✉)
Leibniz University Hannover, Hanover, Germany
e-mail: hannah.elmer@hist.uni-hannover.de

J. Muller (ed.), *Automata, Cyborgs, and Mutants*, Palgrave Studies in Medieval and Early Modern Medicine,
https://doi.org/10.1007/978-3-031-93201-4_12

contexts in which it might happen, the actors involved, and the purposes it could serve—we can see how later medieval people sought to determine the boundaries between life and death and what their interventions (or lack thereof) into death suggest about human agency and the ability of communities to affect the fates of individuals. As some reanimations were not for the sake of prolonging earthly life but rather for the purpose of securing the soul's well-being, they prompt us to consider the relationship between life and death (and definitions of life) from a spiritual rather than physiological angle. Along these lines, certain controversies around reanimation reveal how theological positions helped create conditions in which life/death were not merely understood but also made visible. In this context, as a number of scholars have shown, ideas about the General Resurrection at the end of time were critical, but so were other ideas about the sacraments and their role in salvation.

This chapter is foremost based on research I conducted on approximately twenty holy people and shrines in central Europe (predominately from Polish lands), starting at around 1200 with the earliest bids for papal canonization and extending into the seventeenth century, with the bulk coming from the thirteenth through fifteenth centuries. For reasons of space, I have limited the evidence I cite here to the collections of Hedwig of Silesia, Elizabeth of Hungary/Thuringia, Stanislaw of Szczepanów (Bishop of Kraków), Hyacinth of Poland, and Caesarius of Heisterbach's exempla collection, the *Dialogus miraculorum*; these sources were mostly composed in the thirteenth century (Hyacinth's collection is early fourteenth), and they all remained important well into the early modern period.[2] Nevertheless, my observations and arguments hold true for the other sources as well.[3] Because the collections for Hedwig, Hyacinth, Stanislaw, and Elizabeth were all closely associated with their canonization campaigns, for simplicity's sake I term them "canonization collections."[4] As these sources are closely tied to a clerical-elite culture, I make no claims to present a complete picture of reanimation in later medieval culture, were such a picture even feasible. That said, the sources provide a good starting point for investigations into reanimation and raise fundamental questions about the boundaries around human life and embodiment.

General Resurrection

If later medieval Christians did not pass under sculpted or painted images of the Last Judgment when they entered their local church (and certainly cathedral), they would likely encounter them somewhere inside, perhaps as a wall painting or as part of an altarpiece. In these images not only would they confront the separation of the good from the bad by Christ as Judge, but they would encounter the good and the bad as fully embodied—re-embodied—humans. This day of judgment is also the day of the General Resurrection that marks the end of time.[5] These images would serve as a reminder that all activities undertaken here on earth were ultimately important because of their merit (or lack thereof) for each individual's eternal fate, a fate that would be experienced in a fully physical, albeit perfected, state. This life was framed by the next.

The ultimate orientation towards eternal life, be it in heaven or hell, meant that bodily death did not mark the end of existence, and the idea of an embodied eternity meant that the body's fate after death, and not just the soul's, was a subject of concern, and had been so since the earliest years of Christianity.[6] As the centuries continued to pass after Christ's incarnation without his Second Coming, people increasingly speculated about the location and activity of the soul before this day of judgment, and how, exactly, the soul would be reunited with a flesh that had since decayed into dust.[7] Thus, the period between bodily death and the Last Judgment became a time of suspense; increasingly, it also became a time during which eternal fates could change.[8] Against this background, death is perhaps better viewed as a phase than as a line, and a phase during which the body and soul underwent radically different experiences.[9] The flesh decayed and disintegrated, while the soul either remained relatively constant (suffering in hell or resting in peace in heaven) or underwent purification in purgatory (en route to heaven). These ideas about the afterlife and end times necessarily also frame ideas about possible contact between the living and the dead.

For the most part, souls were removed from sight, but not infrequently, some would visit the living through dreams and visions, and some (though less frequently) would return in more corporeal form as revenants. Reanimation was also a possibility but occurred under different (and limited) conditions. Scholars have examined the dreams, ghosts, and revenants, but hardly the actual reanimations.[10] While the focus of this chapter is ultimately the reanimations, it is important to briefly situate them among these other forms of contact between the living and the dead.

Dream-Visions

Most ephemerally, people returned in dreams or visions.[11] This is a broad category in terms of both purpose and the type of person who might appear—very often the spectral visitor would be someone clearly saintly but anonymous; however, sometimes known, "normal" people would appear to friends or family.[12] Often the sources distinguish between someone receiving such a vision while awake or dreaming, making clear that the visitor was not fully embodied and that his or her presence could happen in a different mode: the absence of body did not preclude interaction, and such interactions could even have corporeal effects. Both dream and wakeful visions occur in all the sources, though in slightly different capacities: Caesarius's exempla feature more "normal" people, while the canonization collections present the "saintly" types almost exclusively. This moment from the miracle collection of St Stanislaw gives a sense of these kinds of encounters:

> Later, after she had closed her eyes in order to rest, but was still unable to sleep, an aged and venerable man came to her and said, "distressed and troubled woman, through your prayers to St Stanislaw, it has happened that your son will be returned to you this night, since your prayers and tears and the vow you vowed to that saint on whose day the boy first became sick have been heard." When the mother opened her eyes, seeing no one, she returned to the house where the boy lay.[13]

As this is the least "embodied" form of return, and one that has received extensive treatment by scholars in several disciplines, we will turn away from it now, but it is worth bearing in mind that the returning spirits could come at any given time, and that often they appeared of their own volition, by a previous agreement, or in response to petition, as the above case illustrates. As in this case, a dream-vision might accompany a reanimation.

Revenants

The next category to briefly consider is that of the revenants, a group that has also received substantial treatment by scholars.[14] These people are understood to be dead, but they possess a corporeality not found in the visions; they are most prevalent in twelfth- and earlier thirteenth-century sources and again in texts from the sixteenth and seventeenth centuries.[15] Their embodiment is ambiguous: sometimes they appear in a seemingly commonplace human guise, other times they manifest in a more rarefied form. Nevertheless, they have an animate, bodily aspect but are not properly living: it is evident that their souls belong somewhere beyond this earthly world and their presence is subject to rules different from those governing ordinary human life. This is clearest in the *Dialogus*, also the main source in which such revenants appear, which tells of several people seen suffering purgatorial punishments on earth and who beg the living for aid. For example, a certain dead knight Frederick was seen burdened by sheepskins and earth, which he said was the weight of all he had stolen during his life. He begged an anonymous man to ask his sons to help ease his burdens by returning unlawfully claimed land, which they ultimately refused to do. Interestingly, Frederick did not return directly to his sons, suggesting his ability to act on earth was limited—rather than coming directly to a specific person, he was left roving, only able to ask those he encountered by chance for help.[16] In contrast to the dream-visions, revenants do not appear in the canonization collections. Yet similar to the visions, the stories of revenants show how the souls of the deceased could physically appear and interact in the quotidian world; they also show how the deceased could even still effect spiritual change, often by describing their purgatorial torments to the living, encouraging them to use their still-living bodies to do spiritual work on the sufferers' behalf.

As shown above in the brief vision narrative from St Stanislaw's collection, the dead could also perform spiritual work on behalf of the living, but the people able to do such work could only do so because of a degree of holiness. Some of them would become officially canonized as saints, but others might have unofficial, local cults. Although these "good" people were not encountered as ambiguously embodied spirits, roving moors or haunting villages, they could still be accessed physically through contact with the bodies they left behind at death—through their relics.[17] Thus, independent of moral status, the body after death maintained a connection to its soul, even if the precise configuration was contested.[18] This is doubly significant because corporeal reanimation was often achieved through saintly help—through the help of the deceased holy. Saints could facilitate physical healing and bestow spiritual solace to the living, and the living could aid those suffering in purgatory. Thus, these cases do not only indicate that change happens throughout the period of death, but they also demonstrate the transferability of spiritual deeds, that a person could alter someone else's post-death experience.

Corporeal Reanimation

People who returned to the living as flesh and blood, animate human beings could elicit different attitudes and approaches in various sources. Caesarius's *Dialogus miraculorum* not only reinforces the observation that death was not medicalized but also shows—with few exceptions[19]—the community surrounding the dying/dead person to have accepted the (impending) death. The community members are active insofar as they fulfill the prescribed deathbed rituals, but they do not otherwise seek to interfere with death. Reanimation in these cases happens without external, bodily intervention. Instead, if the person were good, divine forces might bring him or her back to life to make amends and thereby avoid suffering in the afterlife. We see this in the following anecdote told by Caesarius:

> Not long after [a lay brother] fell ill and when everyone present believed him to be dead, his soul left him and wished to ascend to its rest, but saw before its eyes, that half-penny, which he had forgotten to mention in his confession. And it grew to such a size that it seemed bigger than all the world. And as it stood continually in his way when he pressed upwards, with nothing else to stop his soul, at the prayer of the angels he was allowed to return to the body. He told his vision to the astonishment of all and the whole penny was sent by the abbot to the ferryman with all speed. As soon as he received it, in the very same moment according to their calculations the lay brother expired.[20]

Here, remarkably, angels intercede on the lay brother's behalf, and it is through their prayers that he returns to life. His reanimation allows him to explain the situation to his abbot, who then assumes the responsibility to repay the debt and secure the man's peaceful death.[21] What is significant is that God is persuaded not by the man's actions or by community's works but rather by the angels, who witnessed the ascent of the man's soul to heaven to be blocked by a single, trivial sin. Two layers of divine power intersect in this reanimation. But more importantly, the angels intercede *after* the man died, showing the soul's fate to be in flux directly after death.

Reanimation here is thus entirely in service to the soul's well-being, having surprisingly little to do with the body, and nothing to do with extending the man's life on earth. But it does show how earthly life and the hereafter intermesh and reciprocally affect one another. As Katharine Park mentions, these monks and lay brothers in the *Dialogus* come back to fix something minor so that they can die a "good" death.[22] But Caesarius also tells of a few sinful souls who return briefly to describe the horrors of their fates. Those few words are the extent of the action, however, making these reanimations impotent to alter the afterlife. Whether positive or negative, Caesarius's cases fulfill the important didactic function of instructing the living on what awaits them after death. Thus, reanimation in the *Dialogus* is actually focused on death rather than life, but it is so because of how death serves to link earthly life with eternal life. Death is the bridge between the two, rather than a caesura. For the "good" souls, though their healing is ultimately for their long-term spiritual well-being and not for the sake of the body, it can only occur through this (living) body.

The miracles in this exempla collection unsurprisingly emphasize elements that make the cases useful for didactic purposes.[23] The *Dialogus* presents reanimation from a soteriological angle, explaining why the soul was permitted to return to the body, how long the two were reunited, and what that miracle enabled. Although all of the reanimated people lived long enough for there to be no doubt that they did, indeed, come back to life, none remains alive for more than a few hours or days.[24] They come back in order to change their eternal fate or to warn the living. This makes them fundamentally different from almost all of those reanimated in the canonization collections.

The canonization collections provide us more information about how death-life was discerned, and they show a wider range of responses by the community to the death of one of their members. The contrast with Caesarius's exempla should serve as a warning against drawing conclusions about medieval attitudes and practices based on a single source genre. Because in most of the canonization cases the success of the reanimation was clear, the challenge facing those trying to secure miracle status for their event was to prove that death had, in fact, occurred in the first place. Laura Smoller has made this point with evidence from the miracle collection of St Vincent Ferrer.[25] Whereas in Caesarius's cases the reanimated people described their post-mortem experiences, which helped explain why they were allowed to return to their bodies, the canonization collections prove death by other means, but not by such first-person evidence. Instead, witnesses testify to the death of the person in question and to the use other forms of proof, such as duration of death or physical appearance of the corpse, to do so. A miracle in the collection of Elizabeth of Thuringia/Hungary shows the common strategy of presenting the duration of death:

> "The boy remained dead up to the beginning of the next day, while these same two women incessantly invoked both God and blessed Elizabeth. Finally, in the early morning, the boy moved his hand, drew a breath with some difficulty, and came back to life, having been cured. In the evening of the following day he began to speak."[26]

Other cases from Elizabeth's collection emphasize the deceased's physical condition:

> "The two of them began shouting and some people from that village came, all of whom judged the boy to be dead. He lay there with his mouth agape and his eyes open in an awful way. His skin was dark and wrinkled all over, as if he had been boiled in water, and he had a very tight stomach. His arms and legs, that is to say all his extremities, had gotten stiff."[27]

Sometimes the physical condition could be combined with a temporal indicator to prove death, as in this case of St Hedwig of Silesia: A boy was caught by a mill wheel and drowned. Once he was finally pulled out, he appeared "blueish gray, cold and stiff, and as rigid as wood. Nor was this a surprise, since the boy who had died had been submerged underwater for as long as it takes to walk a Lombard mile."[28]

In addition to providing criteria that was considered compelling for proving death, the canonization collections provide glimpses into strategies people employed to bring someone back to life. Since it was necessary to show that recovery did not come about by natural means in order for a case to achieve miracle status, we often hear of medical or other interventions people may have performed when trying to bring the person back to life, if only to discount them. Hedwig's reanimation of the drowned boy mentioned above not only reverses the order of the vow and the physical/medical manipulation of the body, but it also shows the ineffectiveness of the interventions by highlighting the length of time the child remained dead.[29] *After* vowing to Hedwig, the father of the boy expelled the water from the boy's body and placed him over a fire, but the boy remained motionless from midday to sundown. The faithful kept invoking God through the merits of Hedwig, and finally the boy began to breathe and by the following morning, he rose healed, suffering no pain, but marked with a bruise like a black cloth.[30] We see physical manipulation of the bodies; people didn't *just* pray. And while they were careful to assign responsibility for success to divine intervention, human efforts could be part of the picture. The father's actions did not invalidate the case's miracle status—any return from the dead was understood to be the work of the divine—but they blur the line between human and supernatural agency.

Although these cases are concerned to prove death, they still describe the return to life, providing physical descriptions of the processes. Another of Elizabeth's miracles illustrates this clearly: "Once the invocation had been made, they saw the natural color of the boy return all at once and began to feel the pulse in his veins. [...] And we, who heard this testimony, saw the boy alive and healthy."[31] Through such cases we can get a sense of what kinds of signs people associated with a return to life and perhaps the order in which they might appear. Compared to Caesarius's exempla, the canonization collections thus present death in more physical terms and focus on the transformations the body would undergo in dying and reanimating.

Another key difference between the canonization collections and the *Dialogus* concerns the initiation of reanimation miracles. In brief, the canonization collections show far more involvement by the living community. Whereas Caesarius presents the communities as passive and restricted to ritual roles—they tend to be present only to undertake the standard deathbed rituals and make no effort to hinder, let alone reverse, the death—the canonization collections often show an extremely active community. The desperation and the efforts people took to reverse death might at first glance seem self-explanatory. Yet the resignation seen in the *Dialogus* reminds us that there are other ways people could approach death. Seeking a miracle was not a given.[32]

The above-mentioned case of the drowned boy that Hedwig reanimated displays a number of these elements: alongside the physical interventions (expulsion of water, heating by the fire) is a vow by the father to visit Hedwig's tomb and copious prayers by others. In these miracles, the works of others, both the "normal" community and the holy saint, are crucial for bringing people back to life. A vow to the saint generally initiates the divine involvement. Someone promises to do something—usually visit the saint's tomb, perhaps with a gift or to have a mass performed—and after the words are spoken, the miracle occurs (though sometimes with delay).[33] This is by far the most common mode for miracles of every type. Sometimes supplication without a formal vow suffices. Most of the cases involve someone making a vow on behalf of the dead (child). These vows vary in their goal (baptism, expunging sin/negligence), but all promise to somehow honor the saint, usually by visiting the tomb. But sometimes other actions are recognized as pivotal and no vow or supplication is mentioned. For instance, concerning two adults, Hedwig's merits seem to act on their own without her or anyone else doing anything extra. A deceased woman is plucked out of punishing flames, presumably in purgatory, by Hedwig, despite never having heard of the saint, let alone having prayed to her. The woman lives for an additional four years, during which time she publicizes Hedwig's power.[34] A man dies and is buried near Hedwig's body; proximity to the saint is seen as the reason for his reanimation two days after having died.[35] These resuscitations happen in slightly odd ways—at the desire of Hedwig without any bidding. While both adults end up publicizing their miracles, which can be seen as a kind of "repayment," they do so after the fact, and their expressions of gratitude are not presented as part of a deal with the saint to secure reanimation.

Although in Hedwig's collection we find a clear divide between the reanimation of children and adults, with the adult cases rather reminiscent of those described by Caesarius because of the impetus being spontaneous divine action, other saints have adults returning to life after intense emotional outpouring and beseeching by surviving family members.[36] These devotional interventions show two things: that the revivification was due to the soul's activity (rather than a mechanical understanding of life-body), and that agency in reanimation was distributed and *not* in the hands of the person benefitting from the miracle (in contrast to many other healing miracles).

Unbaptized Children

All of the cases discussed here, and indeed all of the cases that were included in such "official" collections, are unambiguous with respect to the life that returns to the body. As mentioned above, it is the intervening death that needs to be proven. But there is a subset of reanimation cases that can, and sometimes do, bring the question of *life* more sharply into focus. These are the children who died before they could be baptized. Like Caesarius's "good" people, such children, almost always infants, would not need to return to life for very long for the wrong of dying in an unbaptized state to be righted. The soteriological requirement of baptism contributed to creating conditions in which life would even be sought on such bodies. According to medieval Christian and early modern Roman Catholic doctrine, anyone who died without baptism would be damned.[37] This belief derived from the early fifth-century writings of St Augustine against the Pelagians, in which the Church Father unambiguously asserted the necessity of baptism for the salvation of all souls. Because of Augustine's authority in the Church, his statements that this apply even to infants defined the standard theological position; in short, without the sacrament even the smallest children would suffer the flames of hell for eternity.[38] This idea was, in turn, rooted in Augustine's anthropology and the relationship between human nature and original sin: Adam and Eve's disobedience rendered human nature corrupt and thus unfit to be in the presence of God. But God in his mercy provided a way out of this predicament: through the sacrifice of the crucifixion, humankind was redeemed, and through the sacrament of baptism, human nature could be cleansed of its original sin. To restore the soul, however, baptism had to be performed on the living body.

Baptism thus became a means of protecting a person from the dangers of dying in a corrupted state.[39] This increasingly meant baptizing the person before they reached an age when they could understand the basic tenets of Christianity. Over the course of the Middle Ages, infants, rather than adults, became the main recipients of the sacrament.[40] An additional response by the highest Church officials to this baptismal imperative was to permit anyone to perform the sacrament in emergency situations. While it was still preferable to have the full ritual performed by a priest, as long as the person baptizing spoke the correct words, used water, and desired the sacrament to work, the emergency version would be effective.[41] Pastoral manuals and ecclesiastical legislation describing such procedures make clear that the body need to be alive for the sacrament to work, but they do not address potential issues in discerning life and death.[42]

For the lay practitioners faced with the emergency of a dead, unbaptized infant, however, things might not be so clear. Is the child truly dead? Is there some way to make it not dead? Efforts to reanimate unbaptized infants appear already in the earliest canonization collections.[43] Since an emergency baptism could be performed very quickly, the infant did not need to return to life for very long to receive the sacrament. Indeed, the canonization collections show a number of such cases, with the infants living usually for a few hours, perhaps up to a few days. St Stanislaw was credited with reanimating an infant who died shortly after birth and lay dead for several hours; the child remained alive for the rest of the day and night.[44] These time frames, while short, are nonetheless long enough that we can imagine the returned state of life to be fairly evident. Controversies erupting in the fifteenth century and then throughout the early modern period around extremely temporary reanimations, however, raise the questions of how long is long enough and what signs are clear enough for a return to life to be certain. We hear of bodies flickering with life for mere minutes. Bishops complained that people were mistaking for life the slight bodily movements or other changes that would be better explained by natural causes, such as the heat of nearby fire.[45] While some groups recognized certain ephemeral signs as indicating life, others denied the sufficiency of such signs for proving reanimation.

The pressure to baptize thus pushes the issue of determining the line between life and death, which ordinarily does not appear very ambiguous. Although baptism is ultimately what is at stake, it is not what becomes contested. A consequence of these baptismal practices were efforts to define life more precisely, efforts that could also raise questions about expertise, agency, and relationships between the human, the natural, and the divine. Who determines which signs count as life and which refer to some other cause? Who can do what to a body, and what do the resulting changes mean? How can we recognize the work of God? Questions that we encountered above in discerning death return here, but now for pinpointing life. As controversies continued to flare up throughout the early modern period, it is clear that this was an area where ambiguity could not be reliably resolved.[46] Indeed, although such controversies waned by the later eighteenth century, it was not better definitions or more precise measurements of life that finally brought them to an end another two centuries later, but rather shifts in the Church's soteriological stance. In the second half of the twentieth century, space was made for hope that God might find ways other than baptism to save the souls of such children.[47] While other scholars have discussed how such theological changes meant that people no longer needed to try for miraculous reanimations and emergency baptisms, they have not emphasized that such changes also changed how and where people sought to locate the margins life. People no longer needed to look for temporary life in these souls-bodies: not only the afterlife, but *this* life had changed its shape.

Notes

1. Katharine Park, "Birth and Death," in *A Cultural History of the Human Body in the Medieval Age*, ed. Linda Kalof (Bloomsbury Publishing, 2010), 36.
2. Aleksander Semkowicz, ed., "Vita sanctae Hedwigis ducissae Silesiae. (Vita maior, vita minor, genealogia)," in *Monumenta Poloniae historica. Pomniki dziejowe Polski*, vol. 4 (Lwów: Księgarnia Gubrynowicza i Schmidta, 1884), 501–655; Kenneth Baxter Wolf, ed., *The Life and Afterlife of St. Elizabeth of Hungary: Testimony from Her Canonization Hearings* (New York: Oxford University Press, 2011); Wojciech Kętrzyński, "Vita sancti Stanislai Cracoviensis episcopi (Vita maior)," in *Monumenta Poloniae historica. Pomniki dziejowe Polski*, vol. 4. (Lwów: Księgarnia Gubrynowicza i Schmidta, 1884), 319–438; Ludwik Ćwikliński, ed., "De vita et miraculis sancti Iacchonis (Hyacinthi) ordinis fratrum predicatorum. Auctore Stanislao lectore Cracoviensi eiusdem ordinis," in *Monumenta Poloniae historica. Pomniki dziejowe Polski*, vol. 4, 6 vols. (Lwów: Księgarnia Gubrynowicza i Schmidta, 1884), 818–903; Caesarius of Heisterbach, *Dialogus Miraculorum*, ed. and trans. Nikolaus Nösges and Horst Schneider (Turnhout: Brepols, 2009).
3. For other collections, see Hannah Elmer, "Alive Enough? A Conflict over Divine Presence and Natural Power in the Reanimation of Dead Infants, 1400–1545" (New York, Columbia University, 2019), chap. 2.
4. There is much diversity here; for some of the ways that politics and chronology could affect the content of these collections see Gábor Klaniczay, *Holy Rulers and Blessed Princesses: Dynastic Cults in*

Medieval Central Europe (New York: Cambridge University Press, 2002); Laura A. Smoller, "Defining Boundaries of the Natural in Fifteenth-Century Brittany: The Inquest into the Miracles of Saint Vincent Ferrer (d. 1419)," *Viator* 28 (1997): 333–59.

5. For a brief sketch of the tremendous variety and different emphases that perhaps characterize medieval ideas about death and the afterlife, see Caroline Walker Bynum and Paul Freedman, "Introduction," in *Last Things: Death and the Apocalypse in the Middle Ages* (Philadelphia: University of Pennsylvania Press, 2000), 2–10.
6. For an extensive and nuanced analysis of a variety of ways in which relationships between body and soul were expressed and understood over the course of late Antiquity and the Middle Ages, see Caroline Walker Bynum, *The Resurrection of the Body in Western Christianity, 200–1336* (New York: Columbia University Press, 1995).
7. Caroline Walker Bynum and Paul Freedman, "Introduction", op. cit.; Caroline Walker Bynum, *The Resurrection of the Body in Western Christianity, 200–1336,* op. cit.
8. The dynamism of this time after death is especially evident in ideas of purgatory, an invention of the High Middle Ages. See the classic treatment by Jacques Le Goff, *The Birth of Purgatory* (Chicago: University of Chicago Press, 1984); and more recently, Chiara Franceschini, *Storia del limbo* (Milan: Feltrinelli, 2017).
9. Katharine Park, "Birth and Death," op.cit. talks about this through the terms of the different life stages.
10. Of the few pieces to consider reanimation, see Christian Krötzl, "Evidentissima signa mortis. Zu Tod und Todesfeststellung in mittelalterlichen Mirakelberichten," in *Symbole des Alltags, Alltag der Symbole: Festschrift für Harry Kühnel zum 65. Geburtstag*, ed. Blaschitz et al. (Graz: Akademische Druck- u. Verlagsanstalt, 1992), 765–75., and to some degree Laura A. Smoller, "Defining Boundaries of the Natural in Fifteenth-Century Brittany," op. cit.
11. Such visions are ubiquitous in medieval writings. Vision visitations have their own literary topoi and are marshaled in a number of genres to various ends. For a recent treatment with a useful overview of the scholarship, see Michael D. Barbezat, "Converse with the Dead as a Technology of the Self: Agreements to Return from the Other-World in Peter of Cornwall's Book of Revelations,"

Journal of Medieval History 48, no. 1 (2022): 32–56, https://doi.org/10.1080/03044181.2021.2018022.

12. Barbezat examines such vision-visitations in these more "mundane" relationships Michael D. Barbezat, "Converse with the Dead as a Technology of the Self," op cit.
13. "et cum clausisset oculos causa quiescendi eodem die circua occasum solis, nec tamen dormiret, apparuit ei quidam uir canus et reuerendus, qui dixit ei: Mulier anxia et tribulata, obtentum est hoc per preces sancti Stanislai, quod filius tuus reddetur tibi in hac nocte, quia exaudite sunt preces tue et lacrime et uoueas uotum illi sancto, in cuius die incepit puer infirmari. Et mater apertis oculis et neminem uidens reducta est in domum, ubi iacebat puer." Wojciech Kętrzyński, "Miracula sancti Stanislai," in *Monumenta Poloniae historica. Pomniki dziejowe Polski*, vol. 4 (Lwów: Księgarnia Gubrynowicza i Schmidta, 1884), 302, op. cit.
14. Among others, see Nancy Caciola, *Afterlives: The Return of the Dead in the Middle Ages* (Ithaca, NY: Cornell University Press, 2016); Jean-Claude Schmitt, *Ghosts in the Middle Ages: The Living and the Dead in Medieval Society* (Chicago: University of Chicago Press, 1998); Winston Black, "Animated Corpses and Bodies with Power in the Scholastic Age," in *Death in Medieval Europe: Death Scripted and Death Choreographed*, ed. Joëlle Rollo-Koster (New York: Routledge, 2017); Katharine Park, "Birth and Death," op. cit.
15. For an argument linking the decline of revenants in favor of ghosts with the rise in scholastic-university culture, see Winston Black, "Animated Corpses," op. cit.
16. See Book 12, Chapter 14, Caesarius of Heisterbach, *Dialogus Miraculorum*, 2208–10.
17. Of the vast literature on saints, see Caroline Walker Bynum, *Christian Materiality: An Essay on Religion in Late Medieval Europe* (Cambridge, MA: MIT Press, 2011); Robert Bartlett, *Why Can the Dead Do Such Great Things? Saints and Worshippers from the Martyrs to the Reformation* (Princeton: Princeton University Press, 2013); André Vauchez, *Sainthood in the Later Middle Ages* (Cambridge: Cambridge University Press, 1997).
18. For some of the different positions, see Winston Black, "Animated Corpses," 88ff.

19. One of the most prominent exceptions is the story of a nun who attempted suicide and was rescued. “Tunc considerantes quod supradicta foret sanctimonialis, iam pene suffocata, territi sunt omnes, et curam ei adhibentes, postquam aquam evomuit et loqui potuit, interrogaverunt eam dicentes: ‘Quare tam crudeliter egisti, soror?’ Respondit illa: ‘Dominus iste,’ digito Priorem ostendens, ‘minatus est mihi, quia me mortuam sepelire deberet in campo. Unde magis elegi gurgitem hung supernatando descendere, quam bestialiter in campo sepeliri.’…” Caesarius of Heisterbach, *Dialogus Miraculorum*, 772–74.
20. Caesarius of Heisterbach, *Dialogue*, 2:267.
21. “Capitulum 35 - De converso de Cynna qui propter obulum reversus est ad corpus. Cumque sursum nitenti semper obsisteret, neque aliud esset quod animam impediret, ad angelorum petitionem permissa est redire ad corpus. Mirantibus cunctis, visionem retulit, et missus est ab Abbate nautae integer denarius sub omni celeritate. Quem ut ille recepit, eadem hora secundum considerationem conversus exspiravit…” Caesarius, *Dialogus miraculorum*, 2122-2124.
22. Katherine Park, “Birth and Death,” op cit.
23. For more on exempla, see Jacques Le Goff, *In Search of Sacred Time: Jacobus de Voragine and the Golden Legend* (Princeton, New Jersey: Princeton University Press, 2014); Eamon Duffy, “Introduction,” in *The Golden Legend: Readings on the Saints* (Princeton: Princeton University Press, 2012), xi–xx.
24. I have counted roughly sixteen reanimations in the entire collection. For an analysis of several of these cases and how they are situated in the collection as a whole, see Hannah Elmer, “Alive Enough?,” op. cit. chap. 2.
25. Laura A. Smoller, “Defining Boundaries of the Natural in Fifteenth-Century Brittany,” op. cit.
26. Kenneth Baxter Wolf, *The Life and Afterlife of St. Elizabeth of Hungary*, op. cit., 99–100.
27. Ibid., 106–7.
28. Aleksander Semkowicz, “Vita sanctae Hedwigis,” op. cit., 622.
29. Smoller also discusses a few cases of Vincent Ferrer where the typical order or argumentation is different. Laura A. Smoller, “Defining Boundaries of the Natural in Fifteenth-Century Brittany,” op. cit.

30. Et licet nulla spes vite haberetur de ipso, prefatus tamen Waltherus de beate Hedwigis confisus suffragio, eam sepius invocando dicebat: "beata Hedwigis, suscita puerum istum, quem si suscitaveris, tuum cum ipso pedibus meis adibo sepulcrum." Excussa quoque de corpore pueri, quantum excuti potuit, aqua, posuit ipsum ad ignem. Iacuit autem inmobilis, ut procul dubio mortuus, a meridie usque ad solis occasum diei festi predicti. Cum vero non desisterent fideles dominum invocare per beate Hedwigis merita pro puero revocando ad vitam, eodem sero respirare incipiens, sequenti mane surrexit puer incolumis, nichil paciens omnino doloris, in dorso ipsius ad instar panni nigri ex compressione predicta livore tantummodo remanente. Aleksander Semkowicz, "Vita sanctae Hedwigis," op. cit., 622.
31. Kenneth Baxter Wolf, *The Life and Afterlife of St. Elizabeth of Hungary*, op. cit, 106–7.
32. Of course, that many of the reanimated subjects in the canonization collections are small children and not the adults presented by Caesarius is also an important factor.
33. In addition to the case just mentioned, see also the reanimation of Petruscha by Hedwig in Semkowicz, "Vita sanctae Hedwigis," 623–24.
34. In villa hospitalariorum Lissocitz nuncupata, Olomuncensis dyocesis, quedam mulier finem vivendi accepit. Que dum per diem mortua teneretur, pro delictis, a quibus se vivens non plene per satisfactionem purgaverat, penis addicta flammarum incendia sustinebat. Deus autem secundum prophetam mirabilis in sanctis suis, volens beate Hedwigis ammirabilem, eciam in remotis manifestam fieri sanctitatem, sic disposuit, ut prefate mulieri defuncte et in flammis existenti ipsa iam dicta sancta visibiliter appareret et eam potenter ab incendio traheret ipsamque divina virtute pristinam revoaceret ad vitam. Hec femina prius in vita sua numquam audierat de sancte preconio sanctitatis aut de ipsius miraculis, sed per ipsam de flamme cruciatibus eruta et a mortuis ad statum pristinum revocata, per quatuor annos, quibus vixit post resurrectionem iam dictam, non cessavit eam extollere laudibus, sed enarravit omnibus, quibus poterat, quod factum fuerat circa ipsam ostendensque multis quedam signa adustionis, que in eius remanserant corpore ad miraculum divineque pietatis beneficium

plenius comprobandum. Aleksander Semkowicz, "Vita sanctae Hedwigis," op. cit., 624.

35. In opido Prusicz Wratizlaviensis dyocesis quidam mortuus fuit. Qui dum sepeliri deberet in cimiterio monasterii, in quo sancte Hedwigis est corpus sepultam, postquam duobus iam iacuerat diebus exanimis, a mortuis resurrexit. Istius fama miraculi pervenit ad ducem, qui statim credere noluit, sed hominem resuscitatum coram se adduci precepit, fecit quoque ad suam vocari presenciam dominum ville, in qua manens decesserat homo ille, qui dicebatur revocatus ad vitam, ut et ipse veniens, si quam de hoc facto sciret, interrogatus diceret veritatem. Uterque autem comparens coram ipso duce constanter asserebat et veram fuisse illius hominis mortem et resusciationem iam dictam interventu beate Hedwigis veraciter esse factam. Aleksander Semkowicz, "Vita sanctae Hedwigis," op. cit., 624–25.
36. See Ludwik Ćwikliński, "De vita et miraculis sancti Iacchonis," op. cit., 850–62.
37. Things get somewhat more complicated in the twentieth century with shifts brought about through the Second Vatican Council and other later adjustments that permit the faithful a modicum of hope that the unbaptized infants (including aborted fetuses) might nonetheless be saved through God's mercy. See Elke Pahud de Mortagnes, "Der versperrte Himmel: das Phänomen der sanctuaires à répit aus theologiegeschichtlicher Perspektive," *Schweizerische Zeitschrift für Religions- und Kulturgeschichte* 98 (2004): 31–47.
38. For a very helpful overview of the variety of positions in late antiquity, helping show the contingency of the position ultimately adopted by the Church, see Pier Franco Beatrice, *The Transmission of Sin: Augustine and His Sources* (New York: Oxford University Press, 2013).
39. Maxwell E. Johnson, *The Rites of Christian Initiation: Their Evolution and Interpretation* (Collegeville, MN: Liturgical Press, 2007), 260.
40. Park argues the intensification of the Church's teachings on this contributed to a number of lay responses, and desire for baptism was not limited to salvation. The unbaptized were excluded from consecrated ground, adding another layer of shame for the grieving family, and in some areas baptism was also seen to protect people

from drowning or other types of harm. See Katharine Park, "Birth and Death," op. cit., 23–24.

41. Midwives were increasingly trained to do this. See Kathryn Taglia, "Delivering a Christian Identity: Midwives in Northern French Synodal Legislation, c. 1200–1500," in *Religion and Medicine in the Middle Ages*, ed. Peter Biller and Joseph Ziegler (York: York Medieval Press, 2001), 77–90; Katharine Park, "Birth and Death," op. cit., 26.
42. See, for instance, the 1320 statutes by Bishop Nanker of Kraków in: Udalricus Heyzmann, ed., *Starodawne prawa polskiego pomniki*, vol. 4 (Warsaw: Nakł. Akademii Umiejętności, 1875), 4ff.
43. See, for instance, a case in the collection for the Empress Kunigunda (the second official papal canonization process). "De S. Cvnigvnde Imperatrice, Virgine, Conivge, Vidva, Demvum sanctimoniali benedictina, Confygiae et Bambergae in Germania (BHL 2002, 2006, 2007, 2008)," in *AASS*, vol. Martii 1 (Antwerp: J. Meurisum, 1668), 273–78, here 277C.
44. 8. De filio Richardi a morte suscitato. Cum Richardus Lombardus cum uxore pregnante Cracoviam venisset beati Floriani festivitate invitante, tercia nocte uxor ipsius bene natos gemellos, masculum et feminam, est enixa. Sed masculus sequenti die nativitatis sue sine baptismi gracia est defunctus. Quod cum nunciatum esset patri, dolor cordis apprehendit eum non tam pro defuncta prole, sed pro morte anime. Qui consurgens ad Deum, cum omni fiducia supplicat Deo prece cum lacrimis importuna, ut per beati Stanislai gloriosa merita resuscitare dignaretur filium suum ad percipiendum duntaxat nove regeneracionis sacramentum; unde et mortuum puerum ab hora matutinali usque ad horam diei terciam non permisit sepeliri. Hora vero tercia nunciatum est patri: En filius tuus vivit. Qui gaudio repletus sine mora utramque prolem in ecclesia beate virginis cum sollempnitate debita in Christo ex aqua et spiritu sancto regenerari procuravit. Verum puer, qui resuscitatus fuerat, illa die tota et nocte sequenti supervixit, sed circa auroram temporaliter moriens ad vitam et requiem eternam migravit. O mira Salvatoris clemencia! Mortuus filius ire et sub perpetue salutis relictus discrimine non ob aliud suscitatur a morte, nisi ut renato ex aqua et spiritu sancto per beati Stanislai merita, ut pater pecierat, filiacionem mereretur adopcionis divine. Wojciech Kętrzyński, *"Vita sancti Stanislai,"* op. cit., 400–401.

45. For an extensive treatment of the earliest of such controversies (in the second half of the fifteenth century), see Hannah Elmer, "Alive Enough?," chap. 3, op. cit.
46. For an overview of such reanimation practices and controversies, see Jacques Gélis, *Les enfants des limbes: mort-nés et parents dans l'Europe chétienne* (Paris: Audibert, 2006). Gélis has also published other articles relating to the topic and has been the most prolific scholar in this area. I take issue, however, with some of his methodology and categorization. See Hannah Elmer, "Alive Enough?", introduction, op. cit. A very useful and methodologically sounder treatment can be found in Walter Pötzl, "Die Taufe totgeborener Kinder: Inchenhofen, Hohenwart und Tuntenhausen, Bergatreute und Ursberg, 'Sanctuaires à répit' in Süddeutschland," *Bayerisches Jahrbuch für Volkskunde* (2012): 105–42; Oskar Vasella, "Über die Taufe totgeborener Kinder in der Schweiz," *Zeitschrift für Schweizerische Kirchengeschichte* 60 (1966): 1–75.
47. Elke Pahud de Mortagnes, "Der versperrte Himmel," op. cit.

CHAPTER 13

The Technical Imaginary of Immortality: Being-towards-Death at the Limits of Transhumanism

Jessica Lombard

By discussing transhumanism from a philosophical and phenomenological perspective, this paper examines one of the Trojan horses of transhumanist discourses: the desire for immortality. What all transhumanist movements share is the demand for the right tohuman enhancement through technological artifacts and scientific progress. Enhancement itself encompasses a vast and diverse set of activities, within which the question of im/amortality plays a crucial role.

Although the difficult issues of ageing and death are often treated as psychological, social, or political problems, this paper first argues

J. Lombard (✉)
Centre de Recherche en Histoire des Idées, Université Côte-d'Azur, Nice, France
e-mail: jessicalombard.univ@gmail.com

Institut d'histoire et de philosophie des sciences et des techniques, Université Paris 1 Panthéon-Sorbonne, Paris, France

J. Muller (ed.), *Automata, Cyborgs, and Mutants*, Palgrave Studies in Medieval and Early Modern Medicine,
https://doi.org/10.1007/978-3-031-93201-4_13

that transhumanist movementspresent death as a biological issue. Death is no longer seen as the evolutionary limit of living beings anymore but rather an unnecessary disease to be overcome by technological breakthroughs. Contrary to a common assumption, we suggest that transhumanist movements render human life all the more biological by strictly considering it as an object of natural, biological and medical sciences, in a perspective that recalls the Cartesian animal machine, i.e., an application of automaton thinking to living beings.

To better understand the consequences of this perspective, this paper turns to Heidegger's elucidation of *Sein zum Tode* (being-towards-death) in its *Being and Time*. Heideggerian phenomenology depicts death not only as a biological fact that beffals Cartesian automata but as an ontological structure of human existence, whose life forms in history only because he is essentially facing death.

Moreover, in this approach, this openness to finitude is distinctive of the human being and not of other living beings, which implies that animals relate to death in a way fundamentally different from Dasein. In the framework of the Heideggerian existential analysis, we consider how achieving immortality would abolish the original limit of the human temporal horizon of existence, without pretending to draw hypothesis for other living beings.

Consequently, this chapter claims that transhumanism, by construing death as an external disease that may or may not occur to the human being, reduces the human to his primary biological state—to an automatic animalitas—rather than recognizing Dasein as a being fundamentally shaped by finitude. We argue that by treating death as a strictly biological issue, transhumanist movements reveal how we conceive of an ontological shift in the status of finitude: from an existential disposition to a biological threat and that such a transformation could fundamentally alter our relation to the existential sphere. In this way, transhumanism bears witness to an evolution of our technical imaginary, disclosing how we reimagine our relation to ourselves through technology. It not only proposes concrete strategies for overcoming ageing, disease, or mortality, but also discloses how, today, we conceive of ourselves through technological mediation. By projecting scenarios of rejuvenation, radical longevity, or even immortality, transhumanist discourse reveals a cultural reconfiguration of self-understanding: finitude is no longer envisaged primarily as an existential horizon but as a variable subject to technical intervention. In this sense, transhumanism unveils a new mode of imagining the human condition that ties our relation to ourselves ever more closely to the possibilities (and promises) of scientific and technological developments.

A Journey Towards Immortality

The increase in human life expectancy, i.e. the average time an individual within a population is statistically expected to live, is a significantly documented process, independent of transhumanism. Demographic databases such as Gapminder and WorldData platforms (on which are drawn the following data) show that human life expectancy has significantly evolved: in 1960, the global life expectancy was 50.7 years for men and 54.6 years for women and by the early XXIst century, these figures had risen to 69.6 years and 74.5 years, respectively. It is, in fact, a contingent[1] phenomenon based on various factors such as improved healthcare systems and innovations, improved nutrition and public hygiene and broader sociopolitical changes, which are also due to the development of a variety of technologies.[2] More so, several researchers[3] have noted the difficulty of establishing an a priori limit to human life expectancy, one that could never be exceeded in theory, despite all the efforts made to improve living conditions and access to health care. There is no full-fledged socio-political programme to increase life expectancy in the population, nor is it the subject of specific measures for its enforcement. However, transhumanist movements argue that longevity should be considered as a matter of political concern, one that calls for active transformation through technological progress. Indeed, even if life expectancy has increased independently of transhumanism, a number of scientific books and articles are beginning to put forward how transhumanist perspectives can contribute to the debate surrounding this phenomenon.[4] In general, the ageing of organisms is not considered a disease in itself but an issue that triggers various diseases (like Alzheimer's). It is a complex process consisting of an accumulation of independent deficiencies with little or no causality. No single gene causes ageing; ageing is not the result of a genetic mechanism that ensures its occurrence. It comes from what Linda Partridge defines as "an unregulated side effect of the failure of natural selection to maintain function at the later ages that few individuals reach in nature."[5] That is why it is embedded with the deterministic weight of a natural and local evolution of living beings' organisms, which are precisely alive in that they can meet their end.

Transhumanist movements do not deny that ageing is an inevitable and constitutive biological stage of life, but they wage a war against the biological determination this entails. For transhumanist movements, ageing in itself is a disease that must be cured. As an example, we can mention the *Letter to Mother Nature* of the well-known transhumanist Max More, whose development of the extropianist movement led the way to the rise of transhumanist movements. "With all due respect, we must say that you [Mother Nature] have in many ways done a poor job with the human constitution. You have made us vulnerable to disease and damage. You compel us to age and die—just as we're beginning to attain wisdom."[6] Max More's first Amendment is then that: "We will no longer tolerate the tyranny of ageing and death. Through genetic alterations, cellular manipulations, synthetic organs and any necessary means, we will endow ourselves with enduring vitality and remove our expiration date. We will each decide for ourselves how long we shall live."[7] Indeed, as the French transhumanist Didier Coeurnelle points out,[8] the fact that ageing is universal and inevitable should not prevent us from trying to overcome it. A universal disease is still considered a disease, like the flu. The same goes for diseases which are unavoidable without treatment, like Alzheimer's or AIDS.

Consequently, transhumanism considers ageing and death as essentially biological states that should not determine our evolution nor remain inevitable. They must be postponed with any available technological progress, and immortality is no longer the mythical, illusory fruit of alchemy or spiritual quest, but a scientific aim that should be accessible within the framework of medical research. As transhumanism wants to overcome death during our life, and not afterwards as some religions, it contextualises death as something that happens at the end of our biological life and pleads for overcoming this biological mortality by developing new technologies. If we are living longer and in better shape, if we are deeply involved in the development of medicine, therapy and sometimes enhancement; why would death be the only biological "disease" that we should not try to overcome? Transhumanist authors do not consider death as something that intensifies the human life. For them, it is a border that should be moved away by extending life expectancy as far as possible.

But there is a distinction between amortality and immortality.[9] From an amortalist perspective, the subject can no longer die of old age but remains mortal and vulnerable to accidents, euthanasia or suicide. From an immortalist perspective, the subject is forever immortal. Now, to consider different ways of overcoming death imply that there are different ways of thinking about it, which accounts for the many different transhumanist perspectives on this issue (cryogenic, mind uploading, genetic engineering, etc.) and explains why it is impossible to put them all into one category. As this very diversity precludes us from providing a detailed account of transhumanist discourses in this one article, the dualistic approach we provide nonetheless enables us to consider the critical elements of transhumanist issues related to finiteness. Indeed, this chapter does not intend to classify the various transhumanist movements but rather to propose a philosophical analysis of how transhumanist movements consider death and how it contrasts with a phenomenological perspective in which finitude has an ontological status. To clarify this argument, we mention two major transhumanist perspectives on immortality.

A first transhumanist approach to amortalism is mainly based on the advances in research in the biology of ageing. It explores the links between the ageing of organisms and their cellular senescence (degradation of cell functions), as biologists discover this phenomenon to be increasingly flexible: "Some organisms do not seem to age or age very slowly. This tells us that ageing is a much more plastic and modular phenomenon than we thought before undertaking this research."[10] As an example, the jellyfish *Turritopsis nutricula*,[11] is capable of returning to its polyp stage by cellular transdifferentiation, i.e., by modifying its cells' characters and thus functions. Biological ageing does not seem to be a phenomenon whose temporality is invariant. For this reason, the biology of ageing is a significant field of research now that our knowledge of genetics and genomics is growing. Of course, it is unlikely that a single genetic mutation will have a global impact on cellular degeneration, but it is not a point that transhumanist authors contest. The seven to nine steps, more commonly known as the "hallmarks of ageing,"[12] are well-known to be the Trojan horse of Aubrey de Grey's Strategies for Engineered Negligible Senescence (SENS).[13]

In this perspective, where "anti-ageing does not mean anti-death,"[14] transhumanism centres the overcoming of mortality on a biological approach that would seek a form of *amortality*. The individual would not be able to die of old age, but he would remain subject to the possibility of death by accident, euthanasia, suicide or other contingent means.

The second major way of overcoming death discussed by transhumanist movements entails computer science. It considers the possibility of carnal disembodiment in order to download the *persona* on digital platforms, and suggests that "the mind continues to depend on a substrate to exist and to operate, of course, but there are substrate choice."[15] From then on, it is not only the biological body that becomes optional but any support since only information has any real and consistent value, as the transhumanist Hans Moravec argues: "Though mind is entirely the consequence of interacting matter, the ability to copy it from one storage medium to another would give it an independence and an identity apart from the machinery that runs the program."[16]

Where the previous approach seeks a form of amortality that implies to reduce the organisms' biological degradation, the second one suggests a form of *immortality* because the new substrates are permanent or because the digital mind can be transferred to another one after expiration, being virtually incorruptible and eternal.

But transhumanism does not merely aspire to an ideal of humanity in which death is overcome. It also carries a phenomenological weight in its way of understanding death that a philosophical reading must draw out. To conceive death as a biological phenomenon does not mean assuming that it happens only *to* the body, but rather that the body as that to-*which*-death-happens. If our body is biological, and therefore mortal, then death appears as a biological experience that we are almost doomed to suffer, unless we radically change this biological body or make it un-biological. Thus, when transhumanist movements reduce the ontological issue of finitude to its manifestation in the body, understood biologically, they implicitly reveal a fundamental phenomenological perspective. What is at stake is not death *as* a phenomenon but rather the phenomenon that manifests death; i.e., the body. As the French phenomenologist Jean-Philippe Milet emphasizes, we are confronted "not [with] finiteness as a quality of bodies, but [with] the body as a display of finiteness, that in which it takes consistency or takes shape."[17] It is no longer that the body dies, but that death is embodied, just as any disease is embodied in the body. This means that the question is not on whether the body is made to die, but that it is the embodiment of death in the body that is in question and that, as more and more diseases, could in principle be interrupted.

As we mentioned diseases, that also take shape in the body, it is important to underline that for transhumanist authors, our display of finiteness is not circumscribed by death but also appears in weaknesses, diseases, wounds and any biological limitation that they seek to overcome. In this sense, death is conceived as just another disease; hence transhumanist thinkers wonder why there are no collective efforts to eradicate it, as there are to cure other diseases.

From this perspective, the body's finiteness is expressed in every interaction it maintains with the world. Finiteness is not only about death, but about the broader set of human limitations that are mediated by a dense network of prosthetic artefacts. The human being lives in a very dense technological environment, which functions as a necessary support system to help and repair this vulnerable body that conveys limitations and is marked by weakness, illness, and ultimately death. Transhumanist movements acknowledge that humans are never reconciled with their biological and ontological fragility, as our technological dwelling itself attests. Consequently, they argue that we should not be immortalised in the fact that we are indeed mortals, and confined to mortality as our defining condition. On the contrary, for them, mortality just fuels and intensifies the most ardent immortalist desire. And it is, supposedly, up to us to ensure that death will no longer be embodied in our bodies, just as certain diseases no longer "incorporate" themselves in us once eradicated by technological and medical progress.

This phenomenological stance has an essential impact on how finiteness is understood: it is not the evolutionary limit of living beings anymore, but rather an *unnecessary disease*, hence a strictly biological issue borne by the living body and that should be overcome through technological breakthroughs. In this view, mortality is displaced from an existential horizon to a technical problem to be solved and, as such, transhumanist seems to render human life all the more biological by treating it as an object of the natural, biological and medical sciences. This perspective can be likened to an application of automaton thinking to living beings, like the Cartesian "animal machine"[18] is. What follows from this reduction is a profound shift: it entails a conflicted change in the ontological status of finitude, where the transhumanist construal of our mortal body, and of death, as fundamentally biological issues proves incompatible with a more existential analysis, such as discussed in the phenomenology of Martin Heidegger.

Human Existence and Mortality

First of all, the concept of Dasein needs to be recalled in order to highlight how it is open to the world, a world structured by the horizon of death. In German, *Sein* is the verb "to be," and *Da* carries the a spatial connotation of being "there." Dasein thus literally means "to be there." For Heidegger, it designates, in fact, the structure of the human existence, in contrast to any other being. He invests the term with a new metaphysical meaning: Dasein becomes the name for the ontological structure of the human being. Consequently, the prefix *Da-* does not merely signify being "somewhere" but, combined with the suffix *-sein*, expresses the being of the human as he exists. To be (*Sein*) for the human means to exist as human, and to be human means to be "there," as summarised by the phenomenologist Françoise Dastur: for Heidegger, "existence is as spatial as it is temporal."[19] It is in through this understanding of existence that, for Dasein, *Da-* comes to mean "disclosure."[20]

For Heidegger, disclosure is the way the human appears in the world, or the way the human exists in the world. The human is *Da* in the world, is here; but to be here is to be disclosed to the world. This is the ontological disposition of the human; because only the human is "here in the world" as disclosed to the world, and as the world is disclosed by him. A maybe more accessible rendering of disclosure is openness. For Heidegger, the human opens to the world, and the world is open to the human. This means, quite simply, that for the human, the world is an issue. From a Heideggerian perspective, to "be an issue" means to be "concerned"[21] about something and to care for something.

The example from ethology and biosemiotics that Heidegger draws upon in *The Fundamental Concepts of Metaphysics*[22] is enlightening.[23] Some animals, such as crabs or moths, seek or flee the light. Therefore, they seem to adopt a behaviour implying light consciousness and seem to be aware of light. They move away or towards it. Thus, light can initiate their movement. For those animals, light has a meaning. Some crustaceans, for example, swim in such a way that their eyes are oriented according to the light source, wherever it comes from. Each time, light is a stimulus. But consequently, precisely because this meaning is a stimulus for the animal, according to Heidegger, light never seems to exist per se—as something that first is in itself, then and after is something that can be seen, used, transformed, etc. Neither the crab nor the moth goes towards the light for light itself; but for its warmth, for its brightness, etc. Of course, for the human being, light can also be a category. It can also be a source of warmth or luminosity. But as Dasein opens and discloses to the world, light also gives on light. This light per se, for Dasein, is neither brighter nor more useful. The fact that light can be per se is exactly what the Dasein discloses. That is how Dasein is concerned about light; as he is "there," he is disclosed towards light and anything else in the world.

As such, the world disclosure refers to the disclosure of an already interpreted, symbolically structured world. And it also refers to the disclosure of new horizons of meaning and of previously hidden or unthematised dimensions of meaning of the Being. Precisely because the ontological being of the human is Dasein, is the fact to be here in the world's disclosure, then the question of the *closure* of the human in the world, i.e., the question of human death and possible immortality is of paramount importance.

Heidegger's phenomenological perspective is directly linked to the question of death because Dasein can only be open to the world insofar as he can be closed to it; as he can die. Consequently, the being of death that the human discloses has a very special meaning for him; death is also something that can be comprehend but never grasp. The human is aware that he will die, precisely because death discloses to him.

Consequently, for Heidegger, Dasein is what he calls *Sein zum Tode*.[24] The expression *Sein zum Tode* means the being-towards-death. It implies that the human is the being that exist until death and that has a consciousness of death until the end. To be human means to persist in life, never to arrive at this horizon of death that we have nonetheless in view. Death is a constant becoming that escapes all present.

Then, by making the locution "towards death" a fundamental disposition of Dasein, Heidegger argues that death is an existential structure of the human condition—just as the better-known expression "being-in-the-world." We are human because we are towards death, as much as because we are here in the world. As Françoise Dastur summarises: "to die is not merely an extrinsic determination of existence, an 'accident' of the substance 'human'; but on the contrary an essential attribute of it."[25] Death is co-substantial with the human, because as the human being and finiteness are ontologically linked, the human exists in a temporal flow, in a temporal sphere. Death is the horizon that composes the human time, which both encapsulates and opens our temporality. Not only does the human existence only takes place on the temporal horizon of mortality, but this horizon is also necessary (and not contingent) to the very existence of the human being. Just as with the example of light; without the human, there is no death per se, but also, there is no human without death, which means without time.

Here, we see an inversion of the Cartesian *Cogito*, this determination of the being's substance from the hyperbolic doubt that allows Descartes to characterise the identity of *sum*. The "I think therefore I am" (*cogito ergo sum*) inclines towards a quite different perspective. It is structured in reverse, and goes not from its emergence "therefore I am" (*ergo sum*), but from its finitude "because I die" (*sum moribundus*).[26] "I am because I have the possibility of not being anymore"; or rather "I exist because I carry the possibility of non-existence." This "I am because I am mortal" (*sum moribundus*) does not mean *moribundus* as the ontically dying person, the moribund: but rather as the existentially dying being. A *sum moribundus* that substitutes for the *cogito sum* and that we could also translates by the expression "the being on the horizon of death": by inscribing him on the horizon of time, which is to say on the horizon of existential becoming.

In this philosophical framework, we understand that we exist only because we die, not biologically but ontologically. Heidegger suggests a major phenomenological view on finitude and limits, where they are not obstacles, but also where they should not be accepted only because they are biologically inevitable. On the contrary, they are metaphysical structures through which the human can stand out as a temporal being. Death is a form of *aléthéia*, i.e., a "mode of revealing"[27] that implies a new perspective on reality, as the phenomenologist Etienne Bimbenet underlines: "Because it looks constitutively towards a beyond of itself, [...] the sensible is not only what is revealed, but correlatively what reveals: it allows to see the world."[28] Death is an *aléthéia* itself, which unveils the human as the being that exists in time towards death.

On the Ontological Status of Finitude

This phenomenological way of defining the human being could lead to a paradox. Isn't it possible to refuse to qualify the human as a living being if we describe him as the being that is ontologically structured by death? Can we really say that the human is a living being when we say that death destines him to be *moribundus*, as soon as he enters existence? On the contrary, are not all living beings concerning death, as death is what happens to any organic being? Isn't it even part of the very definition of living beings when we consider them as beings that tries to perpetuate in life?

For Heidegger, Dasein is the only being that sustains an ontological relation to death. Yet, unlike the Cartesian *cogitatio* which is a predicate of human beings, Heidegger does not posit a dualistic view of substances, that would place Dasein on one side and all the other beings on the other. Rather, while openning a debate that would need further developments in the field of ethology, he acknowledges that animals have proper ways of living that human beings may not fully apprehend. Here, to claim that the human has a structural relation to death does not mean that animals have none, but that their relation to death differs from that of Dasein. According to Françoise Dastur, more specifically, this distinction "is by no means absurd in the German language, which expresses death (*Tod*) and dying (*Sterben*) by different words."[29] The distinction thus emerges between death as a biological marker of mortality; and death as the ontological horizon of existence: in this sense, all living beings die, but death is not an existential structure of the being of all living beings.

Without denying the fundamental aspect of this outcome, Heidegger shows that death is for every living being, be it plants, humans or animals, this radical event (*evenire*) that occurs in the world. Furthermore, for Dasein, it is also an essential disposition of the existence towards which it is turned or in which it is taken and seized. Dasein does not only die, but he exists as turned towards or concerned by the horizon of his mortality. This ontological possibility of non-existence differs from the biological death that every living being meet at the end of their life.

So that this depiction is an intent to "highlight the *existential* origin of the concept of death."[30]

The phenomenon that Heidegger tries to unveil and explain is not the human being but death. As such, death is not only the marker of the end of the human life, but it is the temporal structure of what dwells in Dasein's existence itself. It is the canvas or the horizon on which the human's temporality appears and the time of existence detaches, and without which no existence can be detached. According to this existential analysis, the human being exists only because he dies, not biologically but ontologically. The inversion of this relationship to death would entail an ontological change in the being of finitude itself, where it is not anymore a metaphysical standpoint for the human to be opened to, but a biological condition that should be overcome.

The distinction between transhumanist views on immortality and the Heideggerian phenomenological perspective on finiteness here appears irreducible. From a transhumanist standpoint, death is biologically inevitable and therefore theoretically necessary only until it can be overcome by any medical knowledge and technological innovation that would repeatedly extend human life expectancy. In this sense, the necessity of death is purely factual, without any performative or existential scope. However, as drawn from a Heideggerian perspective, once death is understood as structural to the very essence of the human being, it no longer matters whether it is empirically inescapable or technologically surmountable. Death becomes necessary not as a biological fact, but as indispensable to defining what it means to be human. In this light, death is rendered even more necessary than before: not merely as an inevitable fact of biology, but as an essential ontological horizon., which reveals a central point of contention between phenomenology and transhumanism regarding the meaning of finiteness.

As we have discussed, treating death solely as a biological event entails a naturalistic view of the human being. Transhumanism, in this sense, conceives of the human in terms of his primary biological condition, from the very fact of being alive. Consequently, in attempting to overcome death as a biological disease, transhumanist movements "biologize" the human being, by bringing his living organism back mostly as an object of biological sciences. This stance could compare with a Cartesian view on animals[31] that led his opponents to coin the expression "animal machine." In Descartes' dualism, the animal is reduced ontologically to a mere material substance, whereas the human being alone possessess *cogitatio*. In a similar way, transhumanist movements may fail to present the human as an existing being, that is, a being fundamentally structured by disclosure to the world, to time and to finitude.

This perspective is paradoxical in itself, because transhumanist authors strongly insist on biological constraints that are considered harmful (diseases, ageing, death) to put forward the necessity to overcome our biological burden. They are intended to go beyond the biological limits that seem to carry finitude along with the body. They encourage not to be ashamed to consider this body as carrying diseases that need a remedy. As Jean-Yves Goffi shows, the bioethicist James Hughes opens the discussion "by estimating that the real question is to know if we want at all costs to remain human, or if we are ready to assume our status of persons."[32] We understand clearly that the fact of remaining human addresses the biological and mortal life of the human being, while the status of a person refers to a growing autonomy of morphological and psychological freedom.

Consequently, transhumanist movements shape the discussion to overcome our biological limitations. But at the same time, while radically seeking to go beyond it, their approach only focuses on the biological aspects of human life and risks reducing it to diseases, limits and mortality without considering that this temporal structure could have an ontological meaning for the human existence.

To reuse the Heideggerian distinction of his *Letter on Humanism*,[33] this way of thinking portrays the human as enclosed within the framework of his biological *animalitas*, and not as disclosed or opened in his phenomenological *humanitas*. When this transhumanist view of death is set against the Heideggerian existential analysis of finitude, it appears as an application of Cartesian automaton thinking to Dasein. From this framework, we argue that the transhumanist approach to death is not intrinsically axiologically wrong, nor hubristic by essence, but that it fails to apprehend finiteness as an ontological structure of human existence.

Conclusion

Our discussion has shown that defining death as transhumanism does, as an external disease that might or might not afflict the human being, reduces the human back to his primary biological state, to his automatic *animalitas*, rather than recognizing Dasein as an existing being fundamentally shaped by finitude.

Indeed, in the Heideggerian framework, death is not a contingent biological fact but an ontological structure of human existence. Maybe unlike any other "disease," this one constitutes the very becoming of Dasein. In decisive question, therefore, is whether one considers the human to be structurally related to finitude in a way fundamentally different from other living beings. If death is not only a biological event but an ontological determination, then to seek to overcome death in this life (and not only in a hereafter) would mean abolishing the temporal horizon of the existence of Dasein. What would it be, then, to exist without death, not understood as what closes life, but on the contrary, as the canvas from which life is detached and takes shape? To contemplate human life beyond its finitude implies the possibility of losing the measure of our finite time and of dissolving it an infinity that unbinds being and temporality. We may thus conclude that the ancient dream of overcoming mortality, already found in numerous myths and religions, does not bear the same ontological consequences as the prospect of realizing immortality here and now through technological means.

For Heidegger, there is no freedom in the confrontation of death. Accordingly, because transhumanist movements have not freed themselves from the fears and anxieties that death brings, they are never authentically free from it. And in this absence of real freedom, they may confirm that human beings are being-towards-death: for if the post- or trans-human were immortal, they would no longer be burdened by this existential weight.

Thus, transhumanist discourses that aim to overcome death bring us back unquestionably to the phenomenological structure of Dasein that they want to exceed. The being-towards-death never has a strictly biological relation to his end and always experiences existentially what constitutes the very measure of his time.

This invites a final question: does transhumanism in fact remain bound to a traditional conception of time and progress, conceived as a linear succession of states and events culminating in death as the final endpoint? By treating death as a last disease to be cured or postponed, transhumanism could reinforce this linear view of time and experience. In contrast, phenomenology proposes a view of existence as a holistic structure in which the entanglement of human being, world, time, and finitude reveals itself as the very ground of meaning. At the very least, by openning this question, transhumanism should be understood as a witness of the evolution of our technical imaginaries. It does not only argue for human enhancement and later on, human immortality, but reveals how our very relation to ourselves is mediated through technology. By envisioning the possibilities of rejuvenation, radical longevity, or even immortality, transhumanist discourse discloses a shift in how we imagine the human condition. In this sense, what is at stake is less the feasibility of such promises than the way in which the human is already reinterpreted through them. The transhumanist imaginary shows that we no longer conceive of finitude exclusively as an existential horizon, but increasingly as a technical variable, that is something to be studied and measured, in order to be, possibly, prolonged or suspended. It thereby signals an evolution in our cultural self-understanding, in which finitude is refigured through technological mediation rather than accepted as an irreducible ontological structure.

Notes

1. Bapteste, Eric. La vie élastique. Révolutions dans le vieillissement. Belin Education. Un monde qui change. Paris, 2025.
2. Lucie Dalibert, "Façonnement du corps vieillissant par les technologies," *Gérontologie et société* vol. 37, n° 1 (2015): 47–58; Kelly Joyce and Laura Mamo, "Graying the Cyborg. New Directions in Feminist Analyses of Aging, Science, and Technology," in *Age Matters: Re-Aligning Feminist Thinking*, ed. Toni M. Calasanti and Kathleen F. Slevin (New York & London: Routledge, 2007).
3. John R. Wilmoth, "Demography of Longevity: Past, Present, and Future Trends," *Experimental Gerontology* 35, n° 9–10 (2000): 1111–29; Jim Oeppen and James W. Vaupel, "Broken Limits to Life Expectancy," *Science* 296, n° 5570 (2002): 1029–31.
4. Major references include: Eric T. Juengst and al., "Biogerontology, "Anti-aging Medicine," and the Challenges of Human Enhancement," *Hastings Center Report* 33, n° 4 (2003): 21–30; Jennifer Loy, "Centenarian Transhumanism Aging in Place," in *Data-driven Multivalence in the Built Environment*, ed. Nimish Biloria (New York: Springer, S.M.A.R.T. Environments, 2020), 141–56; Gorka Orive and al., "Hacking Human Beings with Machine Biology to Increase Lifespan," *Trends in Biotechnology* 38 (2020), Benjamin Ross, "Living "Forever": Transhumanism and Mortality," in *The Philosophy of Transhumanism: A Critical Analysis*, (West Yorkshire: Emerald Publishing Limited, 2020), 73–97; Christopher Wareham, "The Transhumanist Prospect: Developing Technology to Extend the Human Lifespan," in *The Palgrave Handbook of the Philosophy of Aging*, ed. Geoffrey Scarre (London: Palgrave Macmillan UK, 2016), 517–38; Didier Coeurnelle, "The Death of Death. 143: Longevity, Amortality, Transhumanism, Technoprogressism," *Longévitisme* (2021), https://heales.org/2021/03/02/the-death-of-death-n-143-longevity-amortality-transhumanism-technoprogressism-february-2021/; Dinorah Delfin, "I'm Transhuman," *Immortalists Magazine* 1 (2020); Gregory M. Fahy and al., *The Future of Aging. Pathways to Human Life Extension* (New York: Springer, 2010); Geoffrey Scarre, *The Palgrave Handbook of the Philosophy of Aging* (London: Palgrave Macmillan, 2016); David Wood, *The Abolition of Aging: The Forthcoming*

Radical Extension of Healthy Human Longevity (United Kingdom: Delta Wisdom, 2016).

5. Linda Partridge, "The New Biology of Ageing," *Philosophical Transactions of the Royal Society of London. Series B, Biological Sciences* 365, n° 1537 (2010): 148.
6. Max More, "A Letter to Mother Nature," Blog of Max More, http://strategicphilosophy.blogspot.fr (blog), 1999.
7. Ibid.
8. Didier Coeurnelle, "La mort de la mort. 65: La vieillesse est-elle une maladie ?," *Longévitisme* (2014).
9. Lombard, Jessica. « Ageing and Death: A Focus on How to Transcend Diseases for Transhumanist Movements ». Transgressive Bodies. FORUM: University of Edinburgh Postgraduate Journal of Culture & the Arts 34, no 1 (2023).https://doi.org/10.2218/forum.1.9154.
10. Leonard Guarante, *Ageless Quest: One Scientist's Search for the Genes That Prolong Youth* (New York: Cold Spring Harbor Laboratory Press, 2003).
11. F. Boero and al., "Reversing the Life Cycle: Medusae Transforming into Polyps and Cell Transdifferentiation in Turritopsis nutricula (Cnidaria, Hydrozoa)," *Turritopsis nutricula* 190, n° 3 (1996).
12. Carlos López-Otín and al., «The Hallmarks of Aging», *Cell* 153, n° 6 (6 juin 2013): 1194–1217, https://doi.org/10.1016/j.cell.2013.05.039.
13. Aubrey de Grey and Michael Kope, "A Reimagined Research Strategy for Aging," *Longévitisme*, SENS Research Foundation (2009), https://www.sens.org/our-research/intro-to-sens-research/.
14. Courtney Everts Mykytyn, "Anti-Aging is not Necessarily Anti-Death: Bioethics and the Front Lines of Practice," *Medicine Studies* 1, n° 3 (2009): 209–28.
15. Randal A. Koene, "Uploading to Substrate-Independent Minds," in *The Transhumanist Reader: Classical and Contemporary Essays on the Science, Technology, and Philosophy of the Human Future*, ed. Max More and Natasha Vita-More (New Jersey: Wiley-Blackwell, 2013), 146.
16. Hans Moravec, *M.i.n.d Children. The Future of Robot and Human Intelligence* (Cambridge MA: Harvard University Press, 1988), 119–20.

17. Jean-Philippe Milet, “La Technique de la phénoménologie,” in *Phénoménologie et Techniques(s)*, ed. Pierre-Etienne Schmit and Pierre-Antoine Chardel (Paris: Le Cercle Herméneutique, Phéno, 2008), 54.
18. René Descartes, *Discours de la méthode* (1637; Paris: La Pléiade, reprint 1937). The term “animal machine” was coined by his opponents as an accusation against his views on animality.
19. Françoise Dastur, *Heideggeret la question du temps* (Paris: Presses Universitaires de France, Philosophies, 1990), 47.
20. Martin Heidegger, *Being and Time*, trans. Edward Robinson and John Macquarrie (Oxford: Blackwell Publishers, 1962), 26, §2.
21. Jean Beaufret, *Dialogue avec Heidegger*, vol. 1 (Paris: Éditions de Minuit, 1987), 11.
22. Martin Heidegger, *The Fundamental Concepts of Metaphysics: World, Finitude, Solitude*, trans. William McNeill and Nicholas Walker (Bloomington and Indianapolis: Indiana University Press, 1983; reprint 1995).
23. Lombard, Jessica. « A Foray into Welt and Umwelt: Rereading the Onto-Ethological Discussion between Heidegger and Uexküll ». Springer, Biosemiotics, 31 mai 2024, 273–96. https://doi.org/10.1007/s12304-024-09570-1.
24. Heidegger, *Being and Time*, op. cit., §46.
25. Françoise Dastur, *La mort. Essai sur la finitude* (Paris: Presses Universitaires de France, Épiméthée, 2007), 123.
26. Ibid., 123.
27. Martin Heidegger, “The Question Concerning Technology,” in *The Question Concerning Technology, and Other Essays*, trans. William Lovitt (New York & London: Garland publishing, 1977), 13.
28. Étienne Bimbenet, *Nature et humanité: le problème anthropologique dans l’œuvre de Merleau-Ponty*, (Paris: Librairie philosophique J. Vrin, 2004), 241.
29. Dastur, *La mort. Essai sur la finitude*, op. cit., 128.
30. Ibid., 129.
31. Descartes, *Discours de la méthode*, op. cit., 59, Part V, VI.

32. Jean-Yves Goffi, "Chapitre 1. Nature humaine et amélioration de l'être humain à la lumière du programme transhumaniste," *Journal International de Bioétique*, vol. 22 (2011): 29.
33. Martin Heidegger, *Lettre sur l'humanisme. Lettre à Jean Beaufret*, trans. Roger Munier (Francfort-sur-le-Main: Vittorio Kaustermann, 1946).

PART IV

The Transcendence of the Human: A Posthumanist Dream

CHAPTER 14

Gilles Deleuze's Dream of a Non-human Mankind

Audrey Borowski

Introduction

Even if Gilles Deleuze never refers explicitly to posthumanist thought, much of his work seems to anticipate elements of it in certain crucial respects. Posthumanism first and foremost seeks to relinquish the singularity of the human, conceiving the human as thoroughly intermingled in a network of dynamic relations and as in perpetual becoming.[1] It positions itself explicitly against humanism's claims to exceptionality and separation from nature—claims that more often than not lead to the domination and exploitation of some by others,[2] with devastating consequences for other forms of life.[3] Insofar as much of posthumanism seeks to revisit of the human, Deleuze (1925–1995) and his writing partner Félix Guattari (1930–1992) appear as influential and prescient precursors in this line of thinking. Their work problematizes traditional conceptions of the human being[4]; and Deleuze in particular adumbrates not only

A. Borowski (✉)
University of Cambridge, Cambridge, England
e-mail: ab3204@cam.ac.uk

J. Muller (ed.), *Automata, Cyborgs, and Mutants*, Palgrave Studies in Medieval and Early Modern Medicine,
https://doi.org/10.1007/978-3-031-93201-4_14

new ways of thinking and philosophizing, but even seems to welcome a digitally reconfigured non-human future that could enable humanity to fulfill its potential. While there has been some scholarship on the subject,[5] this article undertakes to probe the connection between Deleuze and posthumanism further and demonstrate that, ultimately for the French philosopher, there has never been such a thing as the "human."

Individual

Breaking from much of mainstream philosophy, of which he had become critical,[6] Deleuze turned toward a radical version of "affirmation," a celebratory metaphysics of difference that circumvents the Kantian transcendental model and its dead-ends in favor of multi-sensory aesthetics.[7] Such a transcendental form of empiricism offers access to the virtualities of real experience—rather than the generalities of a possible experience.[8]

Deleuze construed difference as emphatically affirmative, in that difference is construed in terms of multiplicity and relations rather than in the negative.[9] He locates his ontology beyond essences and categories in order to think the individual in a dynamic and relational manner and in terms of modulation and aleatory becoming, comparing this "freeing of difference" (as Foucault puts it) to a throw of dice.[10] Like a nebula in which all virtualities coexist,[11] life territorializes space according to particular assemblages, activating one possible configuration among many in what amounts to a ballet between actual and virtual. In this manner, the virtual operates as set of potentialities that are always capable of being actualized and combined in an infinite array of possibilities. For Deleuze, the world is composed of virtual possibilities open to nomadic and event-induced differentiation, each actualized world resonating with different possible worlds. It is organized according to decentralized networks called rhizomes, composed of divergent series and precarious states of balance ready to be shattered by the slightest accident and opened into a different world at any moment.[12] These series lay constantly open to coupling with divergent and heterogeneous series through *disjunctive syntheses* that redistribute their relations and singularities.[13] Within this configuration, therefore, the Other emerges as "an expression of a possible world."[14] The "great infinite plane of composition" is thus organized according to a rhizomatic, circuitous model whose lines are perpetually changing.[15]

Deleuze sought to capture the dynamic multiplicity of reality by calling for a refounding of philosophy on the basis of a rhizomatic model that would embrace the "plasticity" of life. Life, like thought, involves networks of non-causally related paths and crossroads. In the service of pure immanence, it unfolds processually, and, freed from "common sense," arises as a series of "problematizations" as it explores "all manner of becomings." Confrontation with the outside engenders zones of indetermination zones of indetermination and *experimentation* and "mad-becoming,"[16] as the mind synthesizes meaningful disjunctions that extent beyond any organism through "irrational" leaps.[17] As a result, thought is governed by a "logic of sense" that precedes any truth-values or Kantian categorical unity and opens itself up to any aberrant movements in such a manner that conceptual terrains metamorphose in the face of events. The elasticity of the concept matches the fluidity of the milieu,[18] with conceptual planes often coexisting with each other: concepts are ontogenetic,[19] and are assembled as easily as they are disassembled, replaced, shifted as they move about on a plane of immanence following no particular predetermined outline. They are "vagabond, indeterminate, nondiscursive"[20] as they course through zones of indiscernibility. Concepts are not derived from essences or predication but from continuous variation and superpositions: "The concept of a bird is found not in its genus or species but in the composition of its postures, colours, and songs: something indiscernible that is not so much synesthetic as syneidetic, an ordering of its components by zones of neighbourhood, an intension," whose "components are neither constants nor variables but pure and simple variations ordered according to their neighbourhood."

The Deleuzian cosmos is thus characterized by a continuous becoming, through the joining of planes, the forming of compounds and sensations, and the constant materialization of new virtualities that bifurcate and enter into relations of dependence.[21] The body exists in a state of "pure potential, pure virtuality," an "infolding of impulses from an aleatory outside" coursed by unfamiliar and non-human impersonal individuation.[22] It is originally nonorganic—"No mouth. No tongue. No teeth. No larynx. No esophagus. No belly. No anus"[23]—and composed of zones and levels whose organs are only acquired according to the event and sensation it encounters. Organs within the body are otherwise indeterminate and can vary: "what is a mouth at one level becomes an anus at another level, or at the same level under the action of different forces. Now this complete series constitutes the hysterical reality of the body."[24] Each body without organs does not define, bounded subject, but resonates with the immensity of the cosmos as it enters in and out of various arrangements. Organisms are transient and only formed in confrontation with their milieu,[25] "imprison[ing]" life[26] according to particular temporary configurations: "No organ is constant as regards either function or position … sex organs sprout anywhere … rectums open, defecate and close … the entire organism changes color and consistency in split-second adjustments." Such fluidity and non-essentiality extend to the face itself—the face is a kind of "grid" or "diagram"—and to sexual difference—bodies are merely territorialized as "man" or "woman"[27] amidst "a thousand tiny sexes."[28]

The organized body enters in composition with various heterogeneous beings, including plants, animals, and even (as we shall later see) artificial devices and technological compounds, in the process radically transforming human beings.[29] "What is real," Deleuze and Guattari argue, "is the becoming itself, the block of becoming, not the supposedly fixed terms through which that which becomes passes. Becoming can and should be qualified as becoming-animal even in the absence of a term that would be the animal become." There exists in this manner a whole range of molecular "becomings" mankind can engage with, becoming-woman, -child, or -animal, all of which entails "a disintegration of oneself."[30]

For Deleuze, assemblages are more socio-historical constructs and the manifestation of particular social, political and economic structures—including of the primitive, despotic and capitalist regimes which he explores in *A Thousand Plateaus*—than the expression of any innate trait.[31] The Deleuzian cosmos is characterized by a continuous becoming, following heterogeneous assemblages whose forces and networks, ranging from social and political arrangements to psychological to biological systems, have the capacity to act on other assemblages and to transform the body into a differential topology that varies according to the affects that course it.[32]

Accordingly, Deleuze dismisses more conventional accounts of genesis, such as evolution by descent and filiation.[33] Deleuze and Guattari's "involutionary" model emphasizes translineal symbiosis in the relentless fluctuations between assemblages and milieus from which individual organisms then develop: "it amounts to the same thing to say that each multiplicity is already composed of heterogeneous terms in symbiosis, and that a multiplicity is continually transforming itself into a string of other multiplicities."[34] In this manner individuation precedes speciation or categorization.[35]

Even in their evolution, organisms are thus indissociable from their milieus, becoming links in a larger structure that encompasses their bodies and actions and those of neighbouring organisms: "An animal is defined less by its genus, its species, its organs, and its functions, than by the assemblages into which it enters."[36] The deterritorialized is thus continually reterritorialized through heterogeneous alliances or "unnatural nuptials," as Deleuze and Guattari describe them, such as contagions, invasions, infections, and generative cross-pollination.[37] The plane of Nature "continually makes and unmakes assemblages, employing every artifice" through such trans-species alliances as the wasp and the orchid or the tick and its host.[38] Life thus unfolds according to a "logic of impersonal individuation," a diagrammatic line in perpetual variation, a heterogenesis that characterizes the logic of sense as divergent series meet.[39] This line of thought accounts perhaps for Deleuze's special affinity for cinema and his insistence on the superiority of flashback technique.

Life thus exceeds organisms as it opens itself onto cosmic forces[40] and enters into a composed chaos that becomes sensory[41] as it becomes other.[42] Deleuze discerns profound similarities between the workings of nature and of art in their pursuit of creation and new affects. He thus postulates an aesthetics of sensation open to creative syntheses in which a nomadic subject's body without organs takes center stage.[43] For Deleuze, art, by creating blocks of sensations, lends itself particularly well to his exploration of the manipulations and transmutations of the body beyond any particular organism.[44] The particular force of art for Deleuze perhaps lies in its unique ability to restore the infinite and make sensible 'the imperceptible forces that populate the world'.[45] The canvas or piece of paper offers the ideal medium to conjure up, modulate, combine, and aggregate virtualities, to stage humanity's 'nonhuman becomings'.[46]

The artistic performs a "a triple liberation here—of the body, of the planes, and of colour," a liberation that "can occur only by passing through the catastrophe; that is, through the diagram and its involuntary irruption: bodies are thrown off balance, they are in a state of perpetual fall; the planes collide with each other; colours become confused and no longer delimit an object."[47] In fact, life alone "creates such zones where living beings whirl around, and only art can reach and penetrate them in its enterprise of co-creation."[48]

A thinker preoccupied with the mechanics of disembodiment and dematerialization and with the becoming of the entities that populate our world,[49] Deleuze aims to dissolve the self and dispel the idea of human exceptionalism, the centrality of human perspective for understanding the world, and the very idea of human essence.

Deleuze envisages an encompassing model which prioritizes becoming over being, and difference and rejects the traditional emphasis placed on the human in favor of life.[50] Each assemblage is coursed by "lines of flight" that deterritorialize it and force it out of itself. Drawing on the immanentist philosophies of Spinoza and Nietzsche,[51] Deleuze's metaphysics of difference thus lays out a non-anthropomorphic vision of the world in which life itself is the ultimate arbiter of value.

Man himself is in the process of becoming, in and with a world made up of 'nothing but quantities of force in mutual "relations of tension,"' constantly open to experimentation with possibilities that are always latent, and hence to destabilization—in fact, these various possible configurations are rooted in chaos and disorder itself.[52] Simply put, to be human, for Deleuze and Guattari, means to undergo non-human becomings.[53]

Ideally, mankind would ultimately become "imperceptible," that is, dilute itself in the pure becoming of the world, becoming everyone and everything: "The imperceptible is the immanent end of becoming, its cosmic formula… By a process of elimination, one is no longer anything more than an abstract line, or a piece in a puzzle that is itself abstract. It is by conjugating, by continuing with other lines, other pieces, that one makes a world that can overlay the first one, like a transparency."[54] A new world, Deleuze and Guattari claim, would thus result from this fusion into the world.

We are evanescent beings in constant mutation, as we affect and are affected by the various configurations we take on.[55] In fact, "There are only haecceities, affects, subjectless individuations that constitute collective assemblages… Nothing subjectifies, but haecceities form according to compositions of non-subjectified powers or affects."[56] Entangled in the world and constantly fluctuating as it entered in relations with other beings, each self is constantly on the the verge of vanishing. The self is "only a threshold, a door, a becoming between two multiplicities," making it indeterminate.[57] We are fleeting beings constantly coursed by various affects: "We are not in the world, we become with the world." Each self is born out of "delirium, chance, indifference," and as such a residuum.[58]

As inheritors of architectonics of ideas and deeply grooved paths of thought-action,[59] it behooves us to liberate ourselves from them as much as possible and "become other" as we move in new directions and form new assemblages.

Deleuze understands the earth as a constantly shifting ground without foundation (*un fond sans fondement*), organized rhizomatically and part and parcel of a broader cosmos whose forces human beings can tap. The flows that make up society can be grasped in terms of continuous processes of de- and re-territorialization, which account for the various social formations, alternating striated spaces associated with the state and smooth spaces with nomadism.[60]

Unlike Husserl, for whom man's attachment to an "originary earth" left no room for alternative modes of perception, Deleuze interprets Nietzsche's call to "remain earthly"[61] as an injunction to deterritorialize the earth so as to reveal its creative potential and help spawn new ways of living, thinking, and inhabiting the earth.[62] New territorial assemblages would thus help instantiate a "new earth" amidst the infinity of virtual earths available and thus help bring forth a "new people": "We do not lack communication. On the contrary, we have too much of it. We lack creation. We lack resistance to the present. The creation of concepts in itself calls for a future form, for a new earth and people that do not yet exist."[63]

Dividual

While Anti-Oedipus greeted capitalism's emancipatory potential in its anarchic tendencies, Deleuze and Guattari's later writings express more wariness of recent capitalist and neoliberal accelerationism. In late neoliberalism, much of life has been imbued with management and marketing's techniques of soft control.[64] With the advent of the digital age, a larger machinic system imposes a new type of relation between humans and technology, "a generalized regime of subjection" in which digital technology's hyperconnectivity facilitates the possible production, exchange, and collection of information in a way that tends to segment and alienate individuals.[65] With great foresight, Deleuze identifies the rise of so-called "control societies"[66] as successors to what Michel Foucault theorized as the society of sovereignty. These control societies oversee the transition from disciplinary enclosures in which individuals are trained and disciplined to decentralized and deterritorialized networks of continuous assessment and monitoring. Individuals are no longer physically confined per se or coerced, but are reduced to "dividuals" as they are replaced with abstract digital aggregates and tracked by ubiquitous control devices and mechanisms that interoperate with each other and are free to deny or grant access.

As control becomes modular and diffuse,[67] pervasive autoregulation takes over as the data mined from us is used to classify us and steer our destinies. Paradoxically, then, the increasing automatization of the world multiplies freeways in such a manner that "people can drive infinitely and 'freely' without being at all confined yet while still being perfectly controlled."[68] Instead of being disciplined, individuals now find themselves determined and funneled into flows they can no longer really. Deleuze fears that the emergent informational economy is ushering in a greatly impoverished horizon by curbing the radical empiricism he extolls. The man of control "surfs" on a series of seamless continuums made up of a homogenized reality,[69] which tolerates little resistance and limits our becomings.

A short detour via Nietzsche is perhaps necessary at this stage. Deleuze had a lengthy and extensive engagement with the German philosopher, notably drawing on his powerfully affirmative concepts of will to power and "possibilities of life" which he felt had the capacity to imbue our lives with an aesthetic dimension.[70] Perhaps most notoriously, Deleuze construes Nietzsche's concept of eternal recurrence not, as we have seen, as the recurrence of the same, but as the assertion of the different.[71] For Deleuze, crucially, Nietzsche's interests lies assertion of the death of God[72] than in that of the death of man—or at least a particular kind of man, the one consecrated by classical metaphysics who, as still described and analyzed within the framework of contemporary philosophy, alternates between entering into relations with "forces of elevation to infinity" and "forces of finitude" which had, in turn according to him, characterized the seventeenth, later nineteenth, and twentieth centuries.[73]

Recent technological advances in particular challenged this dichotomy between finitude and infinity as well as reductive narratives about mankind's exceptionality and isolation in the universe more broadly, and made our embeddedness in and dependence on the universe all the more undeniable. As humanity enters forces and relations with the outside, as Deleuze explains,

> These may be forces within man: the force to imagine, remember, conceive, wish, and so on. One might object that such forces already presuppose man; but in terms of form this is not true. The forces within man presuppose only places, points of industry, a region of the existent. In the same way forces within an animal (mobility, irritability, and so on) do not presuppose any determined form. One needs to know with what other forces the forces within man enter into a relation, in a given historical formation, and what form is created as a result from this compound of forces. We can already foresee that the forces within man do not necessarily contribute to the composition of a Man-form, but may be otherwise invested in another compound or form: even over a short period of time Man has not always existed, and will not exist for ever. For a Man-form to appear to be delineated, the forces within man must enter into a relation with certain very special forces from the outside.[74]

Such humanistic narratives, which restricted conceptions of man and his potentialities, as Foucault argued in *Les Mots et les Choses* (1966), were bound, however, to disappear over time. Commenting on Foucault through the lens of Nietzsche, Deleuze seems to co-opt and extend this line of inquiry. In fact, Deleuze's lifelong engagement with Nietzsche seems to come to a head when he muses openly about the death of man, or rather the end of how man has been conventionally conceived so far: 'The question that continually returns is therefore the following: if the forces within man compose a form only by entering into a relation with forms from the outside, with what new forms do they now risk entering into a relation, and what new form will emerge that is neither God nor Man? This is the correct place for the problem which Nietzsche called "the overman."'[75] The matter of the "death of man" is not, however, addressed entirely pessimistically by Deleuze. In fact, it appears as a prerequisite to overcoming nihilism by allowing subjects to "resonate with and amplify active forces in the world."[76]

Posthuman

Indeed, one of Deleuze's later texts remain striking in this regard by casting a more optimistic glance towards the future in the way it celebrates the reconciliation of mind and matter, being and thought, and nature and technology by envisaging a new form of human and life itself in the shape of new assemblages between man and inorganic elements such as electronic components and the silicon that compose them.

In the appendix to his book on Foucault, entitled "On the Death of Man and the Overman," Deleuze adumbrates the possibility of a more potent form of computation with the advent of third-generation machines, cybernetics, and information technology. In a world in which information saturates the world and substitutes for nature, a form of meta-computing could emerge as a force of creativity and serve as the vehicle for a radical and necessary transvaluation of values, that is, mankind's final overcoming of itself.[77]

Deleuze conjures the prospect of a "superfold" for the twenty-first century, one that would supersede the forces of both finitude and infinity and "free life within man himself, to the profit of another form"[78]*—an* overman, a "new infinite creator" that would be "neither god nor man" but a formal compound of the forces within man and these new forces:

> What would be the forces in play, with which the forces within man would then enter into a relation? It would no longer involve raising to infinity or finitude but an unlimited finity… It would be neither the fold or the unfold that would constitute the active mechanism, but something like the superfold, as borne out by the foldings proper to the chains of the genetic code, and the potential of silicon in third-generation machines… The forces within man enter into a relation with forces from the outside, those of silicon which supersedes carbon, or genetic components which supersede the organism… In each case we must study the operations of the superfold, of which the "double helix" is the best-known example.[79]

This overman, whose infinity would be inscribed in his finitude, would finally overcome barriers that had hitherto appeared insuperable, leaving him "in charge of the animals [as well as] the very rocks, or inorganic matter." Ironically, then, by dissolving itself within this new reality, mankind could hope to break once and for all from its essentially reactive and negative character,[80] free to "discover, and invent, new possibilities of life."[81]

Deleuze conceives of the overman as an artist "capable of the highest affirmation and creation," who inaugurates a new way of thinking, feeling, and being[82]: "To affirm is to unburden: not to load life with the weight of higher values, but *to create* new values… which make life light and active."[83] The "fold of silicon with carbon," the assemblage of man with these new forces, would yield a practically unlimited variety of possible combinations, thus producing new modalities of being and expression, and in the process helping activate potential.[84] Ironically, then, Deleuze's ideal of radical ontology and true philosophy find themselves realized in the posthuman age of the superfold and the overman.

Conclusion

Far from being isolated from his milieu and exceptional, man is embedded in his cosmos and molded by the various forces, transformations, permutations which course him or her. Like much of posthumanist thought, Deleuze contests human exceptionalism and seeks to subvert arborescent and categorical thinking providing fruitful tools to continue problematizing man's constant process of becoming.[85]

Deleuze's philosophy invites us to produce unfamiliar visions of reality, to create a new world removed from conventional dichotomies—artificial/natural, human/non-human.

In fact, he enjoins us perhaps to seize the opportunity to embrace the new technological sphere that has opened up before us, as a platform to push the human form to the limit—to move in new directions and form new assemblages, to become the other that we, at core, already are. He offers a valuable new intellectual framework for thinking about human relationships to the environmental and with technolog, perhaps prefiguring Bruno Latour's well-known assertion that "we have never been modern."[86] Deleuze would perhaps add that we have never been "human" either to start with in the sense of claiming exceptionality.

For Deleuze like several of his fellow philosophers of immanence, posthumanity was always immanent to our condition and the sooner we came to acknowledge this—and lay out the conditions for a future ethics and philosophy instead of falling back on tradition—the sooner we would be able to fulfill mankind's potential.

Notes

1. See Pramod K. Nayar, *Posthumanism* (Cambridge: Polity, 2014); Stefan Herbrechter, *Posthumanism: A Critical Analysis* (London: Bloomsbury, 2013); Cary Wolfe, *What Is Posthumanism?* (Minneapolis: University of Minnesota Press, 2010); N. Katherine Hayles, *How We Became Posthuman* (Chicago: University of Chicago Press, 1999); and Rosi Braidotti, *The Posthuman* (Cambridge: Polity, 2013), and *Posthuman Knowledge*

(Cambridge: Polity. 2019). On the relinquishing of human singularity see Claire Colebrook, "Deleuze after Afro-Pessimism," in *From Deleuze and Guattari to Posthumanism: Philosophies of Immanence*, ed. Christine Daigle and Terrance McDonald (London: Bloomsbury, 2020), 250–67.

2. As Nayar defines it, critical posthumanism operates a "radical decentering of the traditional sovereign, coherent and autonomous human in order to demonstrate how the human is always already evolving with, constituted by and constitutive of multiple forms of life and machines" (*Posthumanism*, op. cit., 9).
3. See Daigle's and McDonald Introduction to *From Deleuze and Guattari to Posthumanism, op cit*.
4. See Alain Beaulieu, "Deleuze, Guattari et le posthumanisme," *Interconnections: Journal of Posthumanism* 1 (2021): 69–70.
5. See, for instance, Edward Butler's contribution, "Hercules of the Surface: Deleuzean Humanism and Deep Ecology," in *An (Un)Likely Alliance: Thinking Environment(s) with Deleuze/Guattari*, ed. Bernd Herzogenrath (Cambridge: Scholars Publishing, 2008), 139–158 and *From Deleuze and Guattari to Posthumanism,* op. cit.
6. See Colin Koopman, "Critical Problematization in Foucault and Deleuze: The Force of Critique without Judgment," in *Between Deleuze and Foucault*, ed. Nicolae Morar, Thomas Nail, and Daniel W. Smith (Edinburgh: University Press, 2016), 87–119.
7. Jean-Clet Martin, *La Philosophie de Gilles Deleuze* (Paris: Editions Payot et Rivages, 2019), 44. Deleuze thus addresses one of philosophy's most enduring problems, which John Rajchman poses in his Introduction to Deleuze's *Pure Immancence: Essays on a Life*, trans. Anne Boyman (New York: Zone Books, 2001), 13: "Can we construct an empiricist or experimental relation to the persistence of this zone or plane of presubjective delirium and pre-individual singularity in our lives and in our relations with others?".
8. See Gilles Deleuze, *Difference and Repetition*, trans. Paul Patton (New York: Columbia University Press, 1994), chap. 3.
9. Deleuze, *Difference and Repetition*, op. cit., 52 ("Difference is the object of affirmation or affirmation itself"), 55 ("Affirmation itself is multiple, that it is a creation but also that it must be created as affirming difference, as being difference in itself").

10. Deleuze, *The Logic of Sense*, trans. Constantin Boundas, Mark Lester, and Charles Stivale (New York: Columbia University Press, 1990), 95, 198.
11. Jean-Clet Martin, *La Philosophie de Gilles Deleuze*, op. cit., 83.
12. Ibid., 33, 271.
13. Deleuze and Félix Guattari, *Anti-Oedipus: Capitalism and Schizophrenia*, trans. Robert Hurley, Mark Seem and Helen R. Lane (Minneapolis: University of Minnesota Press, 1983), 75.
14. Dorian Wiszniewski, "Deleuze's Theory of the Other – A Serial Solipsism in Three moments," *Serial(ity) ... & ..., Symposium*, University of Edinburgh, 2009, https://www.pure.ed.ac.uk/ws/portalfiles/portal/11917241/Deleuze_s_Theory_of_the_Other.pdf; and Deleuze, "Michel Tournier and the World without Others," in *The Logic of Sense*, op. cit, 301–21. The appearance of Friday disrupts Robinson's lived experience and reopens the possibility of other another world and possibilities.
15. Deleuze and Guattari, *What Is Philosophy?*, trans. Hugh Tomlinson and Graham Burchell (New York: Columbia University Press, 1994), 186. See Daigle's and Mcdonald's Introduction to *From Deleuze and Guattari to Posthumanism*, op. cit.
16. Deleuze, *Negotiations, 1972–1990*, trans. Martin Joughin (New York: Columbia University Press, 1995), 110: "Thinking doesn't come from within... It comes from this Outside, and returns to it, it amounts to confronting it."
17. See Deleuze and Guattari, *A Thousand Plateaus. Capitalism and Schizophrenia*, trans. Brian Massumi (Minneapolis: University of Minnesota Press, 1987).
18. Deleuze and Guattari, *What Is Philosophy?*, op. cit., 36.
19. Karen Houle, "Symmetry and Asymmetry in Conceptual and Morphological Formations: The Difference PlantBody Growth can Make to Human Thought," in *From Deleuze and Guattari to Posthumanism: Philosophies of Immanence*, op. cit., 85–105.
20. Deleuze and Guattari, *What Is Philosophy?*, op. cit., 143.
21. Ibid., 20.
22. Brian Massumi, Introduction to Deleuze and Guattari, *A Thousand Plateaus*, op. cit., 38.
23. Deleuze, *Francis Bacon: The Logic of Sensation*, trans. Daniel Smith (London: Continuum, 2003), 45.

24. Ibid., 48. Deleuze explains that "the body without organs is defined thus by an indeterminate organ, while the organism is defined by determinate organs."
25. Jean-Clet Martin, *Deleuze* (Paris: Editions de l'éclat, 2012), 91.
26. Deleuze, *Francis Bacon*, op. cit., 45.
27. Erin Hortle and Hannah Stark, "Framing Sexual Selection: Elizabeth Grosz's Work on Deleuze, Darwin and Feminism," in *Deleuze and Evolutionary Theory*, ed. Michael James Bennett and Tano Posteraro (Edinburgh: Edinburgh University Press, 2022), 62.
28. Deleuze and Guattari, *A Thousand Plateaus*, op. cit., 213.
29. Pascal Michon, "Gilles Deleuze and Félix Guattari and the *Rhuthmoi* of Individuation—Part 2," *Rhuthmos* 15 July 2021, https://rhuthmos.eu/spip.php?article2628.
30. Deleuze and Guattari, *A Thousand Plateaus*, op. cit., 239–40.
31. Erin Hortle and Hannah Stark, "Framing Sexual Selection," op. cit., 62.
32. Levi Bryant, *Onto-cartography: An Ontology of Machines and Media* (Edinburgh: Edinburgh University Press, 2014), 113–14; Jean-Clet Martin, *La Philosophie de Deleuze*, op. cit., 331.
33. For more on this, see Barry Allen, "Unnatural Nuptials," in *Deleuze and Evolutionary Theory*, op. cit., 23–41. On evolution in Deleuze's thought see Keith Ansell-Pearson, *Germinal Life: The Difference and Repetition of Deleuze* (London: Routledge, 1999); Manuel DeLanda, "Immanence and Transcendence in the Genesis of Form," *South Atlantic Quarterly* 96 (1997): 499–514; and Levi Bryant, *Onto-cartography,* op. cit. DeLanda's and Bryant's interpretations of Deleuze place particular emphasis on the complex and machinic structure underpinning his ontology. John Roffe, "The Egg: Deleuze Between Darwin and Ruyer," in *Deleuze and Evolutionary Theory*, op. cit., 42–58, examines the embryological thinking that pervades *Difference and Repetition* and accounts for differentiation.
34. Deleuze and Guattari, *A Thousand Plateaus*, op. cit., 238–4, 263, 238, 249.
35. Deleuze, *Difference and Repetition*, trans. Paul Patton (London: Athlone, 1994) 250; Nathan Eckstrand, "Deleuze, Darwin and the Categorisation of Life", *Deleuze Studies*, 8 (2014), 415–44. As Eckstrand suggests, for Deleuze "species are an illusion patterned onto the appearance of individuals via individuation" (420).

36. Deleuze and Claire Parnet, *Dialogues* (New York: Columbia University Press, 1987), 69.
37. Deleuze and Guattari, *A Thousand Plateaus*, op. cit., 241, 250; and see Rosi Braidotti, "A Theoretical Framework for the Critical Posthumanities," *Theory, Culture & Society* 36 (2019): 31–61.
38. Deleuze and Guattari, *A Thousand Plateaus*, op. cit., 258.
39. Jean-Clet Martin, *La Philosophie de Deleuze*, op. cit., 267.
40. Deleuze and Guattari, *What Is Philosophy?* op. cit., 185.
41. Ibid., 206.
42. Ibid., 177: "Sensory becoming is the action by which something or someone is ceaselessly becoming-other."
43. Jean-Clet Martin, *La Philosophie de Deleuze*, op. cit., 69.
44. Deleuze, *Francis Bacon*, op. cit. For more on the relation between Francis Bacon's paintings and Deleuze's conception of the "body without organs," see Florence Andoka, "Qu'est-ce qu'un corps sans organes?", Philosophiques 16 (2013).
45. Deleuze and Guattari, *What is Philosophy?* op. cit., 173.
46. Ibid., 169.
47. Deleuze, *Francis Bacon*, op. cit., 118.
48. Deleuze and Guattari, *What is Philosophy?* op. cit., 173.
49. As Peter Hallward emphasizes in *Out of this World: Deleuze and the Philosophy of Creation* (London: Verso, 2006).
50. "Deleuze," Todd May writes in *Gilles Deleuze: An Introduction* (New York: Cambridge University Press, 2005), "tries to pry us away from humanism by focusing on a difference that need not be human difference and a one that need not be a person." Humanism commits the "error of believing that the proper perspective for understanding the world is centered on the viewpoint of the human subject" (24).
51. See Alain Beaulieu, "Deleuze, Guattari et le posthumanisme," op. cit.
52. Deleuze, *Nietzsche and Philosophy*, tr. Hugh Tomlinson (London: Athlone Press, 1983), 40; also 83: "the diagram is indeed a chaos, a catastrophe, but it is also the germ of order or rhythm." On Deleuze and Nietzsche see James Mollinson, 437. "Deleuze's Nietzschean Mutations: From the Will to Power and the Overman to Desiring-Production and Nomadism," *Deleuze and Guattari Studies* 16 (2022): 428–53.

53. Alain Beaulieu, "Deleuze, Guattari et le posthumanisme," op. cit. Alain Beaulieu in this regard also accurately notes that while humans have the capacity to enter in composition with non-human entities—including inorganic life—these arrangements never achieve ideal states, thus dooming to failure any attempt to associate Deleuze and Guattari with the transhumanist ideology.
54. Deleuze and Guattari, *A Thousand Plateaus*, op. cit., 279–280 (translation modified).
55. Bruce Baugh, "The Art of Good Encounters: Spinoza, Deleuze and Macherey on Moving from Passive to Active Joy," in *From Deleuze and Guattari to Posthumanism*, 63–84.
56. Deleuze and Guattari, *A Thousand Plateaus*, op. cit., 266.
57. Ibid., 249. In his Introduction, Massumi notes that "to think the body in movement thus means accepting the paradox that there is an incorporeal dimension *of the body*" (65).
58. Christine Daigle, "Deleuzian Traces: The self of the polyp," in *From Deleuze and Guattari to Posthumanism*, op. cit., 41–62, here 43; see also Deleuze and Guattari, *Anti-Oedipus*, op. cit.
59. Karen Houle, "Symmetry and Asymmetry in Conceptual and Morphological Formations," op. cit., 86.
60. Deleuze and Guattari, *A Thousand Plateaus*, op. cit., 372.
61. Friedrich Nietzsche, *Thus Spoke Zarathustra*, trans. Graham Parkes (Oxford: Oxford University Press, 2009), 71: "I beseech you, my brothers, *stay true to the earth* and do not believe those who talk of over-earthly hopes… that the earth may one day belong to the Overman."
62. Alain Beaulieu, "Back to Earth! A Comparative Study between Husserl's and Deleuze's Cosmologies," in *From Deleuze and Guattari to Posthumanism*, op. cit., 106–25, here 115.
63. Deleuze and Guattari, *What Is Philosophy?*, op. cit., 108.
64. Ibid., 19: "Philosophy had not remained unaffected by the general movement that replaced Critique with sales promotion. The simulacrum, the simulation of a packet of noodles, has become the true concept; and the one who packages the product, commodity, or work of art has become the philosopher, conceptual persona, or artist."
65. Deleuze and Guattari, *A Thousand Plateaus*, op. cit., 458.
66. Deleuze, *Negotations*, op. cit., 180.

67. See Alexander R. Galloway, "Computers and the Superfold," *Deleuze and Guatarri Studies* 6 (2012): 513–28.
68. Gilles Deleuze, "Having an Idea in Cinema" (1998), trans. Eleanor Kaufman and Kevin Jon Heller in *Deleuzeand Guattari: New Mappings in Politics, Philosophy and Culture* (Minneapolis: University of Minnesota Press, 1998), 14–19, here 18.
69. Deleuze, *Negotiations*, op. cit., 178. In *La Bienveillance des machines. Comment le numérique nous transforme à notre insu* (Paris: Éditions du Seuil, 2022), Pierre Cassou-Nogues links the unique smoothness of control society to a new synhaptic type of surveillance born out of the new digital assemblages.
70. James Mollinson, "Deleuze's Nietzschean Mutations," op. cit.; Keith Ansell Pearson, "Overcoming the Weight of Man: Nietzsche, Deleuze, and Possibilities of Life", *Revue internationale de philosophie*, 61 (2007), 245–59. Pearson interprets "possibilities of life" according to Nietzsche and argues that Deleuze inherits and refines Nietzsche's insight into human existence and the future of the earth (258).
71. Keith Ansell Pearson, "Living the Eternal Return as the Event: Nietzsche with Deleuze," *Journal of Nietzsche Studies* 14 (1997), 64–97.
72. Deleuze, "On the Death of Man and Superman," in his *Foucault*, trans. Seán Hand (London: Continuum, 1988), 124–32, here 129.
73. Ibid., 124–27.
74. Ibid., 124.
75. Ibid., 130.
76. James Mollinson, "Deleuze's Nietzschean Mutations," op. cit., 441.
77. For Alexander Galloway, the superfold is the new 'active mechanism' for life within computerized control society. See Alexander R. Galloway, "Computers and the Superfold," op. cit., 524.
78. Deleuze, "On the Death of Man and Superman," op. cit., 130.
79. Ibid.,131.
80. Deleuze and Guattari, *What is philosophy?*, op. cit., 58.
81. Deleuze, *Nietzsche and Philosophy,* op. cit., 101.
82. Keith Ansell Pearson, "Overcoming the Weight of Man: Nietzsche, Deleuze, and the Possibilities of Life," *Revue internationale de philosophie* 241 (2007): 245–59, here 248.
83. Deleuze, *Nietzsche and Philosophy*, op. cit., 185.

84. Alexander Galloway notes in this regard that the concluding passage of Deleuze's Foucault reads very much like Donna Haraway's now famous 'CyborgManifesto'. See Alexander R. Galloway, "Computers and the Superfold," op. cit., 524.
85. Deleuze thus anticipates Rosi Braidotti's statement in *Posthuman Knowledge*, "The posthuman is a work in progress. It is a working hypothesis about the kind of subjects we are becoming" (2).
86. Bruno Latour, *We Have Never Been Modern*, trans. Catherine Porter (Cambridge, MA: Harvard University Press, 1993).

CHAPTER 15

Decentering the Human in Philosophy, Cybernetics and Early Modern Automata

Rahma Khazam

This article explores the connections between posthuman philosophical movements, cybernetics and early modern automata. Within posthumanist philosophy, object-oriented ontology (OOO) and new materialism have been at the forefront of attempts to decenter the human. Yet their approaches are very different. OOO, an offshoot of speculative realism, argues that there is always more to reality than humans are able to grasp. Similarly non-anthropocentric are its other claims—that objects have agency, that the world is to be viewed in terms of objects, that human beings are one kind of object among others, and that in order to grasp the world we need to consider not only the interactions between one human and another or between humans and objects, but also the relationships between one object and another.

R. Khazam (✉)
Institute ACTE, Université Paris 1 Panthéon-Sorbonne, Paris, France

EnsadLab, Paris, France

J. Muller (ed.), *Automata, Cyborgs, and Mutants*, Palgrave Studies in Medieval and Early Modern Medicine,
https://doi.org/10.1007/978-3-031-93201-4_15

Likewise seeking to decenter the human, the new materialists view the world in terms of continuities and entanglements between humans and nonhumans. Karen Barad proposes a posthumanist performativity that questions the boundaries between human and nonhuman,[1] while Jane Bennett stresses that agency is not just the prerogative of humans but is distributed between human and nonhuman forces.[2] Both OOO and new materialism put humans and nonhumans on a par, but whereas OOO upholds the autonomy of all objects, new materialists focus on relationality and entanglement.

This essay will highlight the differences between these two philosophical movements, showing how they play out in a number of cybernetic artefacts dating back to the 1940s, as well as in late twentieth-century automated artworks that appear to possess agency. These artefacts and artworks, which include Grey Walter's cybernetic tortoises and Jean Tinguely's *Homage to New York*, 1960, explore the relation between-human and object, human and nonhuman, organic and inorganic. Blurring the limits between life and nonlife, they call to mind the early modern automata, which also appeared to possess what we would today call agency.

In Part I of this essay, I will locate what we might today call nonhuman agency in Vaucanson's automata. In Part II I will show how the idea of nonhuman agency has taken on new and more expansive forms in contemporary posthuman philosophical movements, whether OOO or new materialism. In Part III, I will look at the divergences between these two philosophical positions in terms of how they understand nonhuman agency and in Part IV, I will focus on how these divergent views of nonhuman agency can be seen to play out in twentieth-century Posthumanism. As will become clear, the discourse about agency is central to these two alternative approaches to posthumanism, one of which emphasizes the autonomy of its agents and the other their interdependence.

That said, the distinction between these two approaches is starting to blur: recent developments in critical theory have given rise to a new term, neomaterialism, which has to a large extent absorbed OOO while confirming the ascendancy of new materialism, albeit, as Suhail Malik notes, without paying attention to their significant differences.[3] This essay will seek to reinstate some of these differences, focusing in particular on what I take to be the main bone of contention between the two movements, relationality vs. autonomy. This text does not go back over existing discussions as to the respective merits/failings of OOO and new materialism,[4] but should rather be viewed as taking a new tangent that complements these discussions. Crucially, it shows that what is really at stake in the relationality vs. autonomy dispute is our past, current and future relationship with the nonhuman agents that surround us.

Vaucanson's Automata

In her book *The Restless Clock*, Jessica Riskin explores the long and convoluted history of a much-debated question—whether the material world is "self-moving" and displays agency[5]—tracing it back to early automata such as those of Jacques Vaucanson. The latter were capable of playing the flute or even defecating, these being functions "that their designers took to epitomize the animate and the organic."[6] They were part of a long and still ongoing dialectic between life and machinery, in which the dominating principle "that a scientific explanation must not attribute will or agency to natural phenomena" was opposed across the centuries by a steady stream of objectors who maintained that agency was inherent in nature.[7]

Agency is thus the central term in Riskin's book, yet what is meant by it is not so much agency in the classic sense, whereby an agent is an entity capable of action, and agency is the exercise of the ability to act,[8] as a far more complex definition of the term. Riskin points to various kinds of agency—"living forces, sensitive capacities, vital fluids, and self-organizing tendencies"—whose unifying principle is that the activity is generated from "within," the agency of things being defined in her book as "an activity that appears to originate within themselves."[9] Indeed, it is the very fact that agency is what "appears" to originate within something, in other words, its ambivalence, that accounts for the automaton's fascination. The question that hangs over the automaton—whether agency originates within it or merely appears to do so—means that it cannot be reduced to the mere movement of mechanical parts, but instead exploits that ambivalence. Its apparent agency provokes a range of untoward effects, posing questions that are not raised by ordinary machines.

First and foremost among these questions are far-reaching interrogations about the nature of life. As Riskin notes, eighteenth-century automata articulated "certain differences between natural and artificial life that triggered the invention of machines that undermined those differences. But these machines in turn led people to rethink what constituted life."[10] Indeed, articulating such distinctions seems to have been one of the main functions of these kinds of artefacts. As Aurélia Gaillard points out, "the difference between all the simulacra of the classical age, whether 'machines made of earth', statues, dolls, automata, puppets, tableaux vivants—was one of degree rather than type, in that each simulacrum represented a transitional stage between the inanimate and the animate."[11]

The automaton was a puzzle on many accounts, thereby testifying to its makers' creativity.[12] As Gaillard shows, it drew on the technique of *mise en abyme*: inasmuch as Vaucanson's duck was meant to simulate the mechanics of digestion, it was a "representation of a machine representing a living being," that was "modeled on nature itself, viewed to a large extent as a machine." In other words, it was something of a monstrosity, triggering feelings of unease in the viewer.[13] There was also a connection with the divine: for just as Vaucanson's devices were made of mechanical parts, so did "God create his beasts from matter alone."[14] The automata did indeed provoke a kind of religious fervour in some spectators, although this did not mean that they "took God's place," but rather that they elicited "the feeling of the divine."[15]

In many cases, the automata elicited contradictory responses in viewers. Gilles Bertrand and Gilles Montègre evoke Italian artefacts that provoked not only enchantment and wonder but also the opposite, horror and repugnance. The gardens of Bomarzo for instance, displayed grotesque, misshapen statues,[16] showing how wonder can shade into repulsion and horror. Viewers of Vaucanson's automata may also have veered from fascination to repugnance and back again on seeing machines caricature and ape human actions with such skill. Bertrand and Montègre further point out that *commedia dell'arte* and other forms of popular entertainment had similar effects to the automata—generating illusory or optical effects, provoking wonder or unease—yet were still a far cry from machines such as Vaucanson's snake, "whose uniqueness resided in its capacity to move 'of its own accord' once it was set in motion."[17] It was thus essential that the movement or activity, whether fascinating or repelling, appeared to come from within the device, as stipulated in Jessica Riskin's definition of agency. This was what distinguished these machines from other life-like artefacts and prompted viewers to attribute agency to them.[18]

Finally, Jean-Yves Goffi comments on how Vaucanson himself questioned the borders between nature and artifice, most notably in a letter dated 1738, where he referred to his mechanical duck as an animal that drank and even quacked like a real duck.[19] Goffi writes: "He [Vaucanson] reaches the point where he talks about his automaton as if it were an animal several of whose actions are in every way similar to those of a 'real' duck. In other words, he no longer describes his invention as an automaton that closely imitates the mechanics of the actions of a living being; he describes it as if it were a living being, an artificial animal capable of behaving like a real animal."[20]

These accounts of Vaucanson's automata all highlight their ambivalence and theatricality, testifying to the permeability they achieved between technology and life. If we take as our starting-point Riskin's redefinition of agency as the ability to act in a way that appears to originate within the object itself, then machines possessing such agency have effects that are more than just purely mechanistic: depending on the context in which they were constructed, their ability to deploy agency makes them come across as grotesque, or god-like, or as redefining the meaning of life.

As Riskin shows in her book, such an approach to artificial life was unique to the second half of the eighteenth century—people in the seventeenth and nineteenth centuries were more interested in analogy than in simulating life.[21] However the second half of the twentieth century did come to witness a renewed interest in physiological simulation, thanks to the likes of roboticist Rodney Brooks, who gave robots the ability to see and touch, and William Grey Walter, inventor of the first autonomous robots, known as "tortoises."[22] These robots reacted to visual and tactile stimuli, and were thus capable of implementing autonomous actions such as circumventing obstacles.[23] They thereby appeared to exhibit agency—that is, they acted in a way that appeared to originate within themselves.

Nonhuman agency and the uncanny feeling it often elicits is thus to be found in the gap between a real duck or tortoise and a mechanical one, a gap instantiated by Vaucanson's duck, Grey Walter's tortoises and other automata. It is a feeling that calls to mind the Freudian uncanny, but also the more recent term "uncanny valley," an effect produced by certain machines. Coined by roboticist Masahiro Mori in the 1970s, the uncanny valley designates the process whereby robots become increasingly congenial to us, the more humanlike they appear—yet once they pass a certain threshold of resemblance, they provoke a strange sense of unease.[24] This feeling of fascination mixed with unease raises questions about the similarities between humans and nonhumans that resonate not only with lifelike machines but also, as I will now show, with contemporary philosophical movements addressing the posthuman.

Posthumanist Philosophy

If we define posthumanism as contesting the primacy of humans,[25] then the construction of automata that simulate and could eventually take the place of humans is one way of opposing that primacy. Another way, which I shall turn to now, is through new materialism and an offshoot of speculative realism known as object-oriented ontology (OOO), both of which have been at the forefront of attempts to challenge anthropocentrism, either by paying increased attention to nonhuman agency and putting it on a par with human agency, or by adopting a realist or anti-realist perspective. It is surely more than just a coincidence that the display of Vaucanson's automata in 1738 – and of the particular blend of artifice and reality they deployed—was more or less concurrent with Diderot's reading of Berkeley in the late 1740s, a reading which, according to Isabelle Thomas-Fogiel, marked the inception of the idealism-realism debate between "those who upheld the existence of the external physical world and those who rejected it."[26] Realism and idealism argued about the existence of such an external reality beyond the human, while the automata tried to instantiate that external reality by exhibiting an agency that likewise exceeded the human. The discourses on nonhuman agency and realism seemed to converge at that time and they appear to be converging once again in the early twenty-first century, in the realist philosophical movements known as OOO and new materialism, the first of which adopts a realist perspective and the second a realist or idealist perspective depending on the philosopher: not unlike the automata, these realist movements break down the distinctions between human and nonhuman agents, embedding them in the context of the realism-antirealism-idealism debate. In what follows, I will explore the differences and similarities between OOO and new materialism.

OOO's approach to nonhuman agency consists of treating everything as an object—humans being one kind of object and events or things another—while considering that interactions between nonhuman agents are as important as those between humans or those between humans and nonhuman agents. According to OOO founder Graham Harman, humans can only grasp interactions with or between nonhuman agents from the human point of view, but never from the perspective of nonhuman agents themselves. This is because all objects have their own existence that cannot be experienced by others, in other words, they are autonomous and withdraw from one another.[27] As Ian Bogost, another proponent of OOO, points out, "we never understand the alien experience, we only ever reach for it metaphorically."[28]

Whereas OOO's focus is on autonomous withdrawn objects, all of which are on a par, new materialism highlights the vitality and multiple entanglements of matter, challenging the notions of withdrawal and separation. As Diana Coole writes: "New materialism relocates humans by emphasizing their own materiality and through exploring their dependence on fragile or robust material systems and entities on which they leave more or less indelible traces."[29] New materialism thereby challenges the divisions between human and nonhuman agencies, as did the automata. It questions the hierarchies running from human down through animal, vegetable and mineral to the inorganic, by ascribing self-transformative and agential capacities to matter.[30] Jane Bennett argues in favor of ascribing agency to matter in her discussion of a blackout caused by the failure of an electrical power grid: "What is this power it wields? Can it be described as a kind of agency, despite the fact that the term is usually restricted to intentional, human acts? What happens to the idea of an agent once nonhuman materialities are figured less as social constructions and more as actors and once humans are themselves assessed as members of human-nonhuman assemblages?"[31] For Bennett, agency is distributive and agents can be human, biological or viral.[32]

There are thus affinities between OOO and new materialism in terms of questioning the separation between human and nonhuman agents, but there are also many differences, which emerge in their approaches to the key issues of nonhuman agency and realism that I am exploring here. For a start, objects in OOO, which include events as well as nonhuman agents as such, are inherently autonomous, whereas for new materialist philosophers such as Jane Bennett, nonhuman agents are manifestations of energy that do not exist until the human observer is there to perceive them and thus cannot be said to be autonomous.[33] For Harman, on the other hand, objects are unified enduring entities that cannot be dismissed, whether by "reducing them downward to their pieces, or reducing them upward to their effects."[34] Nonhuman agents are thus either autonomous or else entangled with their surroundings, and each of these two alternatives has very different implications and effects.

A second difference between OOO and new materialism resides in their respective approaches to the realism/antirealism/idealism debate. Realism for OOO, and as it is used in this essay, should be understood as implying a mind-independent world. As Cecilia Åsberg, Kathrin Thiele and Iris van der Tuin write: "Where speculative realists strive for an unmediated, wholly a-subjective real, feminist immanence ontology (in the singular plural) insists on the co-constitutive role of the embedded observer, the perspective and the rich agentiality (multi-subjectivity) of context itself."[35] In other words, theirs is not a human-indifferent realism, that is, a realism based on the belief that the world is autonomous and mind-independent, as in the case of OOO, but includes the perspective of the observer. Indeed, new materialist philosopher Diana Coole advocates "a transformed mode of being-in-the-world, one that escapes the limitations of conventional politics, power struggles, and runaway material consumption in pursuit of novel ways of being," while seeking to rebuild and reinvent existing structures.[36] Her brand of new materialism is thus premised on the observer's perspective, and so on mind-dependent interrelations between human and nonhuman agents, and could be construed as a form of idealism.

The different combinations of realism, materialism and antirealism in contemporary philosophy have been described by Graham Harman as follows: OOO is realist but not materialist, the "German Idealist and the Lacanian moves" of Alain Badiou and Slavoj Žižek are clearly antirealist, while Barad's new materialist approach may be dubbed "materialism without realism," inasmuch as it maintains that "objects have no reality apart from their interactions with the mind," and only Manuel DeLanda's version of new materialism is both realist and materialist.[37] Realism and new materialism are thus for the most part antithetical in this respect, with the latter in some cases verging on idealism.

To sum up, OOO and new materialism both disavow anthropocentrism by making room for nonhuman agencies. For Harman, the fact that for OOO everything is an object, whether human, nonhuman, imaginary or immaterial, implies that more attention should be paid to nonhumans, since they outnumber humans. As for new materialism, it seeks to redistribute agency in such a way that the nonhuman is no longer subordinate to the human, but can itself have an impact—as in the case of Jane Bennett, for whom the world is made up of assemblages comprising different kinds of actants, whether human, nonhuman, natural, cultural or technological, none of which can direct the assemblage's actions without the aid of the others.[38] Whereas Bennett speaks of dependence and entanglement between humans and nonhumans, Harman refers to distinct unified objects of many kinds, all autonomous from one another. Yet they both acknowledge nonhuman agency. If we take as our starting-point Riskin's redefinition of agency as the ability to act in a way that appears to originate within the object itself, rather than in our imaginations, in other words that corresponds to a realist rather than the idealist perspective, then the automata can be regarded as instantiating the dispute between OOO and new materialism. For OOO, objects are real and autonomous and can therefore possess an agency that originates within themselves. For the new materialists, objects exist when the human is there to perceive them, and they are therefore dependent on the human mind. A similar question hangs over the automaton—whether agency does indeed originate within it or whether it is a product of our imagination and/or minds.

Confronting the Nonhuman

It is at this point that we might ask how we want to envisage the relation between human and nonhuman—in terms of continuity and entanglement between humans and nonhuman agents, that is, by blurring the boundaries between them? Or should humans renounce their anthropocentric standpoint and acknowledge the nonhuman's agency and thus autonomy? The automata led the way by posing the question as to where agency originates, in the machine or in our minds. OOO and new materialism broaden the discussion, asking us to choose between autonomy and relationality, between an external reality, or a world that includes the human observer and thus stems from the activity of our minds. In what follows, I explore the notions of relationality and autonomy from the perspectives of new materialism and OOO and come to the conclusion that both movements are problematic.

Relationality

The notion of relationality is actively endorsed by new materialism, as is continuity between humans and nonhuman agents. One way of establishing continuities is to reverse the dualisms traversing contemporary thought: Rick Dolphijn and Iris van der Tuin highlight, for instance, new materialism's attentiveness to "materiality [and] processes of materialization," as a means of reversing the process whereby matter "has been so much neglected by dualist thought."[39] Another method is to reconceptualize divisive concepts. Commenting on Rosi Braidotti's version of new materialism, they note: "The exterior and the interior, the subject(ive) and the object(ive), the individual, the social, and the symbolic are conceptualized as co-constitutive instead of pre-determined levels or layers."[40] In her own writing, Rosi Braidotti lays particular emphasis on relationality. As she points out: "I emphasize the embodied, embedded and transversal selves that we are, bonded by ontological relationality. Embodied and embedded because we are deeply steeped in the material world. Transversal because we connect but also differ from each other. And yet we are structurally related to one another, to the human and nonhuman world that we live in [...] The key term is indeed relationality, operating in an embedded and embodied and grounded, multi-directional and multi-scalar manner."[41] Relationality also underlies what Braidotti calls posthuman subjectivity, a merging of human and nonhuman agents that she believes has already occurred and that opens up opportunities for new and fruitful interrelationships. She writes: "Posthuman subjectivity is a transversal alliance that nowadays involves non-human agents. This means that the posthuman subject relates at the same time to the Earth—land, water, plants, animals, bacteria—and to technological agents—plastic, wires, cells, codes, algorithms [...] so that the frame and scope of epistemological and ethical subjectivity is enlarged."[42] Such transversality and embeddedness in the material world contrasts sharply with OOO, inasmuch as objects are no longer viewed as autonomous, separate and irreducible to one another, but as increasingly hybrid and interchangeable, to the point where there are no longer any precise limits as to where stable identities begin and end.

Yet the loss of stable identities can pose problems. First of all, it encourages passivity or even resignation, for if all entities are entangled with one another, they no longer have full control over their actions but are exposed to the whims, influence and invasiveness of others. Furthermore, entanglement can lead to its opposite, withdrawal and protectionism. As Paul Rekret points out, many new materialists feel the need to shield matter from critique, because critique implies human superiority: "Indeed critique [...] is rejected through a directly antonymic set of ethical metaphors, as 'reductive', 'resentment', 'distancing', 'conceit', or 'hubris' found to maintain the priority of the human in some way. In maintaining the priority of the human in some way, the new materialist argument goes, theory fails to 'register' the world in itself."[43]

Turning now to OOO, we may note that it offers a solution to the question of relationality that avoids the failings of new materialism. Harman's goal is to eliminate anthropocentrism and in doing so, he puts all things on a par. Indeed, as Miguel Penas López observes, there are parallels between Harman's thinking and the work of Gilbert Simondon, even though their methods differ:

> Simondon's philosophy eludes the epistemological opposition between subject and object as being the starting point of philosophy by way of raising a more primary ontological question, i.e., the question about the genesis of both the subject and the object. This ontogenetical inquiry strips the subject and its relation with the world of all ontological privilege. On the other hand, Harman looks for the metaphysical framework of every object, which makes it possible to put all kinds of objects on equal footing. Instead of a basic rift between humans and objects, Harman's philosophy points out a general rift between objects.[44]

Objects thus remain distinct despite being on an equal footing, and they can also enter into relations, albeit only up to a certain point.[45] This also applies to the automata: even though we can relate to them on different levels, they retain their specificity.

Autonomy

Moving on to the question as to whether humans should renounce their anthropocentric outlook and acknowledge the nonhuman's full autonomy, we may note that new materialism engages with subjectalism rather than autonomy, meaning that thought can "access the absolute, at the price of identifying this absolute with some dimension of subjective existence itself."[46] In other words, although certain forms of new materialism claim to grasp nature non-subjectively, in reality they do not. They are merely replacing the subject by some other term such as matter, which grasps nature in the subject's stead.

Slavoj Žižek likewise offers a critical assessment of new materialism's surrogate subjects. He notes that "selves or multiple agents are everywhere in different guises," which makes it all the more unclear whether it is we who are ascribing vitality to matter, or whether vitality is inherent in matter.[47] Barad's notion of entanglement does not leave much room for autonomous nonhuman agency either. As Harman observes: "When two things are entangled, it means that they are *not* autonomous, but rather that they mutually co-constitute each other."[48]

OOO, on the other hand, explicitly upholds the autonomy of all objects, inasmuch as Harman adheres to the realist thesis that the world is mind-independent.[49] It should be specified that the mind-independence asserted here does not imply a "world *without* humans," as is often assumed, but only a "world independent of humans."[50] Yet it could be claimed that ascribing independence, autonomy, or withdrawal, or anything else for that matter, to objects, is a subjectalism, as mentioned earlier in connection with new materialism. This would be true if the subject/object distinction still existed in OOO, but if we accept that OOO has eliminated that particular distinction, as it claims to do by putting all objects on a par, then there can no longer be any opposition between subject and object, nor any subjectalist hierarchy between them. For OOO, all objects are "equally *objects*."[51] The automata also challenge the subject-object distinction inasmuch as their lifelikeness puts them on a par with humans. All in all, OOO's insistence on autonomy appears to point towards an equitable and balanced approach that is more in line with the ethos of the automata.

The Humanization of Matter

On this reading, autonomy would appear to have won the day. However, there is a serious objection to OOO and new materialism that has yet to be taken into account: the fact that it is becoming increasingly difficult to speak of matter in isolation from the human. As Suhail Malik observes, matter is increasingly the product of rationalized anthropogenic processes, which "presents an insurmountable theoretical and practical obstacle for naturalized materialisms which typically look to subordinate rational cognition and fabrication to material and organizational processes that precede or are external to anthropogenic involvement."[52] The fact that matter is anthropogenically constructed, or less and less "natural," means that human rationality can no longer be subordinated to material processes, as new materialism would have it. This also poses a problem for OOO, which ascribes autonomy from humans to nonhuman agents that are no longer as nonhuman as they once were. For Malik, failing to recognize that matter is pre-anthropogenic, as in the case of the new materialists and OOO, is in any case a step backwards in view of the anthropogenic interventions that will be necessary to meet the demands of climate change. The latter appear to call for accelerationist-style solutions "to advance the artificing of both anthropic conditions and nature via [...] discourse, technoscience, and other praxes."[53] Seen from this angle, the only possible option would appear to be more human intervention at the expense of the nonhuman. This option would seriously undermine not only new materialism but also OOO, since they both seek to decenter the human in favor of the nonhuman.

However, it would not resolve the question of the divide between "objects, individuals and substances on the one hand, and processes, flux and becoming on the other." More than just an unresolved quarrel between two philosophical movements, it is an ongoing dilemma: as López notes, it is inscribed in an ongoing and still unresolved debate, namely whether "the intrinsic nature of reality is static or dynamic, whether we can isolate in it immutable traits or encounter nothing but perpetual change and transience."[54] This is a dispute that goes back to pre-Socratic philosophy, on the one hand to Parmenides, who held that there was only one eternal reality, and on the other hand to Heraclitus, who favored ongoing change.[55] It is a testimony to the staying-power of this dispute that it is resurfacing today not only in continental philosophy, but also in art, as I will show in Part IV.

Art and Nonhuman Agency

Like the early modern automata, robotic artworks are particularly suited to exploring such notions as nonhuman agency, autonomy and relationality. An example is Jean Tinguely's *Homage to New York*, 1960,[56] which was made of bicycle wheels, motor parts, a piano, metal drums and other discarded objects and measured 23 ft. long and 27 ft. high. As stated in the press release, activation took place in MoMA New York's sculpture garden, whereupon 15 motors controlled by 8 timers or delay relays set in motion its different parts. The piano was played by mechanical arms, while metal parts struck each other, making percussive sounds. Saws and hammers began to move, gradually destroying the work's principal components, and causing it to disintegrate within half an hour.[57] Tinguely's self-destructing machine raises the question of nonhuman agency and intentionality, while highlighting the interrelations between things and the impact they can have on each other—aligning itself with the tenets of new materialism.

Another example is Chris Burden's *Samson*, 1985. It consists of a jack connected to a gearbox and a turnstile, which pushes two pieces of wood into the walls of the exhibition space. When visitors enter through the turnstile, the jack puts pressure on the wooden beams. Over time, it could cause the walls to collapse, thereby destroying the building.[58] The suggestion is that the installation operates autonomously without our being aware of it, and that, rather like in OOO, the nonhuman will always elude our control. At the same time however, these works are ambivalent: after all, *Samson* could also be viewed as an assemblage of different interconnected parts whose movements affect each other, while *Homage to New York* could equally well be about the autonomy of machines that self-destruct without human assistance. By exploring both options, these works seem to be asking us to make our choice as to how we want to envisage the relation between human and nonhuman, as we enter the era of the posthuman.

The notion of nonhuman agency harks back to Vaucanson's automata—as noted above, the construction of automata that appear to possess agency is one way of questioning the primacy of humans and engaging with posthumanism. The notion of nonhuman agency has considerably evolved over the centuries, resurfacing in different contexts, not least OOO and new materialism. At a time when the way forward for posthumanism is widely thought to be relationality, the possibility of an alternative option foregrounding autonomy—as prefigured in OOO and certain robotic artworks—may help to redefine our future relationship with nonhuman agents and put it on a more balanced and less invasive footing.

As I have attempted to show, Vaucanson's automata posed questions that anticipated the development of posthumanist philosophy. Only now, three hundred years on, are the implications of these questions finally starting to be understood.

Notes

1. Karen Barad, "Posthumanist Performativity: Toward an Understanding of How Matter Comes to Matter," *Signs: Journal of Women in Culture and Society* 28, no. 3 (2003).

2. Jane Bennett, *Vibrant Matter: A Political Ecology of Things* (Durham: Duke University Press, 2010).
3. Suhail Malik, "Dispossessed and Watered-Down: Hegemonic Neomaterialism and its Limitations," in *Springerin*, no. 1 (2016): par. 2–3, https://www.springerin.at/en/2016/1/enteignet-und-verwassert/ (accessed July 22, 2022).
4. See, for instance, Thomas Lemke, "Materialism without matter: the recurrence of subjectivism in object-oriented ontology," *Distinktion: Journal of Social Theory* 18, no. 2 (2017), and Paul Rekret, "A critique of new materialism: ethics and ontology," *Subjectivity* 9, no.3 (2016), https://www.researchgate.net/publication/305079128_A_critique_of_new_materialism_Ethics_and_ontology (accessed July 21, 2022).
5. Jessica Riskin, *The Restless Clock: A History of the Centuries-Long Argument over What Makes Living Things Tick*, (Chicago: The University of Chicago Press, 2016), 3, 7.
6. Jessica Riskin, "Eighteenth-Century Wetware," *Representations* 83, no. 1 (2003): 104.
7. Riskin, *The Restless Clock*, 2.
8. Stanford Encyclopedia of Philosophy, https://plato.stanford.edu/entries/agency/ (accessed July 11, 2022).
9. Riskin, *The RestlessClock*, 3.
10. Riskin, "Wetware," *Representations*, 115.
11. Aurélia Gaillard, "De 'la possibilité du flûteur automate' (Helvétius): les automates du XVIIIe siècle comme merveilles de substitution," in *L'automate: modèle métaphore machinemerveille*, eds. Aurélia Gaillard, Jean-Yves Goffi, Bernard Roukhomovsky, Sophie Roux (Pessac: Presses Universitaires de Bordeaux, 2012), 364.
12. Gaillard, "De 'la possibilité du flûteur automate'", 364.
13. Gaillard, "De 'la possibilité du flûteur automate'", 369, 370, 371.
14. Gaillard, "De 'la possibilité du flûteur automate'", 373.
15. Gaillard, "De 'la possibilité du flûteur automate'", 374.
16. Gilles Bertrand and Gilles Montègre, "De la curiosité pour les automates dans l'Europe des Lumières. Machines et mécanismes en Italie au prisme du regard des étrangers (1680–1820)," in *L'automate: modèle métaphore machinemerveille*, eds. Aurélia Gaillard, Jean-Yves Goffi, Bernard Roukhomovsky, Sophie Roux (Pessac: Presses Universitaires de Bordeaux, 2012), 404.

17. Bertrand and Montègre, “De la curiosité”, 406.
18. For an in-depth discussion of the importance of movement in automata and robots and the question of their lifelikeness, see *Behavioral Objects 1, A Case Study: Céleste Boursier-Mougenot*, eds. Samuel Bianchini and Emanuele Quinz (Berlin: Sternberg Press, 2016) and *Behavioral Objets Behavioral Matter*, eds. Samuel Bianchini and Emanuele Quinz, forthcoming.
19. Jean-Yves Goffi, “Nature et artifice,” in *L'automate: modèle métaphore machinemerveille*, eds. Aurélia Gaillard, Jean-Yves Goffi, Bernard Roukhomovsky, Sophie Roux (Pessac: Presses Universitaires de Bordeaux, 2012), 24.
20. Jean-Yves Goffi, “Nature et artifice,” 24–25.
21. Riskin, “Wetware,” *Representations*, 101.
22. Riskin, “Wetware,” *Representations*, 116, 118.
23. Bristol Robotics Laboratory, “Grey Walter and his Tortoises,” University of Bristol News and features, http://www.bristol.ac.uk/news/2008/212017945378.html (accessed July 16, 2022).
24. Rina Diane Caballar, “What is the Uncanny Valley?” IEEE Spectrum, https://spectrum.ieee.org/what-is-the-uncanny-valley (accessed July 16, 2022).
25. See Jay David Bolter, “Posthumanism,” Wiley Online Library, https://onlinelibrary.wiley.com/doi/pdf/10.1002/9781118766804.wbiect220 (accessed July 18, 2022).
26. Isabelle Thomas-Fogiel, “L'opposition entre réalisme et idéalisme?: Genèse et structure d'un contresens,” Revue de Métaphysique et de Morale 95, no. 3 (2017), 9–10.
27. Ian Bogost, *Alien Phenomenology, or What It's Like to Be a Thing* (Minneapolis/London: University of Minnesota Press, 2012), 65.
28. *Ibid.*, 66.
29. Diana Coole, “From Within the Midst of Things: New Sensibility, New Alchemy, and the Renewal of Critical Theory,” in *Realism Materialism Art*, eds. Christoph Cox, Jenny Jaskey, Suhail Malik (Berlin: Sternberg Press, 2015), 43.
30. Coole, “From Within the Midst of Things,” *Realism Materialism Art*, 43.
31. Jane Bennett, “The Agency of Assemblages and the North American Blackout,” *Public Culture* 17, no.3 (2005): 446.
32. Bennett, “The Agency of Assemblages,” *Public Culture*, 446.

33. Manuel DeLanda and Graham Harman, *The Rise of Realism* (Cambridge/Malden: Polity Press, 2017), 58.
34. DeLanda and Harman, *The Rise of Realism*, 81.
35. Cecilia Åsberg, Kathrin Thiele, Iris van der Tuin, "Speculative Before the Turn: Reintroducing Feminist Materialist Performativity," *Cultural Studies Review* 21, no.2 (2015): 151.
36. Coole, "From Within the Midst of Things," *Realism Materialism Art*, 44–45.
37. DeLanda and Harman, *The Rise of Realism*, 3–4.
38. Bennett, "The Agency of Assemblages," *Public Culture*, 445.
39. Rick Dolphijn and Iris van der Tuin, New Materialism: Interviews & Cartographies (University of Michigan Library: Open Humanities Press, 2012), 93.
40. Ibid., 106.
41. Rosi Braidotti, Posthuman Knowledge (Cambridge/Medford: Polity Press, 2019), 45.
42. Ibid., 46.
43. Rekret, "A critique of new materialism," *Subjectivity*, par. 2.
44. Miguel Penas López, "Speculative Experiments: What if Simondon and Harman Individuate Together?" in *Aesthetics in the 21st Century: Speculations V*, eds. Ridvan Askin, Paul J. Ennis, Andreas Hägler, Philipp Schweighauser (Brooklyn: Punctum Books, 2014), 230.
45. Graham Harman, *The Quadruple Object* (Winchester/Washington: Zero Books, 2010), 75.
46. Daniel Sacilotto, "Subjectalism," in The Meillassoux Dictionary, eds. Peter Gratton and Paul J. Ennis (Edinburgh, Edinburgh University Press, 2015), 161.
47. Slavoj Žižek, "Materialism, old and new," *Strike!*, Issue 8, Nov-Dec 2014, p.13.
48. Graham Harman, "Agential and Speculative Realism: Remarks on Barad's Ontology," *Rhizomes*, Issue 30 (2016), http://www.rhizomes.net/issue30/harman.html (accessed July 22, 2022).
49. DeLanda and Harman, *The Rise of Realism*, 3–4, 28.
50. DeLanda and Harman, *The Rise of Realism*, 39–40.
51. Harman, *The Quadruple Object*, 5.
52. Malik, "Dispossessed and Watered-Down," Springerin, par. 7.
53. Ibid., par. 8.

54. López, "Speculative Experiments," *Aesthetics in the 21st Century: Speculations V*, 225–226.
55. López, "Speculative Experiments," *Aesthetics in the 21st Century: Speculations V*, 225.
56. This example and the following one are discussed in Behavioral Objets Behavioral Matter, eds. Samuel Bianchini and Emanuele Quinz, forthcoming, along with many other examples of robotic artworks that appear to demonstrate agency.
57. The Museum of Modern Art, "Homage to New York: A Self-Constructing and Self-Destroying Work of Art," For Release Friday, March 18, 1960, https://www.moma.org/documents/moma_press-release_326185.pdf (accessed July 23, 2022).
58. Google Arts & Culture, "Samson," https://artsandculture.google.com/asset/samson/7gH2V-pE_LzBWw?hl=en (accessed July 23, 2022).

CHAPTER 16

From Automaton to Cyborg: in Search of the Autonomy of the Living

Emanuele Clarizio

Life Between Epistemology and Metaphysics

In *The Order of Things*, Michel Foucault exposed his controversial thesis that life did not exist before the end of the twentieth century: "Historians want to write histories of biology in the eighteenth century, but they do not realize that biology did not exist then, and that the pattern of knowledge that has been familiar to us for a hundred and fifty years is not valid for a previous period. And that, if biology was unknown, there was a very simple reason for it: that life itself did not exist".[1] Although historians have actually criticized the overemphasis on discontinuity in the history of sciences and highlighted a more troubled and complex genesis of biology,[2] Foucault's thesis is significant not only for epistemology, but for metaphysics as well. From this perspective, Davide Tarizzo has revived it, asserting that the notion of "life" is the essential category of what he

E. Clarizio (✉)
Center for Medical Ethics (CEM)—Ethics Laboratory, Catholic University of Lille, Lille, France
e-mail: Emanuele.Clarizio@univ-catholille.fr

J. Muller (ed.), *Automata, Cyborgs, and Mutants*, Palgrave Studies in Medieval and Early Modern Medicine,
https://doi.org/10.1007/978-3-031-93201-4_16

calls, in borrowing another Foucauldian expression, an "ontology of our actuality". In other words, we, modern humans, cannot think of ourselves in any other way than as living beings. Furthermore, according to Tarizzo, this genealogy of our ontology has moral rather than epistemological roots. Or in other words, even the epistemology of the living should be understood through a moral genealogy, which provides its metaphysical foundation. For Tarizzo, "The notion of life that Darwin puts forth, when we carefully scrutinize it, looks like a metaphysical anamorphosis of Kant's notion of the will".[3] Following this interpretation, the revolution in biology is not so much epistemological as metaphysical. What is at stake here is not so much the restructuring of knowledge about living beings around the category of "organization", as Foucault thought, but the metaphysical conception of living beings as sustained by a single life-force that transcends them and that has no foundation other than itself, making it indeed a transposition of what Kant's moral philosophy called the will. Life appears therefore as "synthetic unity", "secret force" and "obscure will of the Self" at the same time, that is, as another figure of the sovereign and autonomous subject that dominates modernity: "We, the moderns, are not. We live and we will", as Tarizzo puts it.[4]

Immanuel Kant is obviously the exponent of this transition, for it was he whom enshrined autonomy "as the pure fact of willing, which is the fact of modernity itself".[5] In this light, the link between Kant's second and third *Critiques* is clear, as Tarizzo promptly notes: "If the metaphysical question from which Kant's *Critique of Practical Reason* takes its initial cues is 'what is the will?' the question at the origins of the *Critique of Judgment* is 'what is a purpose?' There is continuity between these two questions".[6] It is in this lexicon of the purpose that the kinship between will and life is to be found, a lexicon that is in fact common to both the second and the third *Critiques*. Moreover, we can stress that not only does Kant investigate the power of judgment in light of the concept of purpose in the third *Critique*, he also sketches out the notion of will in light of the concept of life in the first lines of the second *Critique* ("However dissimilar representations of objects may be [...] the feeling of pleasure by which alone they properly constitute the determining ground of the will [...] is nevertheless of one and the same kind not only insofar [...] as it affects one and the same *vital force* that is manifested in the faculty of desire")[7] and vice versa ("Life is the faculty of a being to act in accordance with laws of the faculty of desire").[8] Thus, according to Tarizzo, the practical concepts of "pure will" and "will of the Self" have their aesthetic-epistemological counterparts in the aesthetic concept of beauty and the teleological concept of the organism, which are respectively the objects of study of the first and second sections of third *Critique*: just as beauty is pure form of finality—pure will—the organism is the concrete paradigm of finality-without-end (or purpose-without-content)—will of the Self -.

The Analogy Between Life and Technology

However, while Foucault observed that the category of organization is the epistemological core of the notion of life, and while Tarizzo noticed the conceptual junction that links life and will to the concept of purpose, they didn't sufficiently remark that both the notions of organization and purpose have a connection to that of technique, which we can assume to be involved in this story, so that in this way of conceiving ourselves as living beings, there is also at stake, to some extent, a certain problematization of the relationship between life and technology. By thinking of itself as a living being, the human has not only thought of life as the anamorphosis of the will, it has also thought of itself in a very intimate, but in some ways paradoxical, as we will see, relationship with technique. Indeed, the notions of purpose and organization not only have a moral genealogy, they also refer to the technological semantics that informs explanations of the functioning of organisms, at least from Aristotle onwards. Indeed, as Georges Canguilhem reminds us on several occasions,[9] Aristotle had already assimilated "the organs of animal motion to *organa*, that is, to the parts of war machines",[10] but he had not yet dared the analogy between the machine and the whole organism. However, if he did not do so, it was only because of the state of advancement of the techniques of his time, and not because of any particular theoretical reluctance, according to Canguilhem: the machines of war known to Aristotle were in fact machines powered by animal or human motive power, whereas Descartes had in mind spring-loaded or hydraulic machines. This difference is crucial according to Canguilhem, because he believes that, for there to be complete assimilation of the organism to the machine, the greater or lesser similarity between the mechanical parts of the machine and the parts of the body is not a sufficient condition, since there must also be an "interval between the storing up and the release of energy by the mechanism that allows one to forget the relationship of dependence between the mechanism's effects and the action of a living being".[11] If

Aristotle does not have a theory of the animal-machine, it is because the machines of his time were powered by animal strength, and the dependence of the machine on the living was so obvious that a conception of the machine as an autonomous entity could not emerge. In Descartes' case, on the other hand, this became possible by virtue of the existence of automata whose operation is largely autonomous from the technical operator (because the driving force comes from elsewhere).

As Foucault recalls, the category of organism is central to the birth of biology, and as Charles T. Wolfe explains, it was in fact established in an ambiguous relationship, at once of analogy and difference, with the notion of machine.[12] Descartes is certainly the apex figure of the twentieth-century mechanism, which Foucault still interprets as opposite to the rationality of biology: "Cartesian mechanism, which subsequently proved an obstacle, was used at first, the historians tell us, as a sort of instrument of transference, and led, rather in spite of itself, from mechanical rationality to the discovery of that other rationality which is that of the living being".[13] Actually, it is easy to see that mechanical rationality is a condition of the rationality of the living being, insofar as the concept of machine served to analogically explain that of organism. Moreover, when referring to Cartesian mechanism, we should distinguish at least two senses of the mechanism: on the one hand, there is the corpuscular theory of the matter, which intends to explain biological phenomena (or all material phenomena), especially in the field of embryology, through a causalistic approach, thus making them fall within the scope of physics.[14] On the other hand, thanks to the analogy between the organism and the machine, we find a further level of explanation, on a physiological level, which can be called *machinistic*, and whose aim is to account for the peculiar organization of the parts of the organism among themselves, as in the organization of the parts in a machine. With the mechanism *stricto* sensu, therefore, Descartes postulates a certain measurability of the phenomena of life, bringing it back into the realm of science, while with machinism, he recognizes a further specificity of the organism, precisely that of organization. In so doing, Descartes instills at the heart of biology an analogy between life and technology that has not ceased to impact the way we have conceived of the organism ever since.

However, as Foucault perhaps indirectly suggests, biology would later be established against this analogy, which would appear to be rather reductive. Of course, this allegation is erroneous, because, as we have seen, it is precisely through the machinic analogy that biology has been able to gain autonomy from physics. But if we consider that the technological metaphor served as a viaticum for biology to disassociate itself from physics, it implies that it ushered in a movement at the end of which biology must also disassociate itself from technology and assert principles of its own. Kant and Leibnitz above all, have in fact specified and radicalized the analogy between the body and the machine,[15] between technique and life, through the metaphor of life as "technique of nature"[16] and that of the body as "natural machine" respectively.[17] Following them, however, biology had the need to free itself from this analogy as well. As sophisticated as these metaphors may be, the main reason for their failure is not reductionism—after all, it is easy to show that no one has ever claimed perfect equivalence between a machine and an organism[18]—but the latent and unascertainable finalism of any analogy between life and technology. Although the analogy body-machine remains always, in fact, an analogy and not an identity, it refers to a transcendent and teleological dimension: while it is true that automata are organized in a manner similar to organisms, and can therefore function as models to study them, it is also true that they are, above all, always created by an intelligence that transcends them, and cannot move by themselves. This was in fact one of the major limitations of Cartesian biology, or proto-biology if one prefers: embryology, which was purely mechanistic, could avoid transcendent causes, but at the same time could not account for the organic functioning of the body, that is, for real dynamic organization and for the articulation between different organs. To explain this property, Descartes needed a machinic physiology, which, however, was teleological. In a sense, it is as if Kant and Leibniz tried to inscribe the transcendence of technique in the immanence of the organism, in a titanic effort to eliminate finalism from biology, but of course what happened was the opposite: when you put technology into the body, it becomes difficult to avoid the idea of a great engineer conceiving/creating the body. Used to explain the autonomy of the living with respect to the rest of the natural world, the automaton fails in its main task: in reducing autonomy to an autonomy of functioning (i.e. automation), the automaton no longer possesses any ontological autonomy, since it is created by a superior intelligence. Although its functioning is automatic, its formation is not autonomous.

From Mechanism to Organology

It seems then that the renunciation of the analogy between life and technology is motivated by the attempt to abandon finalism, rather than reductionism, and to think of the living as an autonomous being. Tarizzo's hypothesis thus appears to be confirmed: the birth of biology, which coincides with the abandonment of mechanistic teleological models, responds to a need to think of life as "will-of-its-Self", as an ultimate and irreducible normative source. Of course, when speaking of the autonomy of the living, it is difficult to separate the moral relevance of this notion from the epistemological one. Not surprisingly, Tarizzo sees in Canguilhem one of the contemporary exponents of this metaphysics: it is hard not to see that the stakes of his vitalism are not merely epistemological, but also moral. Affirming the "normativity of the living" is not a purely epistemological stance, but serves to ground an ethics of the living that subtracts it from other forms of normativity.[19] But if teleology can no longer be an epistemological option, how can vitalism explain technique itself, and the relationship between technique and life? On this point, Canguilhem's project consists in overturning the mechanistic hypothesis that organisms and automata are analogous because they are created by two similar intelligences, so that the machine created by man can function as a model to understand the body—which remains finalistic, into the thesis of the priority of the living over technique. This is precisely the perspective he adopts in his essay *Machine and Organism*:

> Indeed, the problem of the relations between machine and organism has generally been studied only in one direction: almost always, the attempt has been to explain the structure and function of the organism on the basis of the structure and function of an already-constructed machine. Only rarely has anyone sought to understand the very construction of the machine on the basis of the structure and function of the organism.[20]

To overcome this deficiency, he believes that the mechanical theory of life should be resituated within a biological theory of technique. It is because life is capable of creating machines relatively independent from the biological operator that they can then be used analogously as models for the knowledge of organisms. It is clear that what is at stake here is not an epistemological definition of life, but a metaphysical one, as something which is capable of creation - as the only entity that is capable of creation, in fact! It is because Canguilhem needs to think about the human as a free being that he must conceive of it as an autonomous living being, an entity whose purposes are immanent to life and autonomously determined. Unlike the machine, the living is not only organized, it is also autonomous, and this is ultimately the main criterion for the differentiation between machine and organism. It then becomes clear that the strategy with which Canguilhem dismantles mechanism responds to an attempt to eliminate finalism, rather than to reductionism in its attempt to re-assign technology and life to two distinct, though intimately related spheres. Technique is seen in this perspective as the creation of life, and although their relationship remains close, it is no longer played out on an epistemological level, but on a metaphysical one. Canguilhem has thus created the condition for the birth of what he called a "biological philosophy of technique".[21] It is precisely because the living being is conceived of as autonomous, and therefore normative, that it possesses technique, to the extent that the human specializes as *Homo Faber*, maker of tools, as Henri Bergson had already stressed.[22] On the basis of this vitalist vision, it is possible to overturn the relationship between technique and life: if the technological analogy is effective in the epistemology of the living, it is not because the body actually functions as a machine, but because the living being is capable of creating inorganic instruments in the image of its own organs, and it can therefore use the former as a model to study the latter.

As we have seen, in this overturning of mechanism, the issue at stake is more metaphysical than epistemological: it is not only a question of changing the conceptual framework of our knowledge of life, but most of all of proposing a thesis on the nature—normative and creative—of life itself. The role of the notion of technique, in this transition, is crucial: instead of an epistemological model for the knowledge of organized beings, it is now conceived, first by Bergson and then by Canguilhem, as a creation of the living, and of the human living in particular (*Homo Faber*). By overturning mechanism, therefore, Canguilhem also founded a true anthropology of technology. This was possible because Canguilhem's need to propose a normative and creative notion of life has moral and political, rather than epistemological, roots. What ultimately interests him, through the affirmation of the thesis of the normativity of life, is actually the affirmation of the freedom of the human subject, which finds in its own vital normativity the ultimate source of its values. On the contrary, Descartes' mechanism had no anthropological significance, since the freedom of the subject was located in the thinking substance, the mind, so that the analogy between the machine and the organism did not actually affect his anthropology and did not reveal anything about human nature as such. It is no coincidence that the respectively farthest and closest points of reference of what Canguilhem calls the "biological philosophy of technique" are, according to him, Ernst Kapp and André Leroi-Gourhan, exponents of philosophical anthropology and anthropology of technology respectively.[23] According to this line of interpretation, continuity can be established between Kapp and Leroi-Gourhan through Bergson, as they all conceive of technique as a kind of prolongation or extension of the body. Bergson and Leroi-Gourhan do not exactly think that technical objects are an exteriorization of "organic forms", "organic functions", and "organic proportions" of the body, as Kapp thought[24]; however, they do think that technique is a kind of movement of exteriorization, an extension of the vital impulse through which the living creates inorganic forms. Therefore, their biological philosophy of technology overturns the metaphor between life and technique: it no longer thinks of the organ as based on the model of the artifact; it thinks of the tool as an external organ. Bergson defines for instance the "instrument" as "an artificial organ by which the natural organism is extended",[25] so that Canguilhem can speak of Bergson's philosophy as a "general organology", according to which there exists a sort of continuity between natural organs, which are organic and internal to the organism, and artificial organs, which are inorganic and external. Although he speaks of "general organology" while referring to Bergson's philosophy, he uses the same lens to interpret the Cartesian position on technology, in his earlier text on *Descarteset la technique*.[26] And it is true that in the *Dioptric*, while describing the parts of the eye, Descartes separates the "inner organs, which are the nerves and the brain" from the "external organs, between which I understand all the transparent parts of the eye as well as all the other bodies that can be put between it and the object".[27] The retina and the lens would therefore be, for Descartes, external organs in a similar way. Therefore, despite the overturning of the prominent role between the living and the technical proposed by Canguilhem, his interpretation of the relationship between life and technique aims to establish a certain continuity among these different authors, from Descartes to Bergson and even to Leroi-Gourhan. What is interesting then, is that this proximity between life and technique, that was first established on an epistemological level, turns now not only in a thesis about the origin of technique, it attempts to reveal something not only about biology, but also about technology. Canguilhem then calls the general thesis shared by all these authors "organology", a thesis according to which the tool is in fact an external organ of the living human being.

The Cyborg as a Paradox for Organology

What surprises in Canguilhem's account is that organology postulates the continuity between internal and external organs, while attempting to maintain a clear boundary between the autonomous organism on the one side and the non-autonomous machine on the other. On the one hand, we have an extreme proximity between the living human being and technique; however, for this proximity to exist, a limit, boundary or threshold that clearly separates life from technique needs to be delimited. After all, as Wolfe clearly pointed out, the purpose of the analogy between the body and the machine is precisely to explain the living by comparing it to something other than itself[28]; just as, on the other hand, Bergson's organological thought, according to which life extends its work outside the body, presupposes the ontological priority of life over technique. The philosophy of Bergson and Canguilhem is indeed aimed at establishing the primacy of life, from an ontological as well as moral point of view. However, if technology has been prolonging the human body for millennia, this means, as Bergson states in *The Two Sources of Morality and Religion*,that humanity now possesses an enlarged body that prolongs the natural body.[29] The generative capacity of the living human being, a declination of its ontological autonomy, has rendered the products of human technology ungovernable, putting human moral autonomy at risk. This is exactly the kind of paradoxical relationship between life and technology that exists in the Cyborg. In fact, by making life the source of all creation and all norms, Bergson and Canguilhem naturalized technology, to the extent that they were sometimes interpreted as champions of Cyborg thought.[30] Provocatively interpreting Bergson's appeal to mysticism in *The Two Sources of Morality and Religion*, Paola Marrati even argues that.

> Bergson seems to suggest that the natural cyborgs are, well, *too* natural; that our natural capacity for combining biology and technology - for creating new organs and machines - has not gone far enough [...] He seems to suggest that the natural cyborgs must become supernatural ones if we want to loosen the bonds of *natura naturata*, come closer to *natura naturans*, and thus fulfill - quoting the concluding line of *Two Sources* -

"the essential function of the universe which is a machine for the making of gods".[31]

Along the same line of thought, Ian Hacking speaks about Canguilhem's "vision of life" as something "which breaks down the old barriers between natural and artificial, mind and body, manufactured and created",[32] thus opening the way to a connection between the Automaton and the Cyborg, between Descartes and Haraway.[33] In this continuity laden with ambiguity between life and technique, the figure of the Cyborg arrives on the scene in a provocative way, that is, eliminating the safety threshold that marked the distinction between technique and life. In a sense, the Cyborg extends and radicalizes organology, in that it requires an absolute continuity between organic and inorganic organs. But in so doing, the Cyborg modifies the topology of the relationship between technology and life: this relation no longer goes only in the direction of exteriorization. Rather, the Cyborg invites us to mobilize the paradigms of *hybridization* between technique and the body and of the *internalization* of technique into the body. While it is true that what is at stake, from Descartes to Canguilhem, in the definition of the relationship between life and technology, is the demarcation of a boundary between autonomous organic life and non-autonomous inorganic technique, the Cyborg is provocative because it demonstrates that this boundary in organological thought is in fact too fragile. As pointed out by Thierry Hoquet, then, "Cyborgs represent a paradox for the perspective of organology"[34]: through them, we no longer consider natural the "extension of the body in technique", but must consider "the violence of a graft between heterogeneous entities, and relations of parasitism, if not mutual predation".[35] By questioning the threshold that separates technique from life, the Cyborg questions the autonomy of the living with respect to technique. If initially Bergson and Canguilhem had thought of technique as a natural extension of the body, a natural product of life, in order to affirm the autonomy of the living with respect to the normativity of technique, organology ultimately turns against itself and undermines the very integrity of the organism that was considered the material basis of the moral autonomy of the human being. If instruments are merely externalized organs (i.e. if there is no essential difference between a natural organ and an artificial one), then this means that it is possible to hybridize the organism through the incorporation of artificial organs, which thus replace the natural ones. As such, however, the living organism loses its autonomy and finds itself, on the contrary, normalized by technology. What in fact is the autonomy of the living, according to Canguilhem, if not its ability to impose its own norms on its environment? The living being is normative insofar as its adaptation to the environment passes through the modification of this same environment, a modification that is also a technical activity. Technique is in this sense an extension of life in prolonging its normativity; it is a means of appropriating, modifying and metabolizing the environment. However, if artificial tools are assimilated with natural organs to the point of blending in with them, technique no longer merely prolongs the organism, but penetrates it in a movement of internalization that perverts the normativity of the living. It is precisely this insight that gives rise to the notion of the Cyborg, a neologism created by Manfred E. Clynes and Nathan S. Kline as a contraction of the words "cybernetic organism".[36] The Cyborg, unlike other living beings, does not technically modify the environment to make it habitable, but technically modifies itself in order to be adaptable to different environments. The concept was born precisely to imagine the possibility of leading an extraterrestrial life: "Altering man's bodily functions to meet the requirements of extraterrestrial environments would be more logical than providing an earthly environment for him in space... Artifact-organism systems which would extend man's unconscious, self-regulatory controls are one possibility".[37] In this sense, the Cyborg is rooted more on the value of resilience than on that of autonomy; indeed, it could be said that autonomy is no longer a value for the Cyborg, insofar as it promotes the constant alteration of the self, the loss of physical integrity and the personal identity attached to it, and the renunciation of a normative approach to the environment. Taking to the extreme the idea of a proximity between technique and life, entertained by organological thought, the Cyborg thus also reveals the limits of organology. Thinking of technique as a product of the living does not guarantee that we can maintain governance over it, for hybridization may be so extreme that we ourselves become its prostheses, as Samuel Butler suggests in *Darwin among the Machines*, which Hoquet quotes. When the living human becomes an organ of the machine, it completely loses its moral autonomy as well as its ontological priority over the machine.

Conclusion

It therefore seems that organology, understood not only in the strict sense, as a philosophy of technology that thinks of the latter as a creation of life, but in a broader sense, as a philosophy of the analogy between technique and life, which places natural and artificial organs on the same level, finds itself in an impasse, as does mechanism. In fact, both the automaton and the Cyborg fail in their attempt to think of the living as autonomous, in an absolute (i.e. moral) sense, and in a relative (i.e. epistemological) sense. On the one hand, the Cartesian automaton merely illustrates the functioning of the organism by comparing it to that of the machine, but fails to provide a causal explanation in the sense of modern science (as we have seen, only embryology resorts to causal explanations, but this has nothing to do with machinism and automaton thinking, and is insufficient in itself to account for the whole of the interrogations posed by knowledge of the living). Epistemological failure of the Cartesian automaton, then. As well as ontological and moral failure: by comparing the organism to the automaton, Descartes implicitly refers to the figure of the God-engineer or craftsman who builds the body as a highly perfected technical object, infinitely more perfect than any possible construction by man. If the analogy is based on this teleological fiction, then the automaton cannot function as a foundation for the autonomy of the living being (while it is true that the problem of the autonomy of the living is not a Cartesian problem, but arises later with Kant and Leibniz, their technological analogies do not escape the intrinsically teleological nature, as we have seen). Reciprocally, the figure of the Cyborg, which arises from a reversal of the Cartesian posture, results in a similar impasse, both epistemologically and morally: from the epistemological point of view, by postulating the total reversibility of organs and tools, body and machine, the Cyborg denies the specificity of the living (indeed, this is one of the postulates of cybernetics). Consequently, from a moral point of view, since the artefactual device is indiscernible from the natural organ and thus tends to replace it, the subject tends to lose its autonomy, since its regulations, even homeostatic ones, are imposed from the outside, as in a machine, and thus respond to normative systems that transcend those of the organism itself. In the end, therefore, it seems that the biological philosophy of technology, which first arose to emancipate the living from an epistemological point of view and developed to emancipate it on a moral level, fails in both these attempts.

It seems that the impasse to which the figures of the automaton and the Cyborg lead, invites us to abandon the analogy between life and technology if we want to preserve both the epistemological and moral autonomy of the living. From an epistemological point of view, it is therefore a question of elaborating a comprehensive theory of the living that theoretical biology is probably still seeking.[38] From a moral point of view, perhaps the autonomy at stake should be understood less as a metaphysical category than as a category of pragmatic anthropology: that would imply considering autonomy not as an essential property that we possess once and for all as living beings, but as a relational characteristic, situated in a specific context, that is to say a social, political, institutional, and/or gendered context. Autonomy then becomes ever more precarious and relative, a state that we can lose and that perhaps we never, in fact, entirely possess. This therefore implies the need to elaborate an anthropology of technology, rather than an organology, according to which technique is not an organ of the body, but rather its instrument[39] detached from the body and that we can use reflexively, not an entity with any organic or immediate relation to us.

Notes

1. M. Foucault, *The Order of Things. An Archeology of Human Sciences*, trans. Alan Sheridan Smith (London and New York: Routledge, 2005 [1966]), 139.
2. See for instance Ch. T. Wolfe, *La philosophie de la biologie avant la biologie. Une histoire du vitalisme* [*The Philosophy of Biology Before Biology. A History of Vitalism*] (Paris: Classiques Garnier, 2019) and C. Bognon-Küss and Ch. T. Wolfe, eds., *Philosophy of Biology Before Biology* (London and New York: Routledge, 2019).
3. D. Tarizzo, *Life. A Modern Invention*, trans. Mark William Epstein (Minneapolis and London: University of Minnesota Press, 2017 [2010]), 129.
4. Ibid., 13.
5. Ibid., 21.
6. Ibid., 54.
7. I. Kant, *Critique of Practical Reason*, trans. Mary Gregor (Cambridge: Cambridge University Press, 2015 [1788]), 20–21, my emphasis.
8. Ibid., 7.

9. See for instance G. Canguilhem, "Le tout et la partie dans la pensée biologique" [The Whole and the Part in Biological Thinking], in Id., *Études d'histoire et de philosophie des sciences* [Studies in History and in the Philosophy of Sciences] (Paris: Vrin, 1975).
10. G. Canguilhem, "Machine and Organism", in Id., *Knowledge of Life*, trans. S. Geroulanos and D. Ginsburg (New York: Fordham University Press, 2008 [1965]), 79.
11. Ibid., 80.
12. C. Wolfe, "La catégorie d' 'organisme' dans la philosophie de la biologie" [The Category of 'Organism' in the Philosophy of Biology], *Multitudes*, 16 (2004): 27–40.
13. M. Foucault, *The Ordering of Things*, op. cit., 137–138.
14. On this point, see S. Roux, "Les explications de la philosophie mécanique" [The Explanations of Mechanical Philosophy], *La Lettre de la Maison Française d'Oxford*, 4 (1996): 125–136.
15. See F. Duchesneau, *Les modèles du vivant de Descartes à Leibniz* [Models of the Living from Descartes to Leibniz] (Paris: Vrin, 1998).
16. I. Kant, *Critique of the Power of Judgment*, trans. Paul Guyer (New York: Cambridge University Press, 2000 [1789]).
17. Gottfried Wilhelm von Leibniz, "New System of Nature: And of the Communication of Substances, as well as of the Union of Soul and Body", *The Journal of Speculative Philosophy*, vol. 5, n° 3 (1871 [1695]): 214.
18. See C. Wolfe, "La catégorie d' 'organisme'", *Multitudes*, op. cit.
19. G. Canguilhem, *The Normal and the Pathological*, trans. C. R. Fawcett (New York: Zone Books, 1998 [1966]), *passim*.
20. G. Canguilhem, "Machine and Organism", *The Knowledge of Life*, op. cit., 76.
21. Ibid., 94. See on this E. Clarizio, *La vie technique. Une philosophie biologique de la technique* [Technical Life. A Biological Philosophy of Technique] (Paris: Hermann, 2021).
22. H. Bergson, *Creative Evolution*, trans. A. Mitchell (New York: The Modern Library, 1944 [1907]): p. 153.
23. Actually, E. Kapp was not a vitalist at all, but Canguilhem nevertheless identifies his theory as the first biological philosophy of technique.

24. E. Kapp, *Elements of Philosophy of Technology*, trans. L. K. Wolfe (Minneapolis and London: University of Minnesota Press, 2018 [1877]).
25. H. Bergson, *Creative Evolution*, op. cit., 156.
26. G. Canguilhem, "Descartes et la technique" [Descartes and Technique], in Id., *Œuvres Complètes, t. I, Écrits philosophiques et politiques 1926–1939* [Complete Works, t. I, Philosophical and Political Writings 1926–1939] (Paris: Vrin, 2011).
27. Descartes, *Dioptrique* [Dioptric], in Id., *Œuvres* [Works], (C. Adam & P. Tannery, eds.), vol. VI (Paris: Vrin, 1986): 81.
28. C. Wolfe, "La catégorie d' 'organisme'", *Multitudes*, op. cit.
29. H. Bergson, *The Two Sources of Morality and Religion*, trans. R. A. Audra (Notre Dame: Notre Dame University Press, 1977 [1932]).
30. On Bergson and the Cyborg, see P. Marrati, "The Natural Cyborg", *The Southern Journal of Philosophy*, vol. 48, Spindel Supplement (2010): 3-17; on Canguilhem and the Cyborg, see I. Hacking, "Canguilhem amid the cyborgs", *Economy and Society*, 27, 2-3, (2006): 202-216.
31. P. Marrati, "The Natural Cyborg", *The Southern Journal of Philosophy*, 17.
32. I. Hacking, "Canguilhem amid the cyborgs", *Economy and Society*, 202.
33. D. Haraway, *Simians, Cyborgs, and Women: The Reinvention of Nature* (London and New York: Routledge, 1991); D. Haraway, "The Cyborg Manifesto," in *The Cybercultures Readers*, ed. D. Bell and B. M. Kennedy (London and New York: Routledge, 2000): 291–324.
34. T. Hoquet, "Cyborgs, between organology and phenomenology: two perspectives on artifacts and life", in *French philosophy of technology*, ed. S. Loeve, X. Guchet & B. Bensaude-Vincent (Springer, 2018): 274.
35. T. Hoquet, *Cyborg philosophie* [Cyborg Philosophy] (Paris: Seuil, 2011): 59.
36. M. E. Clynes and N. S. Kline, "Cyborgs and space", *Astronautics* (1960): 26-27 & 74-75.
37. Ibid., 26.
38. See G. Longo et al., "In search of principles for a theory of organism", *Journal of Biosciences*, Indian Academy of Sciences (2015), 40 (5); A. Soto, G. Longo et al. [2016], "Toward a

theory of organisms: Three founding principles in search of a useful integration", *Progress in Biophysics and Molecular Biology*, 122(1), 77–82, https://doi.org/10.1016/j.pbiomolbio.2016.07.006.

39. See X. Guchet, *Care in technology*, (London and Paris: ISTE, 2021).

CHAPTER 17

The Mutant and the Cyborg, Between Science and Fiction, the Posthuman and the Transhuman

Thierry Hoquet

Introduction

This chapter is an exercise in blurring the boundaries between definite concepts and closed identities. Nevertheless, simultaneously and paradoxically, I want to emphasize the operativity of dualistic distinctions. To understand cyborgs and mutants, we must inquire into their theoretical function in contemporary culture: what are they the symptoms of? In which theoretical framework do they make sense? How do they allow us to define ourselves? Several dualisms must be reframed for this purpose, especially the dualism between fiction and science. In fiction, mutants are prevalent symbols of coolness and icons of difference; they often stand as the natural origins of supernatural powers. However, it should always be reminded that "mutants" is first and foremost a scientific concept,

T. Hoquet (✉)
Department of Philosophy, Paris Nanterre University, Paris, France
e-mail: t.hoquet@parisnanterre.fr

J. Muller (ed.), *Automata, Cyborgs, and Mutants*, Palgrave Studies in Medieval and Early Modern Medicine,
https://doi.org/10.1007/978-3-031-93201-4_17

linked to the experimental work of Hugo de Vries, then to Thomas Hunt Morgan's experiments at Columbia University, and maybe more importantly to the studies on mutagenesis by Hermann J. Muller, whose portrait in 1927 appeared under the title: "Manipulator of genes."[1] Similarly, with Cyborgs, we all know famous characters such as the RoboCop, "part man part machine"—a sentence later revived and modified by Kevin Warwick, who defined cyborgs as entities "part-animal, part-machine."[2] However, the concept of "cyborg" was invented in an entirely different context. In 1960, Manfred Clynes and Nathan Kline coined the term Cyborg to designate a *cybernetic organism*.[3] Nevertheless, the birth of cyborgs is linked to the word *cybernetic* coined by Norbert Wiener (1894–1964) from the Greek *kubernétès*, the pilot who knows how to sail a ship. Cybernetics is famously defined as the science of "control and communication in the animal and the machine."[4] Therefore, its focus is on transmitting information and controlling actions. Furthermore, as Ian Hacking has noticed, the cyborg's power was much amplified when it came out of the laboratory[5]—despite the vigorous protests of its inventors.[6] I will use the dichotomy Mutant/Cyborg to blur the contours of the human species and to pluralize our concepts of possible futures, disentangling various forms often agglomerated by assimilation.

Prefiguring the Cyborg/Mutant Binary

Undoubtedly, mutants and cyborgs are twentieth-century concepts. However, it may be tempting to say that their proliferation in contemporary culture is similar to the eighteenth-century proliferation of monstrous creatures, a diversity that legions of naturalists, physicians and philosophers strove to catalogue, building classifications, tables, lists of varieties and engravings. Cyborgs and mutants occupy nowadays the place previously occupied by wolf-children[7] and bear-children, the *puella campanica* and other wild maidens or lost boys discovered in the woods.[8] As it emerges from the works of thinkers such as La Mettrie or Condillac,[9] such *enfants sauvages* or "almost-human" creatures functioned as a challenge to traditional metaphysics and an instance in (what was called) a system of "experimental metaphysics."[10] Linnaeus called them *Homo ferus* as they seemed deprived of all the traditional attributes of humankind. Instead of bipedalism, language and civilization, they could be described as "*tetrapus, mutus, hirsutus*" (quadruped, mute, isolated).[11] They made philosophers puzzle about the extent to which the human mind or reason could develop when deprived of any social bonds with its kindreds, or when subjected to the sole empire of need.[12] These individuals were used as *repoussoirs*, figures of oppositional definition—but were also used as *purgative* for philosophers: they had a solid cathartic and cleansing effect on definitions of "Man" as a metaphysical entity.

Some ancient engravings from Rétif de le Bretonne's *Découverte australe* (1781) nicely encapsulate the Mutant/Cyborg binary.[13] I think that such engravings, although coming from the heart of the eighteenth-century culture, provide us with something very similar to the Mutant/Cyborg binary: the anticipation of both the Mutant as Organic-Unexpected-Variation and the Cyborg as Technically-Enhanced-Production; a forerunner and an excellent presentation of modern dichotomies such as Nature/Artefact, Nature/Culture, Wild/Civilized, Savage/Urban, Beasts/Humans.

In the nineteenth century, evolutionary theory asked the question of the best conditions for the emergence or the creation of new forms of life. While it seems inevitable that, in the future, every old lifeform will become extinct, with new species replacing the old ones, it is legitimate to inquire into what will happen to humankind. But this question was already raised among Enlightenment thinkers, through the fear that our own species could be doomed and rejected to the category of the unnumerable unfortunate essays and trials of Nature[14]—what Buffon described as "caprices de la Nature."[15]

The idea of enhancing the human species through human activity and selective breeding was often evoked during the eighteenth century. One can find it present, for instance, in Buffon's 1778 *Epochs of nature,* where the French naturalist states that:

> ... man has known but lately the extent of his powers, and he has not yet come to get sufficient knowledge of this fact. It depends entirely on the use of his intelligence. So, the more he will observe, the more he will cultivate Nature, the more he will acquire means to subdue it and opportunities to extract new resources from its breast without diminishing the treasures of its inexhaustible fecundity. Moreover, what could he not on himself, I mean, on his own species, if his will was always guided by intelligence? Who knows the extent to which man could perfect his own nature, whether moral or physical?[16]

Buffon stresses that humans have acquired an impressive ability to tame the great garden of Nature and "to change, modify and perfect the species": "it is even the greatest right they have over Nature." They have "transformed sterile grass into wheat;" likewise, dogs are "a kind of creation."[17] By extension, humans could even transform themselves into something different through their own activity. However, Buffon does not tell us how we should call the wonderful products of such amazing powers: should those exquisite creations be called *anti*-natural or *counter*-natural? "*Antiphysiques*" or rather "*transphysiques*"? The perspective of breeding allows us to revisit the well-worn Cartesian and iatro-mechanician theme of the "machine body" to envision the production of domesticated breeds specialized in performing certain functions. An example could be the "turnspit dog" that Charles Darwin mentions in his *Origin of species*—a book in which Darwin deeply pondered the powers of live-stock breeders that he compared to a "magician's wand."[18] Herbert George Wells gave the theme its literary expression in his *First men in the moon* (1901) with the idea of intense physiological specialization in the world of the Selenites: commissioners, as they have to go for long runs, are endowed with long swift legs, intellectuals have big round heads to think more and better, etc.[19] The general utilitarian mechanism that produces functional bodies is transposed to the human species.

However, in the eighteenth century, Charles-Augustin Vandermonde explicitly applied similar considerations to humankind in his *Essay on how to perfect the human species* (1756). Vandermonde's reflections bear on the production of perfected organisms. In Vandermonde's terms:

> Since we have succeeded in perfecting the breeds of horses, dogs, cats, hens, pigeons, canaries, why should we not make any trials on the human species? Why indulge in the pleasures of nature, and forget its wonders? Let us extend our research further. If marriage had brought together the two most beautiful voices of the century, or if a skilful French singer had been matched with some Italian woman with a distinguished voice, their children would probably have admirable organs. When the male goldfinch is bred with the serine, the resulting mule has a much more pleasant and flattering song than any other. This way of perfecting the organ of the voice would be less reprehensible than the one used by some of our neighbours. The poor unfortunate sacrificed to these humiliating ordeals acquire their voices at the cost of their integrity. Can the public applause they receive compensate for their loss and the regrets that devour them![20]

In the last sentence, Vandermonde refers obviously to castrates: and again, he is opposing two ways of triggering and monitoring human evolution and of perfecting the production of human beings: by technical means or *cyborgically* (the castrates) or *mutation-wise* by natural birth (the hybrids).

French epistemologist Georges Canguilhem showed in his "Machines and Organisms" that we have to distinguish the problem of the *functioning* machine from that of machine *construction* or morphogenesis.[21] Descartes' infamous metaphysical theme of the *animal-machine* is a doctrine of *functioning* machines: with their nuts and bolts, sieves, hammers, wheels, pulleys and springs. The question of the production of organisms is that of machine construction. Nevertheless, Canguilhem neglects a third aspect of the question: that the organic machine of the body is also a device of production whose productivity might be increased at will. Just like bees are described by Reaumur as a factory or manufacture of propolis,[22] Neo-Fordism only deepened this tendency. Not only is the organism composed of tools (organs): it is a production system which can be bettered. Hence, in order to improve any species, including our own, humans should not only pay attention to the art of producing beautiful children but consider the art of increasing the overall usefulness of bodies and the general output that society extracts from them.

Transhuman/Posthuman: Two Projects to Overcome Our Present Form and Condition

Through variation, a new kind of Otherness is ultimately meant to triumph over our present Sameness. This seems to be the main lesson of evolutionary theory—beautifully encapsulated in the idea that *species are individuals*, a conception Buffon envisioned in his *Natural History*.[23] Nevertheless, as the only animal aware of the parameters of evolution, our species is not necessarily doomed to decline and to accept the evolutionary process passively. As we have seen, Buffon and others called for our energetic intervention in this process. This concept was expressed most forcefully by Julian Huxley in a 1957 book suggestively entitled *New Bottles for new wine*. Huxley then pinned his hopes on what he called "transhumanism":

> The human species can, if it wishes, transcend itself—not just sporadically, an individual here in one way, an individual there in another way, but in its entirety, as humanity. We need a name for this new belief. Perhaps *transhumanism* will serve: man, remaining man but transcending himself by realizing new possibilities of and for his human nature.[24]

Another possible human future bears the different name of "posthumanism." Among many different possible definitions,[25] I will recur to Katherine Hayles' influential book "How we became posthuman?" where she writes: "As you gaze at the flickering signifiers scrolling down the computer screens, no matter what identifications you assign to the embodied entities that you cannot see, you have already become posthuman."[26]

In Hayles' view, posthumanism is about the privilege of informational patterns over material instantiation; the body as a manipulable prosthesis; seamless articulation between the organism and the machine, e.g. between humans and computers; no difference between corporeal existence and computer simulation. Working along those lines, we understand that Mutants and Cyborgs are two possible descriptions of humanity and two diverging visions of its future. They have received two different names in contemporary culture: the *transhuman* emphasizes the emergence of *mutants* as the result of natural variation or accidental alteration occurring or being triggered in the genetic programme of the individual; the *posthuman* has taken the shape of a hybrid compound, which is often called the *cyborg*, obtained by the grafting or merging of the flesh with technical prostheses, a combination of the carbon-based fabric of life with the silicon-based material of the informational computing system. Posthumanism is often considered as leading eventually to the obsolescence of the flesh. A gnostic dimension is obvious in many posthuman projects. Gnosticism believed that humans received their share of God's divinity, and that Jesus Christ was only a human-shaped shell housing a divine spark that can be found in all superior humans.[27] In what follows, I will give a closer analysis of Mutants, and Cyborgs, then compare them and finally draw some general conclusions.

Mutants Between Singularities and Monstrosities

In biology, evolution has always taken and still takes place through favourable mutations, which are then naturally selected—"a two-step process," as Ernst Mayr famously put it.[28] Mutants represent the future of the human species as produced by chance or through the matrix of aleatory individual variations.[29] Mutants have often been compared to *monsters*. For instance, the Dutch biologist Armand Marie Leroi once tried to translate the old categories of monstrosity into the language of mutations: conjoined twins, cases of hypertrichosis (also known as the *Werewolf syndrome*), dwarves, etc.[30] Leroi interpreted the whole catalogue of horrific entities previously depicted as *freaks* as *mutants*, their modern equivalents in the more neutral language of biological or medical conditions affecting the embryo's development.

Being a mixture of Sameness and Otherness, Mutants usually produce fecund offspring with any ordinary human, so their singular features can thus be transmitted to future generations. In short, employing natural selection, mutations can spread into the *normal* population and change its statistical composition. A mutant is an exception to the rule that claims to become a new rule. It forms the beginning of a new story, a singularity attempting to be general. An essential difference between mutants and monsters is that monsters have traditionally been signifiers of God's wrath; they have embodied the punishment for their parents' crimes. However, mutants are born in an entirely natural world and are perfectly innocent. They do not imply anything more than natural development and evolutionary processes. Our fear of mutants shares common features with the fear of monsters, such as bewilderment in front of the powers of life. However, mutants raise several specific problems in contemporary culture.

(a) Mutants are deeply rooted in the Darwinian framework of evolution, which takes the individual organism as its core level.[31] They embody the idea that evolution occurs through the success of favoured variant individuals. Each individual is, through generation, a token of its type or species (it is the offspring of other tokens of the type), but, considered as a mutant, the same individual differs radically from its type. Each mutant offers a synthetical point of view between heredity and variation, sameness and difference, species and individual. It is germinally connected to its ancestry and somatically or phenomenologically different from it. The mutant is both a continuity and a disruption: it inherits, and it initiates.

(b) Because they are radical idiosyncrasies, mutants are closely related to individualistic philosophical stances. Mutants are unique individuals, singular variations. When we push this idea to its extreme, we can all be considered, each and every one of us, as mutants, i.e. singular and unique individuals, potentially marginals or outcasts, misfits excluded for our difference and stigmatized for our radical uniqueness. Mutations—i.e. individual differences—are sometimes depicted as an acute form of adolescent crisis. In graphic novels or cartoons, when mutant teenage characters undergo the emergence of their new power, it is like the revelation of their true nature—a radical transformation or metamorphosis which is often achieved through much suffering. Hence, mutants often embrace a motto: "Become who you *really* are." In that sense, mutants may be said to be "anti-axolotls," as axolotls never go through metamorphosis; they remain child-like creatures forever.[32] The dimension of crisis leading to transformation or metamorphosis is also underlined in the X-Men series, for instance, through a character like Magneto, who discovers his powers as he is taken away from his parents in a concentration camp in Nazi Germany. Fictional mutants like the X-Men embody a possible metaphor for the persecution of the Jews (hence they are being renamed "J-Men"), discovering who they are at the expense of living traumatic events.[33] The mutation is vital labour. The mutation is both an event and a process.

(c) Natural or artificial? Although the changes called mutations may often be produced by radiation or other accidental factors (in the wake of H. J. Muller's *mutagenesis*), mutations are still considered to be spontaneous or aleatory, as their outcome is not always predictable. The abandonment of teleologically oriented variation was a significant change of focus in Darwin's theory of variation.[34] In Darwin's time, the laws of mutations were still unknown, and variations were therefore considered as produced "by chance"—a term Darwin himself rebuked.[35] Mutants were deemed to be *biologically* produced and then *naturally selected*. Technically, no human being *creates* mutants: we breed them once they have been brought to life. The birth of the mutant is the outcome of a universal lottery, of which breeders or selectors might take advantage.

Cyborgs Beyond the Nature/Artificial Binary

Also situated between the natural and the artificial, cyborgs make sense when considered with the myth of Epimetheus in the background.[36] Epimetheus, the God-sent administrator, was responsible for distributing natural qualities among living creatures. He gave sharp teeth to the predators, horns to potential prey, endowed some with shields and shells, and others with great speed—finally giving to each species the features appropriated to its survival. However, when he thought he had reached the end of his mission, Epimetheus realized he had forgotten humankind. Our species is therefore cursed with what we might call an *Epimethean complex*: it has no fur to be protected from the cold, no armour or husk as defence, no claws to attack ... Humankind is always forced to acquire protective or offensive devices through its own means and primarily through its technical ingenuity. Humans, in short, are not naturally equipped to survive. In many cases, it might be argued that they are not even endowed with the mental equipment (such as instincts) required to obtain accurate knowledge of their environment. Humans have no guarantee that their minds or logic are adapted to an adequate knowledge of the outside world.

Our Epimethean complex entails a state of complete deprivation, both practical and theoretical, compelling us humans to the expedients of Prometheus: steal the fire from the Gods, the fire being a means to supplying food, a transition from the raw to the cooked and a metaphor of knowledge and civilization. The Promethean solution to the Epimethean complex suggests that our equipment might be complemented to get a better life and a better knowledge of the world. Hence, the Cyborg is always on the side of the artificial, even though it has natural parts.

Henri Bergson's philosophy helps us conceive tools as extensions of our vital activity. Bergson considers that our mental faculties, particularly our intelligence, are an outcome of evolutionary processes. In the second chapter of his *Creative Evolution* (1907), he carefully depicts evolution as an arms race between prey and predators.[37] He defines humans as *Homo faber*, showing that the human production of instruments and tools is a product of their intelligence. Tools are our species' tactics to increase our survival chances; therefore, instruments are essentially only external organs for us. A thesis we can call "organology"—a term used by Georges Canguilhem («organologie générale»).[38] Organology is a radical position on the origin of tools and artefacts, suggesting a strict equivalence between internal or native organs (like the lungs, the kidneys or the liver) and external or acquired organs (the tools and devices we build and use). The hypothesis of general organology is thus crucial to understanding the cyborg's possibility. Cyborgs are simply hybrid creatures endowed with both kinds of organs, or as Emanuele Clarizio puts it, where the internalization of the technology complements the externalization of the organic (the extended phenotype).[39]

If the production of tools is a natural, biological, or even "organological" process, what makes cyborgs *cyborgs* (instead of just humans)? We claim that the awkwardness or oddity of the cyborg comes from three main difficulties.

(a) The first one is *the RoboCop problem*: the birth of the cyborg lies in the bloody demise of our organic bodies. From a RoboCopian perspective, cyborgs seem to imply mutilation, amputation or ablation of our native organs, whether or not individuals engage in cyborgization voluntarily or accidentally. Verhoeven's *RoboCop* (1987) shows that cyborgs are not only the coupling of human flesh with artificial tools or machines. This RoboCop evolves in a world where medics and capitalists can graft mechanical parts upon a human substrate. By doing so on the mutilated human body of police officer Alex Murphy, they cannibalize what is left of him. In this context, steel limbs are not only prostheses: they function not as an *enhancement* but as a *replacement* of the body. If our native organic bodies are a prison for the soul, then tools appear to be a much more terrible cause of imprisonment.[40]

(b) The second problem is interconnectivity or *the impossible bioport*. In Descartes' reflection of the human composition of body and soul, the pineal gland ensures interconnectivity between the two parts—body and soul. In cyborgs, the problem—real and imagined in popular culture—is connecting or wiring organic elements (such as the brain) with electronic devices such as silicon chips. After all, our brains are water-based mediums, where ions ensure communication, whereas a chip works with the circulation of electrons. Between those two systems, no wiring is possible. The problem of wiring has been greatly emphasized by cyberpunk culture, looking for new interfaces between the human brain and the cybernetic network. The term "bioport" comes from David Cronenberg's movie *eXistenZ* (1999), which depicts a universe where video games are directly *plugged* into the body of the players.[41] In order to play, you have to get a *bioport*, a plug-in where an *umbi-cord* is inserted. Your organism is thus connected to a semi-organic videogame console called *biopod*. The connection of the organism to the *biopod* through the *bioport* is not a connection between an organism and a machine since *biopods* are closer to animals than to machines. They are not repaired but instead surgically cured. *Biopods* are animal-like creatures living in symbiosis with human organisms. Given that no actual wiring is possible between the organism and a machine, *biopods* are more akin to parasites. Our ambivalent relationship with our *bioports* shows our fear of losing our individuality and our anxiety of being mentally hijacked by our connected devices; they also suggest the instability of the organism's connection with its mechanical extensions.

(c) The problem of interconnectivity also leads to what can be called *the Icarus complex*. Icarus' artificial wings melted as he flew too close to the sun. Hence, the myth of "Icarus" signifies that any assembling between technical devices and organisms is unstable and eventually doomed to disunity. There can be no definite connection, firmly bounding organisms to their artefacts.

Those three different figures raise the general problem of obsolescence: obsolescence of the flesh as in *RoboCop*, obsolescence of the connection as in Icarus, and obsolescence of the extended material substrate. Let us think, for instance, of the general project of posthumanism as Hans Moravec's dream of *uploading the human brain into silicon*. The Icarus perspective suggests that our soul might be freed from the organic body and that our minds must undergo constant updates and controls, much like any Microsoft product. Enhanced humans will need periodic maintenance of their mechanical bodies. In this context, the cyborg thus has both corporeal and mechanical limits and tends towards inevitable obsolescence. This situation is, for instance, a great cause of anxiety to the characters of the Japanese anime *Ghost in the Shell*. From a technical perspective, it is indeed difficult, albeit not impossible, to imagine a cyborgian future for the human species. Those three difficulties (RoboCop, the bioport and Icarus) challenge the organological, Bergsonian approach of techniques as natural secretions of living organisms. Instead of considering machines as external organs and organs as internal machines, we have to raise the question of the feasibility of our *becoming-machine*, the durability or instability of such a hybrid compound and the desirability of such a future.

Various Organologies: *The Fly* Compared to *Terminator*

As mentioned earlier, "*organology*" describes the general theory of devices through which a goal is fulfilled. It is essentially the study of organs and tools, the various fashions by which an end may be reached. Cyborgs and mutants embody two different conceptions of *organology*, two different ways for individuals to access functions they do not primarily have. In a given situation, cyborgs are endowed with adequate technical instrumentation, whereas mutants embody an organic body modification. In other words, cyborgs represent (and renew) an organology of instruments, whereas mutants provide an organology of organs.

The difference between the two types of *organologies* can be readily perceived by comparing two movies: *Terminator* (James Cameron, 1984) and *The Fly* (David Cronenberg, 1986), whose plots both centre on the question of impregnating a woman.

In the cyborg movie, the machine-like organism wants to prevent a human couple from conceiving a child who will grow up into a hero. In the mutant movie, copulation and insemination occur between the mutant form and the normal female form, producing "illegitimate offspring" (On this phrase, see below, Sect. 17.6). This son to be born embodies not the future of the parent form but an intermediate breed full of promise.

(a) *Terminator* and *The Fly* confront the nature of the body in two scenes, where the human epidermis is cut. Schwarzenegger's character in *Terminator* cuts his dermal envelope to repair his cables and finally pulls out a seemingly organic eye, revealing the mechanical device that ultimately constitutes its visual device. In *The Fly*, Jeff Goldblum's character faces his mirror as he tears his nails one after the other and watches his former human body collapse and fall apart.

(b) The final scene of both movies also unveils the true nature of the two characters. The Terminator is finally revealed as a machine. Its transformation throughout the movie is a process of denudation as the Terminator is progressively deprived of its external human envelope and exposed to its actual (*mechanical*) nature. On the opposite, the mutant fly embodies a process of metamorphosis: the progressive emergence or manifestation of a new being, a form of life still unseen on the surface of the earth. The last scene of *The Fly* does not reveal the true essence of the individual formerly known as "Seth Brundle," but it shows the existence of a new, hybrid living being named "Brundlefly."

(c) The difference between cyborgs and mutants can also be felt in the characters' relationship with their vehicles. In *Terminator*, the characters always find new appropriate vehicles: they are experts in bike- and carjacking, as they are always riding a new motorcycle or driving a new truck at full speed. Their bodies are perfectly and easily enhanced by new mechanical two- or four-wheel devices. On the contrary, Seth Brundle is prone to be constantly *carsick* and *airsick* and even confesses that he used to be "*tricycle-sick.*" Brundle is not appropriately "enhanced" by technical devices: his only way to overcome his faults is through organic modifications. This can be interpreted through the model of the two organologies. In the technical organology of the cyborg, vehicles are like a natural extension of the body. In the natural organology of the mutant, Brundle is ascribed to his organic condition, and he cannot use other organs than those with which he was naturally endowed. Furthermore, it is only after he completes his natural metamorphosis from Brundle to Brundlefly that he can solve his transportation problems: the natural purpose or effective result of his mutation is to endow him with new powers or features.

In other terms, Bergson's organology must be complemented. Bergson considers machines just as external organs, entirely subservient to organisms. For Canguilhem, who took up the idea from Bergson, machines only existed to facilitate an organic species' survival. However, fictions go further: they consider machines not as organs but as organisms, struggling to survive and exploit their human creators. In the latter case, we have to think of evolution as competition. However, the alternative conception of evolution as symbiosis is still open.

Final Considerations

If we now recapitulate some general aspects of the mutant/cyborg dichotomy, we can encapsulate some of its major traits in the following table.

Mutants	*Cyborgs*
Transhuman	Posthuman
Purely organic flesh	Mechanical body parts
Stochastic variation	Designed organs
Production of new forms by chance events only	Production of new capabilities by intentional transformation
Potential/actual	Private/public
Eugenism	Monopolistic capitalism
Well-born individuals	Enhanced capabilities

Mutants are associated with transhumanism, cyborgs with posthumanism. Regarding the organological tradition, mutants, and cyborgs embody two ways to achieve human enhancement: spontaneously generating new functions (mutants) or conscious and intentional design (cyborgs). However, mutants and cyborgs jointly offer a re-reading of the human body, past, present and future. They raise the question of the similarity of living and inert matter: they literally embody the reconciliation of mechanical and organic beings and offer a new ontology. Cyborgs show that our organism is susceptible to being complemented by technical devices, such as a mechanical heart. Mutants show that our natural organs are organic machines performing various functions. Moreover, both cyborgs and mutants imply that both "natural" and "artificial" organs are heavily teleological, designed by various instances (God, Humans, Nature) but all the same designed. The natural world is a technological world just as the technological world is a natural one. As Leibniz suggested in his *Monadologie* (1715, § 64): "*les machines de la nature, c'est-à-dire les corps vivants, sont encore machines dans leurs moindres parties jusqu'à l'infini.*" If life can be prolonged, extended, augmented, complemented by the machine, it is because life itself is mechanical as much as vitalist in its essence.

But we also need to consider the uneasiness, the feeling of the *Uncanny*, or *Unheimlichkeit*, depicted by those characters. If mutants refer to our fear of monsters (*teratogenesis* is the archetypical form of *mutagenesis*), similarly, cyborgs refer to our fear of facing automata.

Donna Haraway once depicted the moral situation of cyborgs as "*illegitimate offspring of militarism and patriarchal capitalism.*"[42] Such a phrase seems to equate cyborgs with Late Capitalism productions, owned by great corporations or companies. Nevertheless, the adjective *illegitimate* also draws our attention to the fact that cyborgs are literally described as bastards: as children that were not recognized by their parents and who therefore are promised a lower social status. Whoever their progenitors were, they refused to acknowledge cyborgs as members of their kin or kind.

Some critics have considered that the phrase *illegitimate offspring* was inappropriate.[43] The idea of *hybridity* implies *disharmony*: hybrids seem doomed because of their alleged *unnaturalness*. Besides, the term *hybrid* suggests a *pure* or non-hybridized state, which strikes back hybrids with connotations of impurity, corruption and degeneration. In reply, Haraway herself later extended her "*Cyborg Manifesto*" into a "*Companion Species Manifesto.*"[44] Cyborgs draw our attention to everything produced but eventually rejected as undesired and alien to the system. Hence, if cyborgs are bastards, they are also prone to revolt and a ferment of contestation, like mutants and any *natural children*.[45] If, on the one hand, cyborgs are the illegitimate fusion of animals and machines, on the other hand, dogs and companion species try to make sense of the intricacies of life with technoscience. While mutants scare us with their potential monstrosity, cyborgs are instead linked to our amazement in front of the powers of the mechanical and the feeling of uncanniness (*Unheimlichkeit*).

Finally, our fears or uneasiness in front of what can be called "almost-human" figures express a sort of nostalgia for the human.[46] The human's shape *trembles* when confronted with both fictions of mutants and cyborgs, transhuman and posthuman futures. From mutants to cyborgs, the human individual is being turned into a mere living tool, a slave deprived of freedom, a monstrous creature imprisoned in its body. Between mutants and cyborgs, beyond and below them, more extreme forms can be envisioned: amorph bodies, such as blobs; putrescent bodies, such as zombies; symbiotic bodies, such as mechas... Defining the human is an impossible task (*le propre de l'Homme*), and numerous prefixes (like *trans* and *post*, but also *in*, *un*, or *almost*) modify and shatter our very conception of what makes a human, *human*.

Notes

1. *The Science News-Letter* 12, no. 340 (Oct. 15, 1927): 249.
2. Kevin Warwick, *I Cyborg* (Urbana: University of Illinois Press, 2004), 61–62.
3. Initially published in *Astronautics* (1960), Republished in Chris Hables Gray, H. J. Figueroa-Sarriera and Steven Mentor, eds., *The Cyborg Handbook* (New York: Routledge, 1995), 30.
4. It is the subtitle of Norbert Wiener's book: *Cybernetics or control and communication in the animal and the machine* (1948), 2nd ed., (Cambridge (MA): MIT Press, 1962).
5. Ian Hacking, "Canguilhem amid the cyborgs," *Economy and society* 27 (2/3) (1998): 202–216. See also T. Hoquet, "Créations ensauvagées. Les productions humaines dans les dynamiques de l'évolution," in *Bionano-éthique. Perspectives critiques sur les bionanotechnologies*, ed. B. Bensaude-Vincent, R. Larrère, V. Nurock (Paris: Vuibert, 2008), 157–170.
6. "An interview with Manfred E. Clynes," in *The Cyborg handbook*, op. cit., 43–53.
7. "Wolf children" turned out to become the international title of a Japanese anime おおかみこどもの雨と雪 (Hosoda Mamoru) [The Wolf-Children Ame and Yuki].
8. *Histoire d'une jeune fille sauvage trouvée dans les bois à l'âge de dix ans*, published by Madame H…t [Hecquet] (1755), text attributed to La Condamine, presented by Franck Tinland, Bordeaux, Ducros, 1970.
9. See for instance Condillac, *Traité des sensations*, Book IV, chapter 7, « D'un homme trouvé dans les forêts de Lituanie»; La Mettrie, *Traité de l'âme,* chapter 15, « Histoire d'un enfant trouvé parmi les ours», in *Œuvres philosophiques* (1751), t. I, (Paris: Fayard, 1987), 236–237.
10. The expression occurs in Diderot's *Bijoux indiscrets*, chapter XXVI (« Métaphysique de Mirzoza»), vol III (Fictions I), ed. H. Dieckmann, J. Proust, J. Varloot (Paris: Hermann),120. See also Francine Markovits, "Althusser et Montesquieu: l'histoire comme philosophie expérimentale," in *Althusser philosophe*, ed. Pierre Raymond (Paris: PUF (Actuel Marx Confrontation), 199), 31–74.

11. Carolus Linnaeus, *Systema Naturæ per regna tria naturæ, secundum classes, ordines, genera, species, cum characteribus, differentiis, synonymis, locis*, vol. I, editio decima reformata, Holmiae, Laurentii Salvii, 1758, 20. See also Gunnar Broberg, *Homo sapiens* L., *in Studier i Carl von Linnés naturuppfattning och människolära*, 153–253.
12. Franck Tinland, *L'homme sauvage*: *Homo ferus et Homo sylvestris: de l'animal à l'homme* (Paris: Payot, 1968; reed. Paris: L'Harmattan, 2003).
13. Nicolas Rétif de le Bretonne, *Découverte australe par un homme volant ou le Dédale français*, 1781. Darwin's theory, when first published, was sometimes compared to Rétif's work. See for instance, Antoine Laurent Apollinaire Fée, *Le Darwinisme ou examen de la théorie relative à l'origine des espèces* (Paris: Masson, 1864), who cites Rétif's *Découverte australe*, as having defended a thesis comparable to Darwin's: imagining an original unique animal and a unique vegetal as the principles of every living thing. On Rétif, see Fabrice Brandly, "Animaux," in *Dictionnaire critique de l'utopie au temps des Lumières*, ed B. Baczko, M. Porret, F. Rosset (Genève: Georg, 2016), 99–127.
14. This is clearly evinced in Jean-Baptiste Robinet, *Considérations philosophiques de la gradation des formes de l'être, ou les essais de la nature qui apprend à faire l'homme* (Paris: C. Saillant, 1768). See also Benoit de Maillet, *Telliamed, ou entretiens d'un philosophe indien avec un missionnaire français sur la diminution de la mer, la formation de la terre, l'origine de l'homme,* etc. (1748), ed. Francine Markovits (Paris: Fayard, 1984). In a famous passage of his *Letter on the blind* (1749, DPV IV, 50–51), Diderot states: "Cela supposé, si le premier homme eût eu le larynx fermé, eût manqué d'aliments convenables, eût péché par les parties de la génération, n'eût point rencontré sa compagne, ou se fût répandu dans une autre espèce, monsieur Holmes, que devenait le genre humain ? il eût été enveloppé dans la dépuration générale de l'univers; et cet être orgueilleux qui s'appelle homme, dissous et dispersé entre les molécules de la matière, serait resté, peut-être pour toujours, au nombre des possibles.".
15. Buffon, "Les Toucans," in *Histoire naturelle, générale et particulière, avec la description du cabinet du roi*, t. XXII (Paris: Imprimerie Royale, 1780), 109: "Les vrais caractères des erreurs

de la Nature, sont la disproportion jointe à l'inutilité; toutes les parties qui dans les animaux sont excessives, surabondantes, placées à contresens, et qui sont en même temps plus nuisibles qu'utiles, ne doivent pas être mises dans le grand plan des vues directes de la Nature, mais dans la petite carte de ses caprices, ou si l'on veut de ses méprises…".

16. Buffon, *Époques de la nature* (1778), ed. J. Roger (Paris: Musée National d'Histoire Naturelle, 1988), 220 (all translations are mine unless otherwise noted).
17. Buffon, *Histoire naturelle*, op. cit., especially t. V, 3–4 ("La Brebis") and 195–196 ("Le Chien avec ses variétés").
18. Charles Darwin, *On the Origin of Species* (London: Murray, 1859), 30–31.
19. H.G. Wells, *The First Men in the Moon: A Critical Text of the 1901 London First Edition, with an Introduction and Appendices*, ed. Leon E. Stover (Jefferson (N.C.): McFarland, 1998). See especially the text by Wells, "An Age of Specialization" (1904), 274: "There is something of the phonograph in all of us, but in the sort of eminent person who makes public speeches about education and reading, and who gives away prizes and opens educational institutions, there seems to be little else but gramophone. These people always say the same things, and say them in the same note. And why should they do that if they are really individuals?".
20. Charles-Augustin Vandermonde, *Essai sur la manière de perfectionner l'espèce humaine* (Paris: Vincent, 1756), 94: "Puisque l'on est parvenu à perfectionner la race des chevaux, des chiens, des chats, des poules, des pigeons, des serins, pourquoi ne ferait-on aucune tentative sur l'espèce humaine? Pourquoi s'amuser aux plaisirs de la nature, et oublier ses merveilles? Étendons plus loin nos recherches." and 95–96: "Si le mariage avait réuni les deux plus belles voix du siècle, ou qu'un habile chanteur français avait été assorti avec quelque italienne qui eut une voix distinguée, leurs enfants auraient vraisemblablement des organes admirables. Quand on fait accoupler le chardonneret mâle avec la serine, le mulet qui en provient a un chant beaucoup plus agréable et plus flatteur que celui de l'un et de l'autre. Cette façon de perfectionner l'organe de la voix serait moins condamnable que celle dont se servent quelques-uns de nos voisins. Les pauvres malheureux que l'on sacrifie à ces épreuves humiliantes achètent leurs voix au prix

de leur être. Les applaudissements qu'ils reçoivent peuvent-ils les dédommager des regrets qui les dévorent!".

21. Georges Canguilhem, "Machine et organisme" (1946–1947), in *La Connaissance de la vie* (Paris: Hachette, 1952).
22. Antoine Ferchault de Réaumur, *Mémoires pour servir à l'Histoire des Insectes* (Paris: Imprimerie Royale, 6 volumes, 1734–1742).
23. As shown by Jean Gayon, "The individuality of the species: A Darwinian theory?—from Buffon to Ghiselin, and back to Darwin," *Biology and Philosophy*, vol. 11 (1996): 215–244.
24. Julian Huxley, *New Bottles for new wine. Essays* (London: Chatto et Windus, 1957), "Transhumanism,"13–17.
25. Cary Wolfe, *What is Posthumanism?* (Minneapolis: University of Minnesota Press, 2010).
26. N. Katherine Hayles, *How we became posthuman. Virtual bodies in cybernetics, literature, and informatics* (Chicago-London: The University of Chicago Press, 1999), xiv.
27. Hans Jonas, *The Gnostic Religion: The Message of the Alien God and the Beginnings of Christianity* (Boston: Beacon Press, 1958).
28. Ernst Mayr, "Accident or design, the paradox of evolution," in *The evolution of living organisms*, ed. Geoffrey Winthrop Leeper (Melbourne: Melbourne U. Press, 1962), 5.
29. See Jean-Michel Salanskis, "Fiction des sujets," *Critique*, n° 709–710 (2006/6–7): 625–636.
30. Armand-Marie Leroi, *Mutants, On the forms, varieties and errors of the human bodies* (New York: Penguin Books, 2005).
31. Jean Gayon, *Darwinism's struggle for survival. Heredity and the hypothesis of natural selection*, trans. Matthew Cobb (Cambridge: Cambridge University Press, 1998).
32. I thank Mateusz Janik for indicating the existence of a Polish novel by Jacek Dukaj called *The Old Axolotl* (Polish: Starość aksolotla, 2015).
33. See Lawrence Baron, "X-Men as J Men: the Jewish subtext of a comic book movie," *Shofar: An Interdisciplinary Journal of Jewish Studies*, 22.1 (2003): 44–52.
34. See Peter J. Bowler, "Darwin's concepts of variation," *Journal of the history of medicine and allied sciences*, 29–2 (1974): 196–212; Rasmus G. Winther, "Darwin on variation and heredity," *Journal of the history of biology*, 33 (2000): 425–455.

35. On the meaning of "chance" (hazard), see Francesca Merlin, *Mutations et aléas: le hasard dans la théorie de l'évolution* (Paris, Hermann, 2013).
36. Plato, *Protagoras*, 320-321c.
37. Cf. Henri Bergson, *L'Évolution créatrice* (Paris: Alcan, 1907), 131 and 140.
38. See my "Cyborgs, Between Organology and Phenomenology: Two perspectives on Artifacts and Life," in *French Philosophy of Technology*, ed. Sacha Loeve, Bernadette Bensaude-Vincent and Xavier Guchet (Dordrecht: Springer Verlag, 2018), 257–277.
39. See Emanuele Clarizio, *La vie technique* (Paris: Hermann, 2021).
40. See for instance the example of people in "half-life" in Philip K. Dick's *Ubik* (1969): these individuals are almost dead and placed in a state of suspended animation.
41. Peggy Cardon, "La mutation fantasmée: le cas Cronenberg," *Critique*, n° 709–710 (2006/6–7): 580–590.
42. Donna Haraway, "A Cyborg Manifesto: Science, Technology, and Socialist-Feminism in the Late Twentieth Century," reprinted in *Simians, Cyborgs and Women: The Reinvention of Nature* (New York: Routledge, 1991), 149–181.
43. See, for instance, Jennifer Gonzàlez rejecting Haraway's definition of the cyborg as *illegitimate breeding* and suggesting that the cyborg is "*what we have always been—regardless of our 'breeding'*", and as such, "*as legitimate as any* other" (J. Gonzalez in Gill Kirkup et al., eds., *The Gendered Cyborg: A Reader* (London: Routledge, 2000), 68).
44. Donna Haraway, *The Companion Species Manifesto. Dogs, people and significant otherness* (New York: Prickly Paradigm Press, 2003).
45. As in Denis Diderot's 1757 play, *Le Fils naturel* (*The Natural Son*), about a young man of unknown parentage.
46. Thierry Hoquet, *Les Presque-Humains. Mutants, Cyborgs, Robots, Zombies, … et nous* (Paris: Le Seuil, Collection « L'Ordre philosophique», 2021).

Name Index

J. Muller (ed.), *Automata, Cyborgs, and Mutants*, Palgrave Studies in Medieval and Early Modern Medicine,
https://doi.org/10.1007/978-3-031-93201-4

Subject Index

J. Muller (ed.), *Automata, Cyborgs, and Mutants*, Palgrave Studies in Medieval and Early Modern Medicine,
https://doi.org/10.1007/978-3-031-93201-4

D

E

F

G

H

N

O

U

V

W

Kolibri 360 GmbH
Ferdinand-Jühlke-Straße 7
99095 Erfurt, Deutschland
produktsicherheit@kolibri360.de